RECENT TREND IN ELECTROCHEMICAL SCIENCE AND TECHNOLOGY

Edited by **Ujjal Kumar Sur**

Recent Trend in Electrochemical Science and Technology
Edited by Ujjal Kumar Sur

Published by InTech

Janeza Trdine 9, 51000 Rijeka, Croatia

Publishing Process Manager Gorana Scerbe

Technical Editor Teodora Smiljanic

Cover Designer InTech Design Team

Additional hard copies can be obtained from orders@intechopen.com

Recent Trend in Electrochemical Science and Technology, Edited by Ujjal Kumar Sur
p. cm.
ISBN 978-953-307-830-4

For EU product safety concerns:
IN TECH d.o.o., Prolaz Marije Krucifikse Kozulić 3, 51000 Rijeka, Croatia,
info@intechopen.com or visit our website at intechopen.com.

Meet the editor

Dr. Ujjal Kumar Sur works as Assistant Professor in the Department of Chemistry, Behala College, Kolkata, India. He is also a visiting scientist in the State Key Laboratory of Physical Chemistry of solid surfaces, Xiamen University, Xiamen, China. Dr. Sur is a well known Scientist in Electrochemistry and Surface Science. His research interests are electrochemistry, self-assembled monolayers of organic molecules, microwave chemistry, nanotechnology and surface-enhanced Raman spectroscopy. He did his Ph.D in Raman Research Institute, Bangalore under the supervision of Professor V. Lakshminarayanan (1998-2003). He did his Postdoctoral research in Loughborough University UK (2003-2005) and in Institute of Atomic & Molecular Sciences, Academia Sinica, Taipei, Taiwan (2006-2008).

Contents

Preface

Electrochemistry is a fast emerging scientific research field connected to both physics and chemistry. It integrates various aspects of the classical electrochemical science and engineering, solid-state chemistry and physics, materials science, heterogeneous catalysis, and other areas of physical chemistry. This field also comprises a variety of practical applications, which include different types of energy storage devices such as batteries, fuel cells, capacitors and accumulators, various sensors and analytical appliances, electrochemical gas pumps and compressors, electrochromic and memory devices, solid-state electrolyzers and electrocatalytic reactors, synthesis of new materials with novel improved properties, and corrosion protection.

This book titled *"Recent Trends in Electrochemical Science and Technology"* contains a selection of chapters focused on advanced methods used in the research area of electrochemical science and technologies; descriptions of electrochemical systems; processing of novel materials and mechanisms relevant for their operation. This book provides an overview on some of the recent development in electrochemical science and technology. Particular emphasis is given both to the theoretical and the experimental aspect of modern electrochemistry. Since it was impossible to cover the rich diversity of electrochemical techniques and applications in a single issue, the focus is on the recent trends and achievements related to electrochemical science and technology. Some of the topics represented in the book are: *study of charge transfer kinetics at interfaces using scanning electrochemical microscope (SECM); electrochemistry of curium in molten salts; application of the negative binomial pascal distribution in probability theory to electrochemical processes; water management and experimental diagnostics in polymer electrolyte fuel cell; Mars electrochemistry; studies of supercapacitor electrodes with high* ***pseudo capacitance****; electrochemical basis of biological activity; nanomaterials preparation by electrochemical methods,* etc.

Ujjal Kumar Sur,
Department of Chemistry,
Behala college, Kolkata,
India

Introduction to Electrochemical Science and Technology and Its Development

Ujjal Kumar Sur
Department of Chemistry, Behala College, Kolkata-60, India

1. Introduction

Electrochemistry is a fast emergent scientific research field in both physical and chemical science which integrates various aspects of the classical electrochemical science and engineering, solid-state chemistry and physics, materials science, heterogeneous catalysis, and other areas of physical chemistry. This field also comprises of a variety of practical applications, which includes many types of energy storage devices such as batteries, fuel cells, capacitors and accumulators, various sensors and analytical appliances, electrochemical gas pumps and compressors, electrochromic and memory devices, solid-state electrolyzers and electrocatalytic reactors, synthesis of new materials with novel improved properties, and corrosion protection.

Electrochemistry is a quite old branch of chemistry that studies chemical reactions which take place in a solution at the interface of an electron conductor (a metal or a semiconductor) and an ionic conductor (the electrolyte), and which involve electron transfer between the electrode and the electrolyte or species in solution. The development of electrochemistry began its journey in the sixteenth century. The first fundamental discoveries considered now as the foundation of electrochemistry were made in the nineteenth and first half of the twentieth centuries by M. Faraday, E. Warburg, W. Nernst, W. Schottky, and other eminent scientists. Their pioneering works provided strong background for the rapid development achieved both in the fundamental understanding of the various electrochemical processes and in various applications during the second half of the twentieth century. As for any other research field, the progress in electrochemistry leads both to new horizons and to new challenges. In particular, the increasing demands for higher performance of the electrochemical devices lead to the necessity to develop novel approaches for the nanoscale optimization of materials and interfaces, for analysis and modeling of highly non-ideal systems.

2. Historical background on the development of electrochemistry

2.1 16th to 18th century developments

- In 1785, Charles-Augustin de Coulomb developed the law of electrostatic attraction.

- In 1791, Italian physician and anatomist Luigi Galvani marked the birth of electrochemistry by establishing a bridge between chemical reactions and electricity on his essay "*De Viribus Electricitatis in Motu Musculari Commentarius*" by proposing a "*nerveo-electrical substance*" on biological life forms.
- In 1800, William Nicholson and Johann Wilhelm Ritter succeeded in decomposing water into hydrogen and oxygen by electrolysis. Later, Ritter discovered the process of electroplating.
- In 1827, the German scientist Georg Ohm expressed his law, which is known as "Ohm's law".
- In 1832, Michael Faraday introduced his two laws of electrochemistry, which is commonly known as "Faraday's laws of Electrolysis".
- In 1836, John Daniell invented a primary cell in which hydrogen was eliminated in the generation of the electricity.
- In 1839, William Grove produced the first fuel cell.
- In 1853, Helmholtz introduced the concept of an electrical double layer at the interface between conducting phases. This is known as the capacitance model of electrical double layer at the electrode|electrolyte interface. This capacitance model was later refined by Gouy and Chapman, and Stern and Geary, who suggested the presence of a diffuse layer in the electrolyte due to the accumulation of ions close to the electrode surface. Figure 1 illustrates the Helmholtz double layer model at the electrode|electrolyte interface.

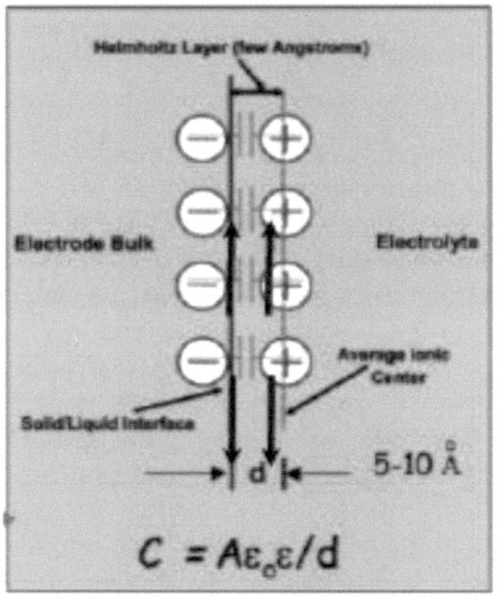

Fig. 1. Schematic diagram of Helmholtz double layer model

- In 1868, Georges Leclanché patented a new cell which eventually became the forerunner to the world's first widely used battery, the zinc carbon cell.
- In 1884, Svante Arrhenius published his thesis on the galvanic conductivity of electrolytes. From his results, he concluded that electrolytes, when dissolved in water,

become to varying degrees split or dissociated into electrically opposite positive and negative ions. He introduced the concept of ionization and classified electrolytes according to the degree of ionization.

- In 1886, Paul Héroult and Charles M. Hall developed an efficient method to obtain aluminium using electrolysis of molten alumina.
- In 1894, Friedrich Ostwald concluded important studies of the conductivity and electrolytic dissociation of organic acids.
- In 1888, Walther Hermann Nernst developed the theory of the electromotive force of the voltaic cell.
- In 1889, he showed how the characteristics of the current produced could be used to calculate the free energy change in the chemical reaction producing the current. He constructed an equation, which is known as Nernst equation, which related the voltage of a cell to its properties.
- In 1898, German scientist, Fritz Haber showed that definite reduction products can result from electrolytic processes by keeping the potential at the cathode constant.

Fig. 2. Pictures of Arrhenius and Nernst

2.2 The 20th century developments

- In 1902, The Electrochemical Society (ECS) of United States of America was founded.
- In 1909, Robert Andrews Millikan began a series of experiments to determine the electric charge carried by a single electron.
- In 1922, Jaroslav Heyrovski invented polarography, a commonly used electroanalytical technique. Later, in 1959, he was awarded Nobel prize for his invention of polarography.
- In 1923, Peter Debye and Erich Huckel proposed a theory to explain the deviation for electrolytic solutions from ideal behaviour.

- In 1923, Johannes Nicolaus Brønsted and Martin Lowry published essentially the same theory about how acids and bases behave.
- In 1937, Arne Tiselius developed the first sophisticated electrophoretic apparatus. Later, in 1948, he was awarded Nobel prize for his pioneering work on the electrophoresis of protein.

Fig. 4. Heyrovsky's polarography instrument

- In 1949, the International Society of Electrochemistry (ISE) was founded.
- In 1960-1970, Revaz Dogonadze and his co-workers developed quantum electrochemistry.
- In 1957, the first patent based on the concept of electrochemical capacitor (EC) was filed by Becker.
- In 1972, Japanese scientists Akira Fujishima and Kenichi Honda carried out electrochemical photolysis of water at a semiconductor electrode and developed photoelectrochemical (PEC) solar cell.
- In 1974, Fleishmann, Hendra and Mcquillan of University of Southampton, UK introduced surface enhanced Raman scattering (SERS) spectroscopy (Fleishmann et al. 1974). It was accidentally discovered by them when they tried to do Raman with an adsorbate of very high Raman cross section, such as pyridine (Py) on the roughened

silver (Ag) electrode. The initial idea was to generate high surface area on the roughened metal surface. The Raman spectrum obtained was of unexpectedly high quality. They initially explained the intense surface Raman signal of Py due to increased surface area. Later, Jeanmaire and Van Duyne (Jeanmaire & Van Duyne, 1977) from Northwestern University, USA, first realized that surface area is not the main point in the above phenomenon in 1977. Albrecht and Creighton of University of Kent, UK, reported a similar result in the same year (Albrecht & Creighton, 1977). These two groups provided strong evidences to demonstrate that the strong surface Raman signal must be generated by a real enhancement of the Raman scattering efficiency (10^5 to 10^6 enhancement). The effect was later named as surface-enhanced Raman scattering and now, it is an universally accepted surface sensitive technique. Although, the first SERS spectra were obtained from an electrochemical system (Py + roughned Ag electrode), all important reactions on surfaces including electrochemical processes can be studied by SERS.

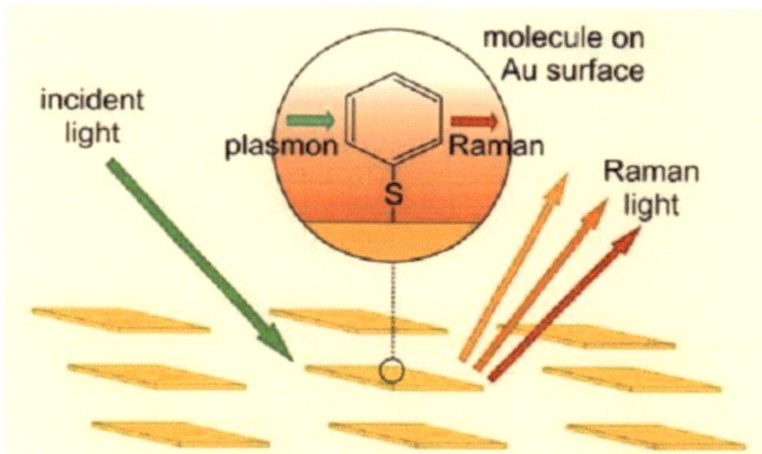

Fig. 5. Schematic diagram to explain the principle of SERS

- In early eighties, Fleischmann and his co-workers at the Southampton Electrochemistry group exploited the versatile properties of microelectrodes in electrochemical studies. The ultramicroelectrodes, due to their extremely small size, have certain unique characteristics which make them ideal for studies involving high resistive media, high speed voltammetry and *in vivo* electrochemistry in biological systems.
- In 1989, A.J.Bard and his group at the University of Texas, Austin, USA developed a new scanning probe technique in electrochemical environment (Bard et al. 1989). This is known as Scanning Electrochemical Microscope (SECM), which is a combination of electrochemical STM and an ultramicroelectrode.

2.3 Recent developments

Development of various electroanalytical techniques such as voltammetry (both linear and cyclic), chrono and pulsed techniques, electrochemical impedance spectroscopy (EIS) as well

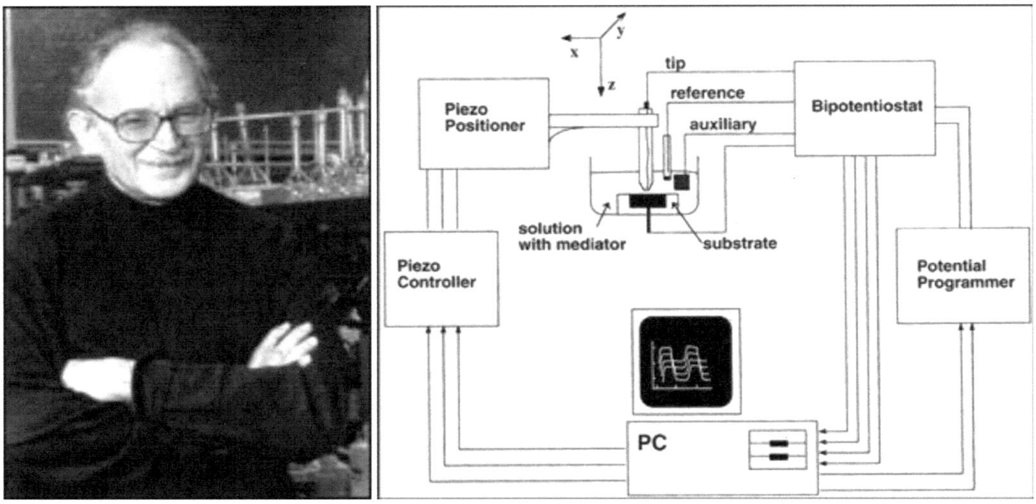

Fig. 6. Picture of A. J. Bard along with the schematic diagram of SECM

as various non-electrochemical surface sensitive techniques such as X-ray diffraction (XRD), X-ray photoelectron spectroscopy (XPS), Infrared (IR) and Raman spectroscopy, SERS, Scanning electron microscopy (SEM), Scanning probe techniques like Scanning tunneling microscope (STM), Atomic force microscope (AFM) and SECM has brought a new dimension in the research of electrochemical science and technology. In the recent time, electrochemical science and technology has become extremely popular not only to electrochemists, but also to material scientists, biologists, physicists, engineers, metallurgists, mathematicians, medical practitioners. The recent advancement in material science and nanoscience & nanotechnology has broadened its practical applications in diversed field such as energy storage devices, sensors and corrosion protection. The invention of fullerenes (Kroto et al. 1985) and carbon nanotubes (Iijima, 1991) (In 1980's and 1990's and the recent invention of graphene made a breakthrough in the development of various energy storage devices with enhanced performance. Graphene was discovered in 2004 by Geim and his co-workers (Novoselov et al. 2004), who experimentally demonstrated the preparation of a single layer of graphite with atomic thickness using a technique called micromechanical cleavage. With inherent properties, such as tunable band gap, extraordinary electronic transport properties, excellent thermal conductivity, great mechanical strength, and large surface area, graphene has been explored for diversed applications ranging from electronic devices to electrode materials. The two dimensional honeycomb structure of carbon atoms in graphene along with the high-resolution transmission electron microscopic (TEM) image are shown in Figure 7. Graphene displays unusual properties making it ideal for applications such as microchips, chemical/biosensors, ultracapacitance devices and flexible displays. It is expected that graphene could eventually replace silicon (Si) as the substance for computer chips, offering the prospect of ultra-fast computers/quantum computers operating at terahertz speeds.

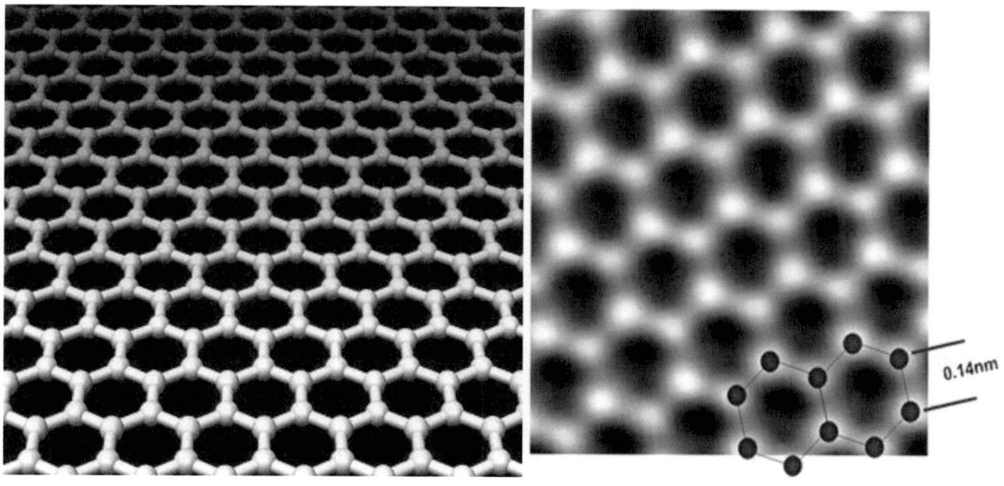

Fig. 7. Two dimensional honeycomb structure of graphene along with the high-resolution TEM image.

3. Conclusion

This book titled "*Recent Trend in Electrochemical Science and Technology*" contains a selection of chapters focused on advanced methods used in the research area of electrochemical science and technologies, description of the electrochemical systems, processing of novel materials and mechanisms relevant for their operation. Since it was impossible to cover the rich diversity of electrochemical techniques and applications in a single issue, emphasis was centered on the recent trends and achievements related to electrochemical science and technology.

4. Acknowledgement

We acknowledge financial support from the project funded by the UGC, New Delhi (grant no. PSW-038/10-11-ERO).

5. References

Albrecht, M.G., & Creighton, J.A. (1977). Anomalously Intense Raman Spectra of Pyridine at a Silver Electrode. *J.Am.Chem.Soc.*, Vol. 99, (June 1977), pp. 5215-5217, ISSN 0002-7863.

Bard, A.J., Fan, F.-R.F., Kwak, J., & Lev, O. (1989). Scanning Electrochemical microscopy. Introduction and principles. *Anal. Chem.*, Vol. 61, (January 1989) pp. 132-138, ISSN 0003-2700.

Fleischmann, M., Hendra, P.J., & McQuillan, A.J. (1974). Raman Spectra of pyridine adsorbed at a silver electrode. *Chem.Phys.Lett.*, Vol. 26, (15 May 1974), pp. 163-166, ISSN 0009-2614.

Iijima, S., (1991). Helical microtubules of graphitic Carbon. *Nature*, Vol. 354, (7 November 1991), pp. 56-58, ISSN 0028-0836.

Jeanmaire, D.L., & Van Duyne, R.P. (1977). Surface Raman Electrochemistry part 1. Heterocyclic, Aromatic and Aliphatic Amines Adsorbed on the Anodized Silver Electrode. *J. Electroanal. Chem.*, Vol. 84, (10 November 1977), pp. 1-20, ISSN 1572-6657.

Kroto, H. W., Heath, J. R., O'Brien, S. C., Curl, R. F., & Smalley, R. E. (1985). C60: Buckminsterfullerene. *Nature*, Vol. 318, (14 November 1985), pp.162–163, ISSN 0028-0836.

Novoselov, K.S., Geim, A.K., Morozov, S.V., Jiang, D., Zhang, Y., Dubonos, S.V., Grigorieva, I.V., & Firsov, A.A. (2004). Electric field effect on atomically thin carbon films. *Science*, Vol. 306, (22 October 2004), pp. 666-669, ISSN 0036-8075

Part 1

Physical Electrochemistry

1

Electrochemistry of Curium in Molten Chlorides

Alexander Osipenko[1], Alexander Mayershin[1], Valeri Smolenski[2,*], Alena Novoselova[2] and Michael Kormilitsyn[1]
[1]Radiochemical Division, Research Institute of Atomic Reactors,
[2]Institute of High-Temperature Electrochemistry, Ural Division, Russian Academy of Science, Russia

1. Introduction

Molten salts and especially fused chlorides are the convenient medium for selective dissolution and deposition of metals. The existence of a wide spectrum of individual salt melts and their mixtures with different cation and anion composition gives the real possibility of use the solvents with the optimum electrochemical and physical-chemical properties, which are necessary for solving specific radiochemistry objects. Also molten alkali metal chlorides have a high radiation resistance and are not the moderator of neutrons as aqua and organic mediums [Uozumi, 2004; Willit, 2005].

Nowadays electrochemical reprocessing in molten salts is applied to the oxide and metal fuel. Partitioning and Transmutation (P&T) concept is one of the strategies for reducing the long-term radiotoxicity of the nuclear waste. For this case pyrochemical reprocessing methods including the recycling and transmutation can be successfully used for conversion more hazardous radionuclides into short-lived or even stable elements. For that first of all it is necessary to separate minor actinides (Np, Am, Cm) from other fission products (FP).

Pyrochemical reprocessing methods are based on a good knowledge of the basic chemical and electrochemical properties of actinides and fission products. This information is necessary for creation the effective technological process [Bermejo et al., 2007, 2008; Castrillejo et al., 2005a, 2005b, 2009; De Cordoba et al., 2004, 2008; Fusselman et al., 1999; Kuznetsov et al., 2006; Morss, 2008; Novoselova & Smolenski, 2010, 2011; Osipenko et al., 2010, 2011; Roy et al., 1996; Sakamura et al., 1998; Serp et al., 2004, 2005a, 2005b, 2006; Serrano & Taxil, 1999; Shirai et al., 2000; Smolenski et al., 2008, 2009].

Curium isotopes in nuclear spent fuel have a large specific thermal flux and a long half-life.

So, they must be effectively separated from highly active waste and then undergo transmutation.

The goal of this work is the investigation of electrochemical and thermodynamic properties of oxide and oxygen free curium compounds in fused chlorides.

2. Experimental

2.1 Preparation of starting materials

The solvents LiCl (Roth, 99.9%), NaCl (Reachim, 99.9%), KCl (Reachim, 99.9%), and CsCl (REP, 99.9%) were purified under vacuum in the temperatures range 293-773 K. Then the reagents were fused under dry argon atmosphere. Afterwards these reagents were purified by the operation of the direct crystallization [Shishkin & Mityaev, 1982]. The calculated amounts of prepared solvents were melted in the cell before any experiment [Korshunov et al., 1979].

Curium trichloride was prepared by using the operation of carbochlorination of curium oxide in fused solvents in vitreous carbon crucibles. Cm^{3+} ions, in the concentration range 10^{-2}-10^{-3} mol kg^{-1} were introduced into the bath in the form of $CmCl_3$ solvent mixture.

The obtained electrolytes were kept into glass ampoules under atmosphere of dry argon in inert glove box.

2.2 Potentiometric method

The investigations were carried out in the cell, containing platinum-oxygen electrode with solid electrolyte membrane which was made from ZrO_2 stabilized by Y_2O_3 supplied by Interbil Spain (inner diameter 4 mm, outer diameter 6 mm). This electrode was used as indicating electrode for measuring the oxygen ions activity in the investigated melt. The measurements were carried out versus classic Cl^-/Cl_2 reference electrode [Smirnov, 1973]. The difference between indicator and reference electrodes in the following galvanic cell

$$Pt_{(s)}, O_{2(g)} \left| ZrO_2\left(Y_2O_3\right) \right| Melt\,under\,test \left\| Solvent\,melt \right| Cl_{2(g)}, C_{(s)} \quad (1)$$

is equal to

$$\varepsilon = \varepsilon^o - \frac{RT}{2F} \ln \frac{a_{O^{2-}} \cdot p_{Cl_2}}{a^2_{Cl^-} \cdot p^{1/2}_{O_2}} \quad (2)$$

where a is the activity of the soluble product in the melt (in mol·kg^{-1}); P is the gas pressure (in atm.); ε^o is the difference of standard electrode potentials of the reaction 3 (in V); T is the absolute temperature (in K); R is the ideal gas constant (in J·mol^{-1}·K^{-1}); n is the number of electrons exchanged and F is the Faraday constant (96500 C·mol^{-1}).

$$2Cl^-_{(l)} + 1/2O_{2(g)} = O^{2-}_{(l)} + Cl_{2(g)} \, . \quad (3)$$

The value ε^o of the reaction (3) *is* the following

$$\varepsilon^o = E^o_{Cl_2/Cl^-} - E^o_{O_2/O^{2-}} = \frac{-\Delta G^o}{2F} \quad (4)$$

where ΔG^o is the change of the standard Gibbs energy of the reaction 3 (in kJ·mol^{-1}·K^{-1}).

$$E_{O_2/O^{2-}} = E^*_{O_2/O^{2-}} - \frac{RT}{2F} \ln\left[m_{eq}\left(O^{2-}\right)\right] \quad (5)$$

where $E_{O_2/O^{2-}}$ is the equilibrium potential of O_2/O^{2-} system (in V); $E^*_{O_2/O^{2-}}$ is an apparent standard potential of the system (in V).

The value of apparent standard potential E^* in contrast to the standard potential E^o describes the dilute solutions, where the activity coefficient $\gamma_{O^{2-}}$ is constant at low concentrations [Smirnov, 1973] and depends from the nature of molten salts. It can be calculated experimentally with high precision according to expression (5). The introducing of oxide ions in the solution was done by dropping calculated amounts of BaO (Merck, 99,999%) which completely dissociates in the melt [Cherginetz, 2004].

All reagents were handled in a glove box to avoid contamination of moisture. The experiments were performed under an inert argon atmosphere.

The potentiometric study was performed with Autolab PGSTAT302 potentiostat/galvanostat (Eco-Chimie) with specific GPES electrochemical software (version 4.9.006).

2.3 Transient electrochemical technique

The experiments were carried out under inert argon atmosphere using a standard electrochemical quartz sealed cell using a three electrodes setup. Different transient electrochemical techniques were used such as linear sweep, cyclic, square wave, differential and semi-integral voltammetry, as well as potentiometry at zero current. The electrochemical measurements were carried out using an Autolab PGSTAT302 potentiostat-galvanostat (Eco-Chimie) with specific GPES electrochemical software (version 4.9.006).

The inert working electrode was prepared using a 1.8 mm metallic W wire (Goodfellow, 99.9%). It was immersed into the molten bath between 3 - 7 mm. The active surface area was determined after each experiment by measuring the immersion depth of the electrode. The counter electrode consisted of a vitreous carbon crucible (SU - 2000). The Cl^-/Cl_2 or Ag/Ag^+ (0.75 mol·kg^{-1} AgCl) electrodes were used as standard reference electrodes. The experiments were carried out in vitreous carbon crucibles; the amount of salt was (40-60 g). The total curium concentrations were determined by taking samples from the melt and then analyzed by ICP-MS.

3. Results and discussion

3.1 Potentiometric investigations

The preliminary investigations of fused 3LiCl-2KCl eutectic and equimolar NaCl-KCl by of O^{2-} ions are present in Table 1. In this case, the potential of the pO^{2-} indicator electrode vs. the concentrations of added O^{2-} ions follows a Nernst behavior (eq. 5). The experiment slope is closed to its theoretical value for a two-electron process, which shows the Nernstian behavior of the system.

To identify curium oxide species and to determine their stability, the titration of Cm^{3+} by O^{2-} ions was performed. To estimate stoichiometric coefficients of reactions that involve initial components, the ligand number "α" was used.

Molten solvent	Temperature, K	$E^*_{O_2/O^{2-}}$ (in V vs. Cl^-/Cl_2)	$\frac{RT}{2F}$ (exp.)	$\frac{RT}{2F}$ (theor.)
3LiCl-2KCl	723	-1.087±0.001	0.072±0.001	0.072
	823	-1.102±0.001	0.082±0.001	0.082
	923	-1.275±0.004	0.091±0.001	0.0911
NaCl-KCl	1023	-1.351±0.001	0.101±0.001	0.101
	1073	-1.448±0.003	0.134±0.002	0.106
	1123	-1.374±0.001	0.111±0.001	0.111
NaCl-2CsCl	823	-0.751±0.001	0.083±0.001	0.083
	923	-0.771±0.001	0.092±0.001	0.092
	1023	-0.985±0.001	0.113±0.009	0.102

Table 1. The parameters of calibration curve for 3LiCl-2KCl, NaCl-KCl and NaCl-2CsCl melts, (molality scale)

$$\alpha = \frac{\left[O^{2-}\right]_{added}}{\left[Cm^{3+}\right]_{initial}} \tag{6}$$

where $\left[O^{2-}\right]_{added}$ is the added concentration of oxide ions in the melt, (in mol·kg^{-1}); $\left[Cm^{3+}\right]_{initial}$ is the initial Cm^{3+} concentration, (in mol·kg^{-1}).

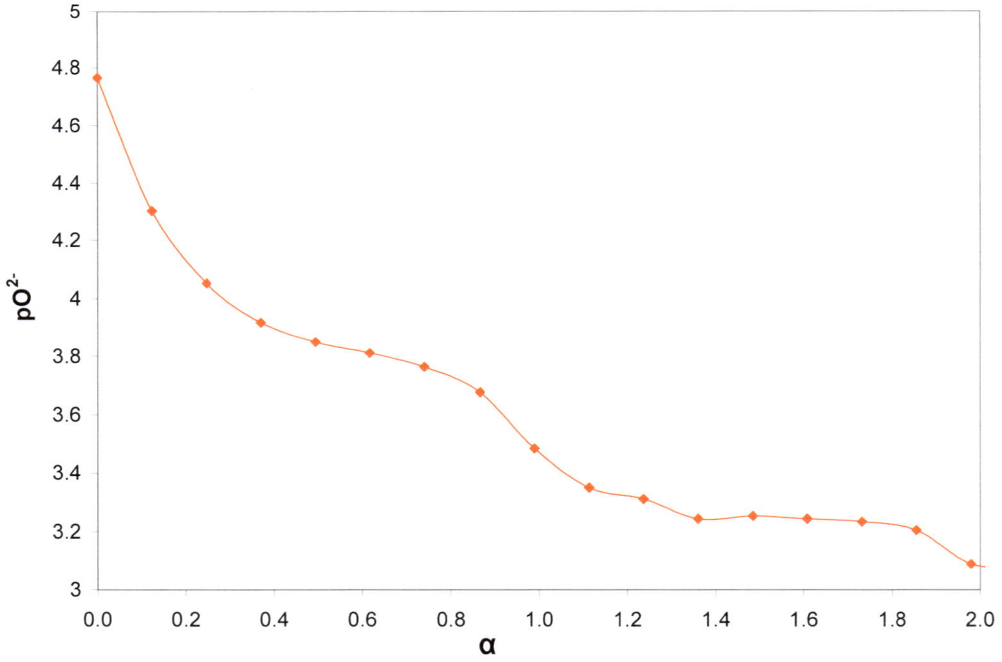

Fig. 1. Potentiometric titration of Cm^{3+} solution by O^{2-} ions in NaCl-2CsCl at 1023 K. $[Cm^{3+}]$ = 1.2·10^{-3} mol·kg^{-1}

The potentiometric titration curve pO^{2-} versus *a* in the NaCl-2CsCl-$CmCl_3$ melt shows one equivalent point for *a* equal to 1, Fig. 1. This can be assigned to the production of solid oxycloride, CmOCl. The shape of an experimental curve shows the possibility of formation of soluble product CmO^+ in the beginning of titration [Cherginetz, 2004]. The precipitation of Cm_2O_3 did not fixed on experimental curves. One of the reasons of these phenomena may be the kinetic predicaments in formation of insoluble compound Cm_2O_3.

Therefore, the titration reactions can be written as:

$$Cm^{3+}_{(l)} + O^{2-}_{(l)} \Leftrightarrow CmO^{+}_{(l)} \qquad (0 < \alpha < 0.5) \qquad (7)$$

$$Cm^{3+}_{(l)} + O^{2-}_{(l)} + Cl^{-}_{(l)} \Leftrightarrow CmOCl_{(s)} \qquad (0.5 < \alpha < 1.0) \qquad (8)$$

$$2CmOCl_{(s)} + O^{2-}_{(l)} \Leftrightarrow Cm_2O_{3(s)} + 2Cl^{-}_{(l)} \qquad (1.0 < \alpha < 1.5) \qquad (9)$$

Combine expressions (8) and (9), $Cm_2O_{3(s)}$ formation is described by (10):

$$2Cm^{3+}_{(l)} + 3O^{2-}_{(l)} \Leftrightarrow Cm_2O_{3(s)} \qquad (10)$$

The chloride ions activity in the melt is one. By applying mass balance equations (11, 12) and the expressions of the equilibrium constant of the reaction (7) and the solubility constants of the reactions (8, 10) it is possible to calculate the concentration of CmO^+ ions and the solubility of CmOCl and Cm_2O_3 in the melt:

$$\left[O^{2-}\right]_{bulk} = \left[O^{2-}\right]_{added} - \left[CmO^{+}\right]_{bulk} - \left[CmOCl\right]_{precipitated} - 3\left[Cm_2O_3\right]_{precipitated} \qquad (11)$$

$$\left[Cm^{3+}\right]_{bulk} = \left[Cm^{3+}\right]_{initial} - \left[CmO^{+}\right]_{bulk} - \left[CmOCl\right]_{precipitated} - 2\left[Cm_2O_3\right]_{precipitated} \qquad (12)$$

where $\left[O^{2-}\right]_{bulk}$ is the equilibrium concentration of oxide ions in the melt, (in mol·kg^{-1}); $\left[Cm^{3+}\right]_{bulk}$ is the equilibrium concentration of curium ions in the melt, (in mol·kg^{-1}); $\left[CmO^{+}\right]_{bulk}$ is the equilibrium concentration of curium oxide ions in the melt, (in mol·kg^{-1}).

$$K_{eq}^{CmO^+} = \frac{\left[CmO^{+}\right]}{\left[Cm^{3+}\right]\cdot\left[O^{2-}\right]} \qquad (13)$$

$$K_{s}^{CmOCl} = \left[Cm^{3+}\right]\cdot\left[O^{2-}\right]\cdot\left[Cl^{-}\right] \qquad (14)$$

$$K_{s}^{Cm_2O_3} = \left[Cm^{3+}\right]^2\cdot\left[O^{2-}\right]^3 \qquad (15)$$

The formation of CmO^+ ions in the range ($0 < \alpha < 0.5$) is described by the following theoretical titration curve:

$$\alpha = \frac{\left[O^{2-}\right]_{bulk} \cdot \left[\frac{K_{eq}^{CmO^+}}{\left[Cm^{3+}\right]_{initial}} + \left[O^{2-}\right]_{bulk} \right]}{\frac{\left[Cm^{3+}\right]+1}{K_{eq} + \left[O^{2-}\right]_{bulk}}} \quad (16)$$

When CmOCl is precipitating ($0.5 < \alpha < 1.0$), the theoretical titration curve can be written as:

$$\alpha = 1 + \frac{1}{\left[Cm^{3+}\right]_{initial}} \left[\left[O^{2-}\right]_{bulk} - \frac{K_s^{CmOCl}}{\left[O^{2-}\right]_{bulk}} \right] \quad (17)$$

In the range ($1.0 < \alpha < 1.5$), where Cm_2O_3 is precipitating, the theoretical titration curve is:

$$\alpha = 1.5 + \frac{1}{\left[Cm^{3+}\right]_{initial}} \left[\left[O^{2-}\right]_{bulk} - \frac{1.5 \cdot \left(K_s^{Cm_2O_3)}\right)^{1/2}}{\left[O^{2-}\right]_{bulk}^{3/2}} \right] \quad (18)$$

Molten solvent	Temperature, K	$pK_{eq}^{CmO^+}$	pK_s^{CmOCl}	$pK_s^{Cm_2O_3}$
3LiCl-2KCl	723 823 923	2.5±0.2 2.4±0.2 0.8±0.1	7.5±0.2 5.7±0.2 5.2±0.2	15.5±0.5 12.7±0.5 12.5±0.5
NaCl-KCl	1023 1073 1123	2.6±0.2 2.4±0.2 1.3±0.1	5.9±0.2 5.8±0.2 5.6±0.2	12.9±0.4 12.6±0.4 12.1±0.4
NaCl-2CsCl	829 923 1023	4.2±0.2 3.4±0.2 3.7±0.2	7.9±0.2 7.5±0.2 6.7±0.2	20.1±0.3 18.5±0.3 16.8±0.3

Table 2. The experimental values of dissociation constants of CmO^+, CmOCl и Cm_2O_3 in fused solvents at different temperatures, (molatility scale)

The best conformity of the experimental and theoretical titration curves at different temperatures is obtained with the constants, offers in Table 2. All results are presented in Tables 3-5. Thermodynamic data allowed us to draw the potential–pO^{2-} diagrams, Fig. 2-4, which summarized the stability areas of curium compounds in different solvents a various temperatures.

The decreasing of the temperature and the shift of the ionic radius of the solvent (in z/r, nm) [Lebedev, 1993] from LiCl up to CsCl mixtures show regular decreasing of the solubility of curium in the solvents [Yamana, 2003].

System	Expression for equilibrium potential	Apparent standard potential (V vs. Cl^-/Cl_2) $[Cm^{3+}] = 1$ mol·kg^{-1}
1. $Cm^{3+} + 3e^- \leftrightarrow Cm$	$E_1 = E_1^* + \frac{2.3RT}{3F}\log\left[Cm^{3+}\right]$	$E^*(1) = -2.924$
2. $CmO^+ + 3e^- \leftrightarrow Cm + O^{2-}$	$E_2 = E_1^* - \frac{2.3RT}{3F} pK_{eq(CmO^+)} +$ $+\frac{2.3RT}{3F}\log\left[CmO^+\right] + \frac{2.3RT}{3F}$	$E^*(2) = -3.055$
3. $CmOCl + 3e^- \leftrightarrow Cm + O^{2-} + Cl^-$	$E_3 = E_1^* - \frac{2.3RT}{3F} pK_{S(CmOCl)} +$ $+\frac{2.3RT}{3F} pO^{2-}$	$E^*(3) = -3.220$
4. $Cm_2O_3 + 6e^- \leftrightarrow 2Cm + 3O^{2-}$	$E_4 = E_1^* - \frac{2.3RT}{6F} pK_{S(Cm_2O_3)} +$ $+\frac{2.3RT}{2F} pO^{2-}$	$E^*(4) = -3.286$
5. $Cm^{3+} + O^{2-} \leftrightarrow CmO^+$	$pK_{eq} = -4.7455 + 5426/T$	$pK_{eq} = 2.5$
6. $Cm^{3+} + O^{2-} + Cl^- \leftrightarrow CmOCl$	$pK_s = 1.5132 + 3394/T$	$pK_s = 7.5$
7. $2Cm^{3+} + 3O^{2-} \leftrightarrow Cm_2O_3$	$pK_s = 0.779 + 10407.5/T$	$pK_s = 15.5$

Table 3. Equilibrium potentials and values of apparent standard potentials of redox system in 3LiCl-2KCl at 723 K. $[Cm^{3+}] = 1$ mol·kg^{-1}. Potentials are given vs. Cl^-/Cl_2 reference electrode

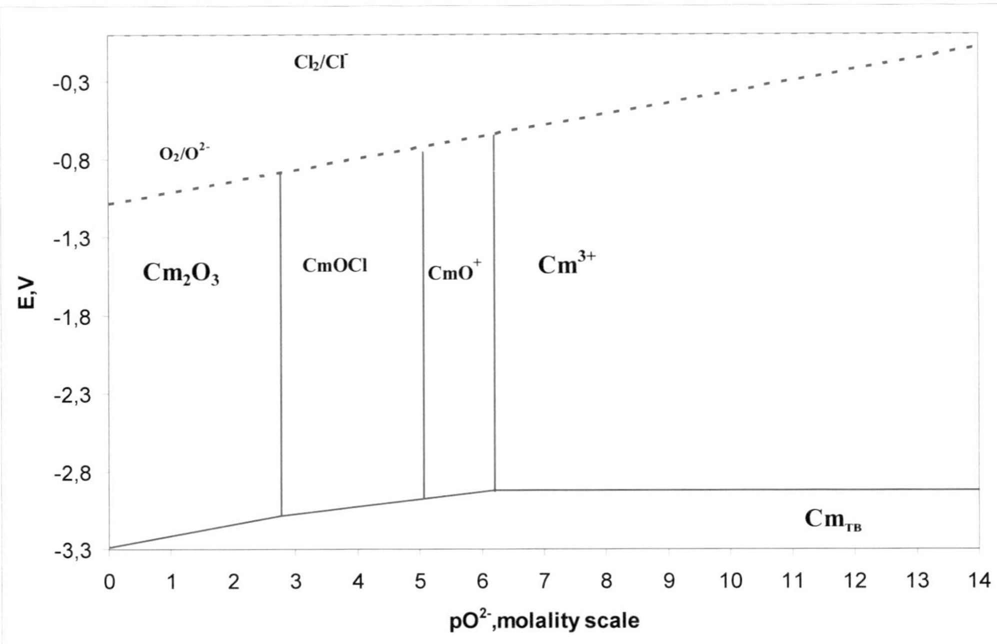

Fig. 2. Potential–pO^{2-} diagram for curium in 3LiCl-2KCl eutectic at 723 K. $[Cm^{3+}] = 1$ mol·kg^{-1}. Potentials are given vs. Cl^-/Cl_2 reference electrode

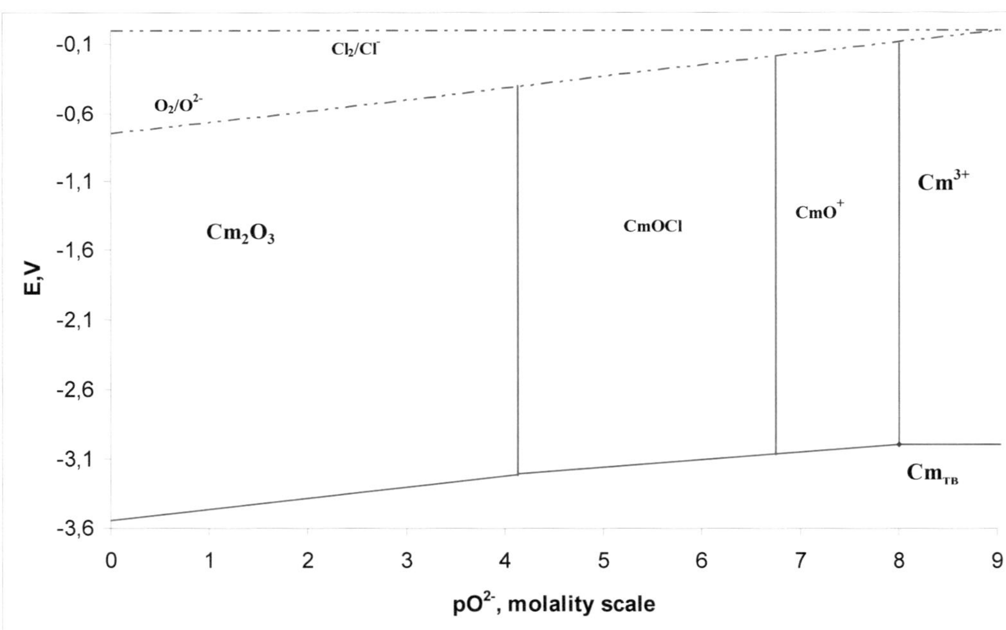

Fig. 3. Potential–pO^{2-} diagram for curium in equimolar NaCl-KCl at 1023 K. $[Cm^{3+}] = 1$ mol·kg^{-1}. Potentials are given vs. Cl^-/Cl_2 reference electrode

System	Expression for equilibrium potential	Apparent standard potential (V vs. Cl^-/Cl_2) $[Cm^{3+}] = 1$ mol·kg $^{-1}$
1. $Cm^{3+} + 3e^- \leftrightarrow Cm$	$E_1 = E_1^* + \frac{2.3RT}{3F}\log\left[Cm^{3+}\right]$	$E^*(1) = -2.727$
2. $CmO^+ + 3e^- \leftrightarrow Cm + O^{2-}$	$E_2 = E_1^* - \frac{2.3RT}{3F}pK_{eq(CmO^+)} + \frac{2.3RT}{3F}\log\left[CmO^+\right] + \frac{2.3RT}{3F}pO$	$E^*(2) = -2.915$
3. $CmOCl + 3e^- \leftrightarrow Cm + O^{2-} + Cl^-$	$E_3 = E_1^* - \frac{2.3RT}{3F}pK_{S(CmOCl)} + \frac{2.3RT}{3F}pO^{2-}$	$E^*(3) = -3.128$
4. $Cm_2O_3 + 6e^- \leftrightarrow 2Cm + 3O^{2-}$	$E_4 = E_1^* - \frac{2.3RT}{6F}pK_{S(Cm_2O_3)} + \frac{2.3RT}{2F}pO^{2-}$	$E^*(4) = -3.165$
5. $Cm^{3+} + O^{2-} \leftrightarrow CmO^+$	$pK_{eq} = -4.7455 + 5426/T$	$pK_{eq} = 2.6$
6. $Cm^{3+} + O^{2-} + Cl^- \leftrightarrow CmOCl$	$pK_s = 1.5132 + 3394/T$	$pK_s = 5.9$
7. $2Cm^{3+} + 3O^{2-} \leftrightarrow Cm_2O_3$	$pK_s = 0.779 + 10407.5/T$	$pK_s = 12.9$

Table 4. Equilibrium potentials and values of apparent standard potentials of redox system in equimolar NaCl-KCl at 1023 K. $[Cm^{3+}] = 1$ mol·kg^{-1}. Potentials are given vs. Cl^-/Cl_2 reference electrode

System	Expression for equilibrium potential	Apparent standard potential (V vs. Cl^-/Cl_2) $[Cm^{3+}] = 1$ mol·kg $^{-1}$
1. $Cm^{3+} + 3e^- \leftrightarrow Cm$	$E_1 = E_1^* + \frac{2.3RT}{3F}\log\left[Cm^{3+}\right]$	$E^*(1) = -2.996$
2. $CmO^+ + 3e^- \leftrightarrow Cm + O^{2-}$	$E_2 = E_1^* - \frac{2.3RT}{3F}pK_{eq(CmO^+)} + \frac{2.3RT}{3F}\log\left[CmO^+\right] + \frac{2.3RT}{3F}$	$E^*(2) = -3.220$
3. $CmOCl + 3e^- \leftrightarrow Cm + O^{2-} + Cl^-$	$E_3 = E_1^* - \frac{2.3RT}{3F}pK_{S(CmOCl)} + \frac{2.3RT}{3F}pO^{2-}$	$E^*(3) = -3.430$
4. $Cm_2O_3 + 6e^- \leftrightarrow 2Cm + 3O^{2-}$	$E_4 = E_1^* - \frac{2.3RT}{6F}pK_{S(Cm_2O_3)} + \frac{2.3RT}{2F}pO^{2-}$	$E^*(4) = -3.546$
5. $Cm^{3+} + O^{2-} \leftrightarrow CmO^+$	$pK_{eq} = -4.7455 + 5426/T$	$pK_{eq} = 4.2$
6. $Cm^{3+} + O^{2-} + Cl^- \leftrightarrow CmOCl$	$pK_s = 1.5132 + 3394/T$	$pK_s = 7.9$
7. $2Cm^{3+} + 3O^{2-} \leftrightarrow Cm_2O_3$	$pK_s = 0.779 + 10407.5/T$	$pK_s = 20.1$

Table 5. Equilibrium potentials and values of apparent standard potentials of redox system in NaCl-2CsCl eutectic at 829 K. $[Cm^{3+}] = 1$ mol·kg^{-1}. Potentials are given vs. Cl^-/Cl_2 reference electrode

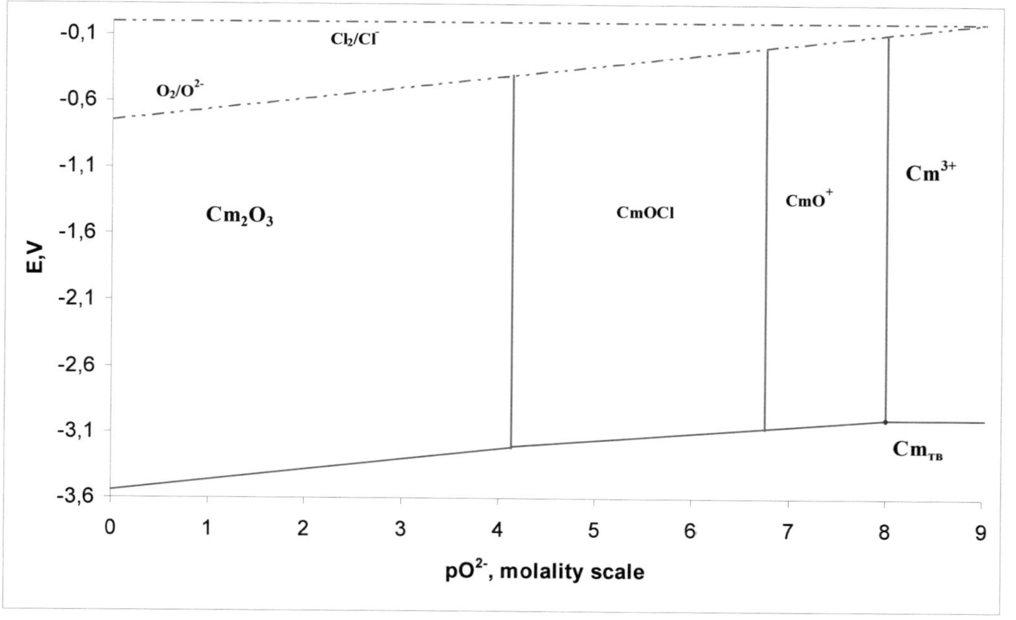

Fig. 4. Potential–pO^{2-} diagram for curium in equimolar NaCl-2CsCl at 829 K. $[Cm^{3+}]$ = 1 mol·kg^{-1}. Potentials are given vs. Cl^-/Cl_2 reference electrode

3.2 Transient electrochemical technique

3.2.1 Voltammetric studies on inert electrodes

The reaction mechanism of the soluble-insoluble Cm(III)/Cm(0) redox system was investigated by analyzing the cyclic voltammetric curves obtained at several scan rates, Fig. 5, 6. It shows that the cathodic peak potential (E_p) is constant from 0.04 V/s up to 0.1 V/s and independent of the potential sweep rate, Fig. 7. It means that at small scan rates the reaction Cm(III)/Cm(0) is reversible. In the range from 0.1 V/s up to 1.0 V/s the dependence is linear and shifts to the negative values with the increasing of the sweep rate. So in this case (scan range > 0.1 V/s) the reaction Cm(III)/Cm(0) is irreversible and controlled by the rate of the charge transfer. On the other hand the cathodic peak current (I_p) is directly proportional to the square root of the polarization rate (υ). According to the theory of the linear sweep voltammetry technique [Bard & Folkner, 1980] the redox system Cm(III)/Cm(0) is reversible and controlled by the rate of the mass transfer at small scan rates and is irreversible and controlled by the rate of the charge transfer at high scan rates.

The number of electrons of the reduction of Cm(III) ions for the reversible system was calculated at scan rates from 0.04 up to 0.1 V/s:

$$E_p - E_{p/2} = -0.77\frac{RT}{nF} \qquad (19)$$

where E_P is a peak potential (V), $E_{P/2}$ is a half-peak potential (V), F is the Faraday constant (96500 C·mol^{-1}), R is the ideal gas constant (J·K^{-1}·mol^{-1}) and T is the absolute temperature (K), n is the number of exchanged electrons. The results are 3.01±0.04.

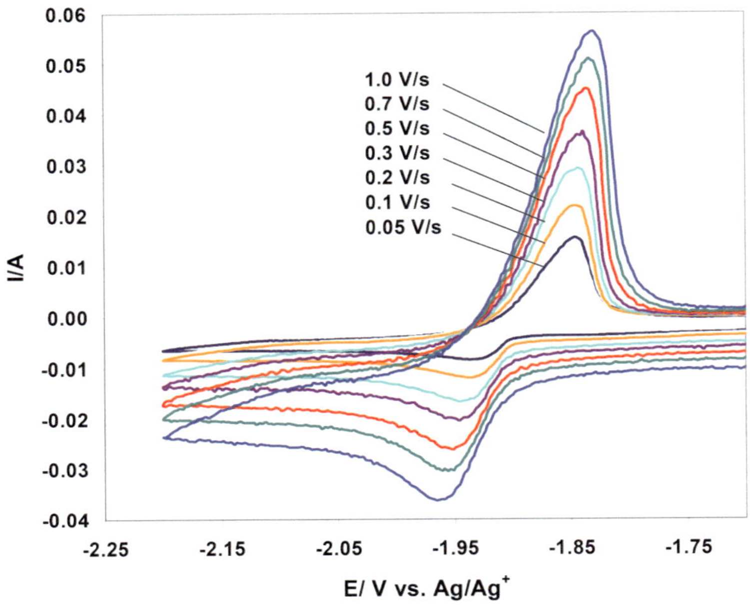

Fig. 5. Cyclic voltammograms of fused 2LiCl-3KCl-$CmCl_3$ salt at different sweep potential rates at 723 K. Working electrode: W (S = 0.36 cm^2). [Cm(III)] = $5.0 \cdot 10^{-2}$ $mol \cdot kg^{-1}$

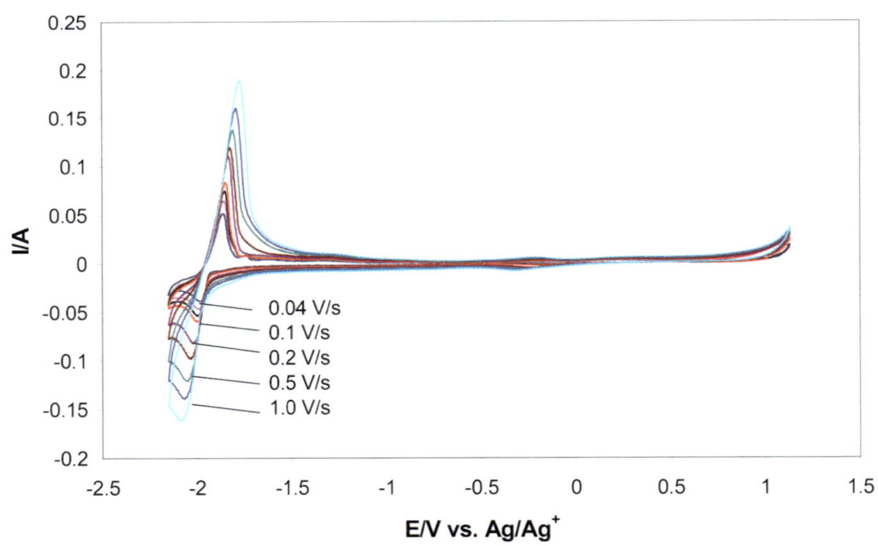

Fig. 6. Cyclic voltammograms of NaCl-2CsCl-$CmCl_3$ at different sweep potential rates at 823 K. Working electrode: W (S = 0.31 cm^2). [Cm(III)] = $4.4 \cdot 10^{-2}$ $mol \cdot kg^{-1}$

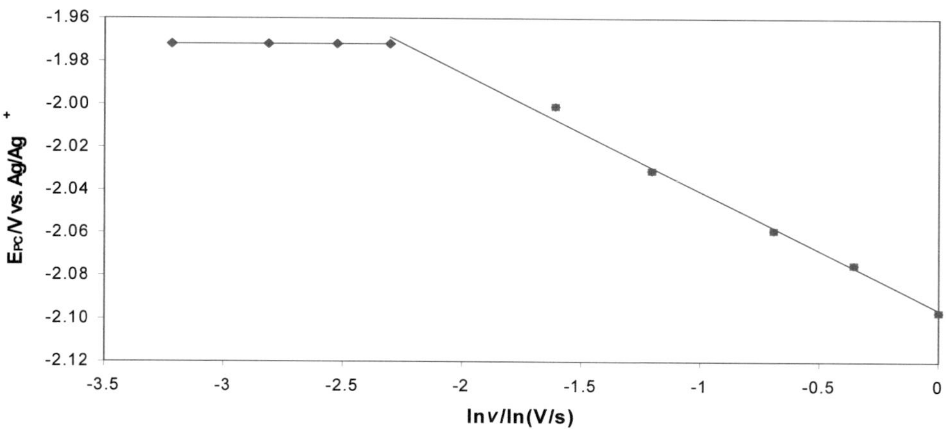

Fig. 7. Variation of the cathodic peak potential as a function Naperian logarithm of the sweep rate in fused NaCl-2CsCl-$CmCl_3$ at 823K. Working electrode: W (S = 0.59 cm^2). [Cm(III)] = $4.4 \cdot 10^{-2}$ mol·kg^{-1}

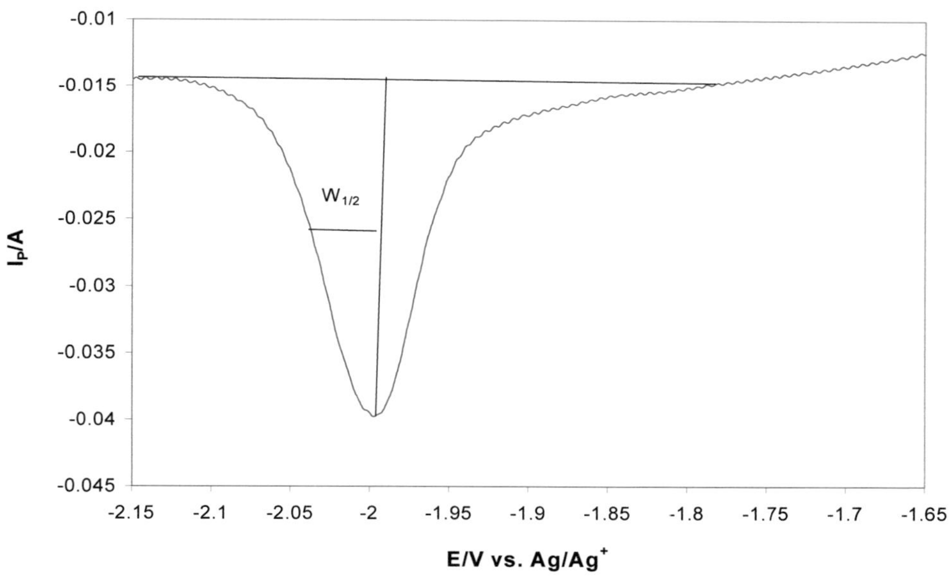

Fig. 8. Square wave voltammogram of NaCl-2CsCl-$CmCl_3$ at 25 Hz at 823 K. Working electrode: W (S = 0.29 cm^2). [Cm(III)] = $9.7 \cdot 10^{-3}$ mol·kg^{-1}

The square wave voltammetry technique was used also to determine the number of electrons exchanged in the reduction of Cm(III) ions in the molten eutectic NaCl-2CsCl. Fig. 8 shows the cathodic wave obtained at 823 K. The number of electrons exchanged is determined by measuring the width at half height of the reduction peak, $W_{1/2}$ (V), registered at different frequencies (6–80 Hz), using the following equation [Bard & Folkner, 1980]:

$$W_{1/2} = 3.52 \frac{RT}{nF} \tag{20}$$

where T is the temperature (in K), R is the ideal gas constant (in $J \cdot K^{-1} \cdot mol^{-1}$), n is the number of electrons exchanged and F is the Faraday constant (in $C \cdot mol^{-1}$).

At middle frequencies (12-30 Hz), a linear relationship between the cathodic peak current and the square root of the frequency was found. The number of electrons exchanged determined this way was close to three (n = 2.99±0.15).

The same results were found in the system $3LiCl\text{-}2KCl\text{-}CmCl_3$ [Osipenko, 2011].

On differentional pulse voltammogram only one peak was fixed at potential range from -1.5 up to -2.2 V vs. Ag/Ag^+ reference electrode, Fig. 9. It means that the curium ions reduction process at the electrode is a single step process.

Potentiostatic electrolysis at potentials of the cathodic peaks shows the formation of the solid phase on tungsten surface after polarization. One plateau on the dependence potential - time curves was obtained, Fig. 10.

So the mechanism of the cathodic reduction of curium (III) ions is the following:

$$Cm(III) + 3\bar{e} \Rightarrow Cm(0) \tag{21}$$

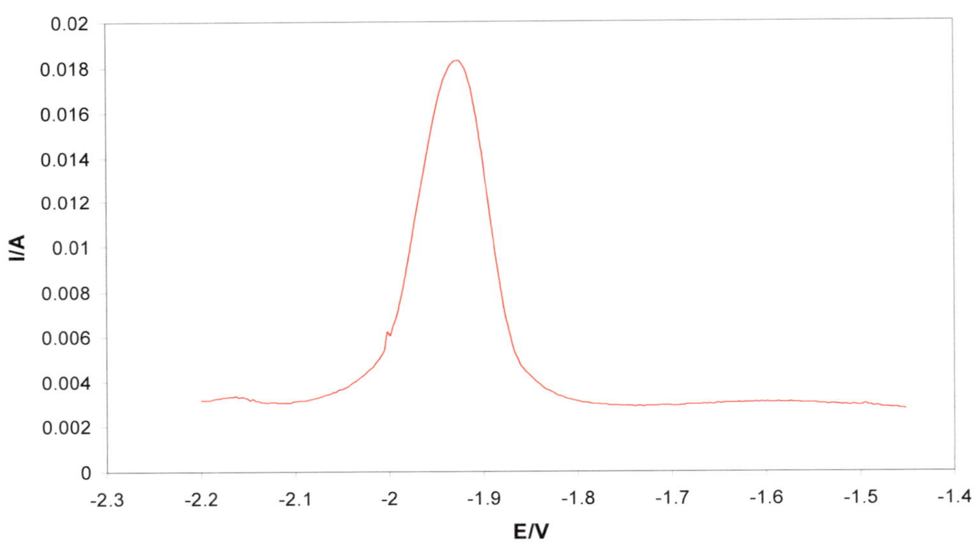

Fig. 9. Differential pulse voltammogram of $NaCl\text{-}2CsCl\text{-}CmCl_3$ melt at 923 K. [Cm(III)] = $4.4 \cdot 10^{-2}$ $mol \cdot kg^{-1}$

3.2.2 Diffusion coefficient of Cm (III) ions

The diffusion coefficient of Cm(III) ions in molten chloride media was determined using the cyclic voltammetry technique and applying Berzins–Delahay equation, valid for reversible soluble-insoluble system at the scan rates 0.04-0.1 V/s [Bard & Faulkner, 1980]:

$$I_p = 0.61(nF)^{3/2} C_0 S \left(\frac{Dv}{RT} \right)^{1/2} \tag{22}$$

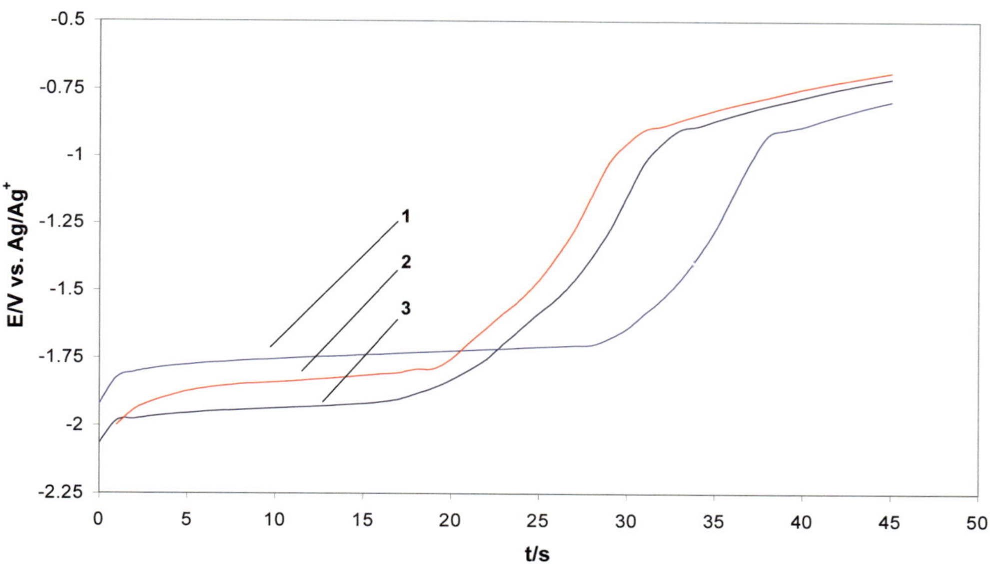

Fig. 10. The potential–time dependences after anodic polarization of W working electrode in NaCl-2CsCl-$CmCl_3$ melt at different temperatures. [Cm(III)] = $4.4 \cdot 10^{-2}$ mol·kg^{-1}. The value of polarization is equal -2.1 ÷ -2.2 V. The time of polarization is equal 5 ÷15 s. 1 – 1023 K; 2 – 923 K; 3 – 823K

where S is the electrode surface area (in cm^2), C_0 is the solute concentration (in mol·cm^{-3}), D is the diffusion coefficient (in cm^2·s^{-1}), F is the Faraday constant (in 96500 C·mol^{-1}), R is the ideal gas constant (in J·K^{-1}·mol^{-1}), n is the number of exchanged electrons, v is the potential sweep rate (in V/s) and T is the absolute temperature (in K).

The values obtained for the different molten chlorides tested at several temperatures are quoted in Table 6.

The diffusion coefficient values have been used to calculate the activation energy for the diffusion process. The influence of the temperature on the diffusion coefficient obeys the Arrhenius's law through the following equation:

$$D = D_o \exp\left(-\frac{E_A}{RT}\right) \pm \Delta \qquad (23)$$

Solvent	T/K	D/cm^2·s^{-1}	-E_A/kJ·mol^{-1}
LiCl-KCl	723	$9.27 \cdot 10^{-6}$	28.2
	823	$1.62 \cdot 10^{-5}$	
	923	$2.57 \cdot 10^{-5}$	
NaCl-2CsCl	873	$6.97 \cdot 10^{-6}$	44.5
	973	$1.33 \cdot 10^{-5}$	
	1023	$2.49 \cdot 10^{-5}$	

Table 6. Diffusion coefficient of Cm(III) ions in molten alkali metal chlorides at several temperatures. Activation energy for the curium ions diffusion process

where E_A is the activation energy for the diffusion process (in $kJ \cdot mol^{-1}$), D_o is the pre-exponential term (in $cm^2 \cdot s^{-1}$) and Δ is the experimental error.

From this expression, the value of the activation energy for the Cm(III) ions diffusion process was calculated in the different melts tested (Table 6).

The average value of the radius of molten mixtures $\left(r_{R^+}\right)$ was calculated by using the following equation [Lebedev, 1993]:

$$r_{R^+} = \sum_{i=1}^{N} c_i r_i \tag{24}$$

where c_i is the mole fraction of i cations; r_i is the radius of i cations in molten mixture, consist of N different alkali chlorides, nm.

The diffusion coefficient of curium (III) ions becomes smaller with the increase of the radius of the cation of alkali metal in the line from Li to Cs (Table 6). Such behaviour takes place due to an increasing on the strength of complex ions and the decrease in contribution of D to the "hopping" mechanism. The increase of temperature leads to the increase of the diffusion coefficients in all the solvents.

3.2.3 Apparent standard potentials of the redox couple Cm(III)/Cm(0)

The apparent standard potential of the redox couple Cm(III)/Cm(0) was determined at several temperatures. For the measurement, the technique of open-circuit chronopotentiometry of a solution containing a $CmCl_3$ was used (e.g. Fig. 10). A short cathodic polarisation was applied, 5-15 seconds, in order to form in situ a metallic deposit of Cm on the W electrode, and then the open circuit potential of the electrode was measured versus time (Fig. 10). The pseudo-equilibrium potential of the redox couple Cm(III)/Cm(0) was measured and the apparent standard potential, E^*, was determined using the Nernst equation:

$$E_{Cm(III)/Cm(0)} = E^*_{Cm(III)/Cm(0)} + \frac{RT}{nF} \ln X_{CmCl_3} \tag{25}$$

being,

$$E^*_{Cm(III)/Cm(0)} = E^\circ_{Cm(III)/Cm(0)} + \frac{RT}{nF} \ln \gamma_{CmCl_3} \tag{26}$$

The apparent standard potential is obtained in the mole fraction scale versus the Ag/AgCl ($0.75\ mol \cdot kg^{-1}$) reference electrode and then transformed into values of potential versus the Cl^-/Cl_2 reference electrode scale or direct versus Cl^-/Cl_2 reference electrode. For this purpose the special measurements were carried out for building the temperature dependence between Ag/AgCl ($0.75\ mol \cdot kg^{-1}$) and Cl^-/Cl_2 reference electrodes. From the experimental data obtained in this work the following empirical equation for the apparent standard potential of the Cm(III)/Cm(0) system versus the Cl^-/Cl_2 reference electrode was obtained using:

$$E^{*}_{Cm(III)/Cm(0)} = -(3.285 \pm 0.004) + (5.48 \pm 0.15) \cdot 10^{-4} T \pm 0.002 \ V \ \text{[3LiCl-2KCl, 723-923 K]} \quad (27)$$

$$E^{*}_{Cm(III)/Cm(0)} = -(3.750 \pm 0.006) + (9.98 \pm 0.16) \cdot 10^{-4} T \pm 0.003 \ V \ \text{[NaCl-KCl, 1023-1123 K]} \quad (28)$$

$$E^{*}_{Cm(III)/Cm(0)} = -(3.407 \pm 0.005) + (5.42 \pm 0.14) \times 10^{-4} T \pm 0.002 \ V \ \text{[NaCl-2CsCl, 823-1023 K]} \quad (29)$$

The relative stability of complex actinides ions increases with the increase of the solvent cation radius, and the apparent standard redox potential shifts to more negative values [Barbanel, 1985]. Our results are in a good agreement with the literature ones [Smirnov, 1973].

3.2.4 Thermodynamics properties

The apparent standard Gibbs energy of formation $\Delta G^{*}_{CmCl_3}$ was calculated according by the following expression:

$$\Delta G^{*}_{CmCl_3} = nFE^{*}_{Cm(III)/Cm(0)} \quad (30)$$

The least square fit of the standard Gibbs energy versus the temperature allowed us to determine the values of ΔH^* and ΔS^* more precisely by the following equation:

$$\Delta G^{*}_{CmCl_3} = \Delta H^{*}_{CmCl_3} - T\Delta S^{*}_{CmCl_3} \quad (31)$$

from which, values of enthalpy and entropy of formation can be obtained:

$$\Delta G^{*}_{CmCl_3} = -950.5 + 0.182 \cdot T \pm 0.6 \ \text{kJ mol}^{-1} \ \text{3LiCl-2KCl} \quad (32)$$

$$\Delta G^{*}_{CmCl_3} = -1085.3 + 0.312 \cdot T \pm 0.8 \ \text{kJ mol}^{-1} \ \text{NaCl-KCl} \quad (33)$$

$$\Delta G^{*}_{CmCl_3} = -986.4 + 0.174 \cdot T \pm 0.6 \ \text{kJ mol}^{-1} \ \text{NaCl-2CsCl} \quad (34)$$

The calculated values are summarized in Table 7. The average value of the radius of these molten mixtures in this line, *pro tanto*, is 0.094 nm for fused 3LiCl-2KCl eutectic; 0.1155 nm for fused equimolar NaCl-KCl and 0.143 nm for fused NaCl-2CsCl eutectic [Lebedev, 1993]. From the data given in Table 7 one can see that the relative stability of curium (III) complexes ions is naturally increased in the line $(3LiCl\text{-}2KCl)_{eut.}$ – $(NaCl\text{-}2CsCl)_{eut.}$.

Thermodynamic properties	3LiCl-2KCl	NaCl-KCl	NaCl-2CsCl
E^*/V	-2.752	-2.779	-2.880
$\Delta G^*/(\text{kJ mol}^{-1})$	-773.4	-781.7	817.1
$\Delta H^*/(\text{kJ mol}^{-1})$	-950.5	-1085.3	-986.4
$\Delta S^*/(\text{J K}^{-1}\text{ mol}^{-1})$	0.182	0.312	0.174

Table 7. The comparison of the base thermodynamic properties of Cm in molten alkali metal chlorides at 973 K. Apparent standard redox potentials are given in the molar fraction scale

The changes of the thermodynamic parameters of curium versus the radius of the solvent cation show the increasing in strength of the Cm-Cl bond in the complex ions $[CmCl_6]^{3-}$ in the line from LiCl to CsCl [Barbanel, 1985].

4. Conclusion

The electrochemical behaviour of $CmCl_3$ in molten alkali metal chlorides has been investigated using inert (*W*) electrode at the temperatures range 723-1123 K. Different behaviour was found for the reduction process. At low scan rates (< 0.1 V/s) Cm(III) ions are reversible reduced to metallic curium in a single step, but at scan rates (>0.1 V/s) this reaction is irreversible.

The diffusion coefficient of Cm(III) ions was determined at different temperatures by cyclic voltammetry. The diffusion coefficient showed temperature dependence according to the Arrhenius law. The activation energy for diffusion process was found.

Potentiostatic electrolysis showed the formation of curium deposits on inert electrodes.

The apparent standard potential and the Gibbs energy of formation of $CmCl_3$ have been measured using the chronopotentiometry at open circuit technique.

The influence of the nature of the solvent (ionic radius) on the thermodynamic properties of curium compound was assessed. It was found that the strength of the Cm–Cl bond increases in the line from Li to Cs cation.

The obtained fundamental data can be subsequently used for feasibility assessment of the curium recovery processes in molten chlorides.

5. Acknowledgement

This work was carried out with the financial support of ISTC project # 3261.

6. References

Barbanel, Ya.A. (1985). *Coordination Chemistry of f-elements in Melts*, Energoatomizdat, Moscow, Russia

Bard, A.J. & Faulkner, L.R. (1980). *Electrochemical Methods. Fundamentals and Applications*, John Wiley & Sons Inc., ISBN 0-471-05542-5, USA

Bermejo, M.R.; de la Rosa, F.; Barrado, E. & Castrillejo, Y. (2007). Cathodic behaviour of europium(III) on glassy carbon, electrochemical formation of Al_4Eu, and oxoacidity reactions in the eutectic LiCl-KCl, In: *Journal of Electroanalitical Chemistry*, Vol. 603, No. 1, (May 2007), pp. 81- 95, ISSN 0022-0728

Bermejo, M.R.; Barrado, E.; Martinez, A.M. & Castrillejo, Y. (2008). Electrodeposition of Lu on W and Al electrodes: Electrochemical formation of Lu-Al alloys and oxoacidity reactions of Lu(III) in eutectic LiCl-KCl, In: *Journal of Electroanalitical Chemistry*, Vol. 617, No. 1, (June 2008), pp. 85- 100, ISSN 0022-0728

Castrillejo, Y.; Bermejo, M.R.; Diaz Arocas, P.; Martinez, A.M. & Barrado, E. (2005). The electrochemical behavior of praseodymium(III) in molten chlorides, In: *Journal of Electroanalitical Chemistry*, Vol. 575, No. 1, (January 1995), pp. 61- 74, ISSN 0022-0728

Castrillejo, Y.; Bermejo, M.R.; Barrado, A.I.; Pardo, R.; Barrado, E. & Martinez, A.M. (2005). Electrochemical behavior of dysprosium in the eutectic LiCl-KCl at W and Al electrodes, In: *Electrochimica Acta,* Vol. 50, No. 10, (March 2005), pp. 2047- 2057, ISSN 0013-4686

Castrillejo, Y.; Fernandes, P.; Bermejo, M.R.; Barrado, A.I. & Martinez, A.M. (2009). Electrochemistry of thulium on inert electrodes and electrochemical formation of a Tm-Al alloy from molten chlorides, In: *Electrochimica Acta,* Vol. 54, No. 26, (November 2009), pp. 6212-6222, ISSN 0013-4686

Cherginetz, V.L. (2004) *Chemistry of Oxocompounds in Ionic Melts,* Institute of Monocrystiles, Kharkov, Ukraina, ISBN 966-02-3244-6

De Cordoba, G. & Caravaca, C. (2004). An electrochemical study of samarium ions in the molten eutectic LiCl+KCl, In: *Journal of Electroanalitical Chemistry,* Vol. 572, No. 1, (October 2004), pp. 145-151, ISSN 0022-0728

De Cordoba, G.; Laplace, A.; Conocar, O.; Lacquement, G. & Caravaca, C. (2008). Determination of the activity coefficients of neodymium in liquid aluminum by potentiometric methords, In: *Electrochimica Acta,* Vol. 54, No. 2, (December 2008), pp. 280-288, ISSN 0013-4686

Korshunov, B.G.; Safonov, V.V. & Drobot, D.V. (1979). *Phase equilibriums in halide systems,* Metallurgiya, Moscow, USSR

Kuznetsov, S.A.; Hayashi, H.; Minato, K. & Gaune-Escard, M. (2006). Electrochemical transient techniques for determination of uranium and rare-earth metal separation coefficients in molten salts, In: *Electrochimica Acta,* Vol. 51, No. 12 (February 2006), pp. 2463-2470, ISSN 0013-4686

Lebedev, V.A. (1993). *Selectivity of Liquid Metal Electrodes in Molten Halide,* Metallurgiya, ISBN 5-229-00962-4, Russia

Novoselova, A. & Smolenski, V. (2010). Thermodynamic properties of thulium and ytterbium in molten caesium chloride, In: *Journal of Chemical Thermodynamics,* Vol. 42, No. 8, (August 2010), pp. 973-977, ISSN 0021-9614

Novoselova, A. & Smolenski, V. (2011). Thermodynamic properties of thulium and ytterbium in fused NaCl-KCl-CsCl eutectic, In: *Journal of Chemical Thermodynamics,* Vol. 43, No. 7, (July 2011), pp. 1063-1067, ISSN 0021-9614

Osipenko, A.; Maershin, A.; Smolenski, V.; Novoselova, A.; Kormilitsyn, M. & Bychkov, A. (2010). Electrochemistry of oxygen-free curium compounds in fused NaCl-2CsCl eutectic, In: *Journal of Nuclear Materials,* Vol. 396, No. 1, (January 2010), pp. 102-1067, ISSN 0022-3115

Osipenko, A.; Maershin, A.; Smolenski, V.; Novoselova, A.; Kormilitsyn, M. & Bychkov, A. (2011). Electrochemical behaviour of curium (III) ions in fused 3LiCl-2KCl eutectic, In: *Journal of Electroanalitical Chemistry,* Vol. 651, No. 1, (January 2011), pp. 67-71, ISSN 0022-0728

Roy, J.J.; Grantham, L.F.; Grimmett, D.L.; Fusselman, S.P.; Krueger, C.L.; Storvick, T.S.; Inoue, T.; Sakamura, Y. & Takahashi, N. (1996). Thermodynamic properties of U, Np, Pu, and Am in molten LiCl-KCl eutectic and liquid cadmium, In: *Journal of The Electrochemical Society,* Vol. 143, No. 8, (August 1996), pp. 2487-2492, ISSN 0013-4651

Sakamura, Y.; Hijikata, T.; Kinoshita, K.; Inoue, T.; Storvick, T.S.; Krueger, C.L.; Roy, J.J.; Grimmett, D.L.; Fusselman, S.P. & Gay, R.L. (1998). Measurement of standard potentials of actinides (U, Np, Pu, Am) in LiCl-KCl eutectic salt and separation of actinides from rare earths by electrorefining, In: *Journal of Alloys and Compounds*, Vol. 271-273, (June 1998), pp. 592-596, ISSN 0925-8388

Serp, J.; Konings, R.J.M.; Malmbeck, R.; Rebizant, J.; Scheppler, C. & Glatz, J-P. (2004). Electrochemical of plutonium ion in LiCl-KCl eutectic melts, In: *Journal of Electroanalitical Chemistry*, Vol. 561, (January 2004), pp. 143-148, ISSN 0022-0728

Serp, J.; Allibert, M.; Terrier, A.L.; Malmbeck, R.; Ougier, M.; Rebizant, J. & Glatz, J-P. (2005). Electroseparation of actinides from lanthanides on solid aluminum electrode in LiCl-KCl eutectic melts, In: *Journal of The Electrochemical Society*, Vol. 152, No. 3, (March 2005), pp. C167-C172, ISSN 0013-4651

Serp, J.; Lefebvre, P.; Malmbeck, R.; Rebizant, J.; Vallet, P. & Glatz, J-P. (2005). Separation of plutonium from lanthanum by electrolysis in LiCl-KCl onto molten bismuth electrode, In: *Journal of Nuclear Materials*, Vol. 340, No. 2-3, (April 2005), pp. 266-270, ISSN 0022-3115

Serp, J.; Chamelot, P.; Fourcaudot, S.; Konings, R.J.M.; Malmbeck, R.; Pernel, C.; Poignet, J.C.; Rebizant, J. & Glatz, J-P. (2006). Electrochemical behavior of americium ions in LiCl-KCl eutectic melt, In: *Electrochimica Acta*, Vol. 51, No. 19, (May 2006), pp. 4024-4032, ISSN 0013-4686

Serrano, K. & Taxil, P. (1999). Electrochemical nucleation of uranium in molten chlorides, In: *Journal of Applied Electrochemistry*, Vol. 29, No. 4, (April 1999), pp. 505-510, ISSN 0021-891X

Shishkin, V.Yu. & Mityaev, V.S. (1982). Purification of alkali chloride metals by direct crystallization. In: Proceedings of the Academy of Sciences. *Journal of Inorganic materials*, Vol. 18, No. 11 (November 1982), pp. 1917-1918, ISSN 0002-337X

Shirai, O.; Iizuka, M.; Iwai, T.; Suzuki, Y. & Arai, Y. (2000). Electrode reaction of plutonium at liquid cadmium in LiCl-KCl eutectic melts, In: *Journal of Electroanalitical Chemistry*, Vol. 490, No. 1-2, (August 2000), pp. 31-36, ISSN 0022-0728

Smirnov, M.V. (1973). *Electrode Potentials in Molten Chlorides*, Nauka, Moscow, USSR

Smolenski, V.; Novoselova, A.; Osipenko, A.; Caravaca, C. & de Cordoda, G. (2008). Electrochemistry of ytterbium(III) in molten alkali chlorides, In: *Electrochimica Acta*, Vol. 54, No. 2, (December 2008), pp. 382-387, ISSN 0013-4686

Smolenski, V.; Novoselova, A.; Osipenko, A. & Kormilitsyn, M. (2009). The influence of electrode material nature on the mechanism of cathodic reduction of ytterbium (III) ions in fused NaCl-KCl-CsCl eutectic, In: *Journal of Electroanalitical Chemistry*, Vol. 633, No. 2, (August 2009), pp. 291-296, ISSN 0022-0728

Uozumi, K.; Iizuka, M.; Kato, T.; Inoue, T.; Shirai, O.; Iwai, T. & Arai, Y. (2004). Electrochemical behaviors of uranium and plutonium at simultaneous recoveries into liquid cadmium cathodes, In: *Journal of Nuclear Materials*, Vol. 325, No. 1, (February 2004), pp. 34-43, ISSN 0022-3115

Willit J. (2005). 7th International Symposium on Molten Salts Chemistry & Technology, *Proceeding of Overview and Status of Pyroprocessing Development at Argonne National Laboratory*, Toulouse, France, August 2005

Yamana, H.; Fujii, T. & Shirai, O. (2003). UV/Vis Adsorption Spectrophotometry of some f-elements in Chloride Melt, *Proceeding of International Symposium on Ionic Liquids on Honor of Marcelle Gaune-Escard*, Carry le Rouet, France, June 2003

2

Application of the Negative Binomial/Pascal Distribution in Probability Theory to Electrochemical Processes

Thomas Z. Fahidy
Department of Chemical Engineering, University of Waterloo, Canada

1. Introduction

The subject of interest here is a set of random observations falling into only two categories: success and failure, their single-event probability being constant and independent of any previous occurrence during a sequence of such events. The probability distribution of their number, known generally as Bernouilli trials, is called the negative binomial distribution (NBD) or the Pascal distribution. In this context, success and failure are completely relative concepts, their physical meaning defined by the experimenter or analyst: the appearance of a substandard product, for instance, among acceptable ones may well be considered success by a quality controller whose objective is the identification of substandards. This distribution serves in general for finding the probability of an exact number, or at least, or at most a certain number of failures observed upon so many successes (or vice versa, depending on the chosen definition of success and failure). Its special form, related to the appearance of the *first* success (or failure) is called the geometric distribution.

Although NBD theory has widely been employed in various technical/technological areas, its utility for the analysis of electrochemical phenomena and electron-transfer processes has not yet been demonstrated to the author's knowledge. The purpose of this chapter, in consequence, is to supply such a demonstration via five specific illustrative examples as a means of stimulating further interest in probabilistic methods among electrochemical scientists and engineers.

2. Brief theory

2.1 Basic concepts and definitions

Let N be the random variable denoting the number of failures occurring in successive Bernouilli trials (Appendix A) prior to the occurrence of K successes (also random). Define $M \equiv N + K$. Then, the probability mass function (pmf) of the NBD, defined (e.g., Doherty, 1990) as

$$\text{pmf (NBD)} = P[N = n; K = k] = C(n+k-1;k-1)p^k q^n \quad 0 < p < 1; n = 0, 1, 2,\ldots; q = 1 - p \qquad (1)$$

or, alternatively [e.g., Walpole et al., 2002] as

$$\text{pmf(NBD)} = P\{M = m; K = k] = C(m-1; k-1)p^k q^{m-k} \quad 0 < p < 1; m = k, k+1, k+2,\ldots; q = 1 - p \qquad (2)$$

yields the probability that the number of independent Bernouilli trials required to achieve n number of failures prior to k number of successes is m = n + k. The cumulative mass function cmf (Weisstein, n.d.)

$$\text{cmf (NBD)} = P[N \le n'; K = k] = p^k\{C(k-1;k-1) + C(k;k-1)q + C(k+1;k-1)q^2 + \ldots + C(n'+k-1;k-1)q^{n'}] = 1 - I_q(n'+1;k) \quad (3)$$

which yields the probability of achieving up to n' (but not more than n') failures prior to achieving k number of successes, is readily computable in terms of the incomplete beta function

$$I_q(n' + 1;k) = \Gamma(n' + k + 1)/[\Gamma(n' + 1)\Gamma(k)]\ \Psi_q(k;n') \quad (4)$$

where

$$\Psi_q(k;n') \equiv \int_0^q [(u^{n'}(1 - u)^{k-1}]\, du$$

requires, in general, numerical integration. Selected values of the Ψ – function are given in Table 1; the k = 5; n' = 0 entry demonstrates insensitivity to small values of success probability, since the value of $\Psi_q(5;0) = (1-p)^5/5$ is essentially 0.2.

Eq.(4) is particularly useful in the case of n'/small k configuration (Appendix B).

k	n'	q = 1 – p			
		0.85	0.90	0.95	0.99
1	2	0.2047	0.2430	0.2858	0.3234
	3	0.1305	0.1640	0.2036	0.2401
	4	0.0887	0.1181	0.1548	0.1902
2	2	0.0742	0.0790	0.0822	0.0833
	3	0.0418	0.0459	0.0489	0.0499
	4	0.0259	0.0259	0.0322	0.0333
3	2	0.0325	0.0330	0.0333	0.0333
	3	0.0159	0.0164	0.0166	0.0167
	4	$8.82\text{x}10^{-3}$	$9.28\text{x}10^{-3}$	$9.49\text{x}10^{-3}$	$9.52\text{x}10^{-3}$
4	2	0.0166	0.0167	0.0167	0.0167
	3	$7.06\text{x}10^{-3}$	$7.12\text{x}10^{-3}$	$7.14\text{x}10^{-3}$	$7.14\text{x}10^{-3}$
	4	$3.49\text{x}10^{-3}$	$3.55\text{x}10^{-3}$	$3.57\text{x}10^{-3}$	$3.57\text{x}10^{-3}$
5	0	0.19998	0.199998	0.1999999	0.2
	1	0.03332	0.03333	0.03333	0.2
	2	$9.50\text{x}10^{-3}$	$9.52\text{x}10^{-3}$	$9.52\text{x}10^{-3}$	0.2

Table 1. A short tabulation of selected values of the Ψ – function in Eq.(4) at small success single-event probabilities and small success numbers

2.2 Important parameters of the NBD

Table 2 contains parameters related to the first four statistical moments, with original notations adjusted to comply with the notation scheme in Doherty (1990) followed in this chapter. The mode (the value of the most frequently occurring random variable) does not exist when k =1 (the case of geometric distribution) and k = 0 (meaningless in NBD context).

Parameter	Expression	Reference
Mean (or expectation)	qk/p	Doherty, 1990
Variance	qk/p^2	Doherty, 1990
Skewness	$(1 + q)/\sqrt{kp}$	Evans, et al., 2000
Kurtosis	$3 + 6/k + p^2/kq$	Evans, et al., 2000
Mode	$q(k - 1)/p; k > 1$	Weisstein, n.d.

Table 2. Fundamental parameters of the negative binomial distribution (NBD)

2.3 The geometric distribution GD: a special case of NBD

Since C(n + k -1; k – 1) reduces to C(n; 0) = 1 when k = 1, the simplified form of Eq.(1) yields the pmf of the geometric distribution as pq: this is the probability of the first success occurring at the n - th Bernouilli trial, (i.e., the probability of (n-1) unsuccessful ("failed") trials prior to the first success on the n-th trial) with mean q/p, variance q/p^2, skewness $(1 + q)/\sqrt{q}$, and kurtosis $9 + p^2/q$. Similarly, from Eq.(2), when k = 1, C(m – 1; 0) = 1 and it follows that

$$\text{pmf (GD)} = P[N = n ; K =1] = pq^{m-1} = pq^n \tag{5}$$

and

$$\text{cmf (GD)} = P [N \le n' ; K = 1] = p (1 + q + q^2 +\ldots+q^{n'}) = 1 - q^{n'+1} \tag{6}$$

2.4 Complementary probabilities

The complement of the cumulative mass function determines the probability that at least (n' + 1) or more failures would occur prior to the appearance of the last success, namely

$$\text{NBD: } 1 - p^k[C(k + 1; k -1) - C(k + 2; k - 1)q - C(k + 3; k - 1)q^2 - \ldots - C(n' + k + 1;q^{n'}] = I_q (n' + 1; k) \tag{7}$$

$$\text{GD: } 1 - p(1 + q + q^2 +\ldots+q^{n'+1}) = q^{n'+1} \tag{8}$$

2.5 Estimation of the single-event success probability from experimental observations

If p is not known a-priori, Evans et al. (2000) recommend the unbiased-method estimator

$$p^*(UB) = (k_0 - 1)/(n_0 + k_0 - 1) \tag{9}$$

and the maximum likelihood-method estimator

$$p^*(ML) = k_0 / (n_0 + k_0) \tag{10}$$

based on available experimental data. The zero subscript refers to anteriority, and the asterisk indicates that Eqs.(9) and (10) are estimators of the unknown ("true") population parameters. When n_0 and k_0 are small, discrepancy between the two estimates can be very large. Conversely, if $k_0 >> 1$, the two estimates are essentially equal.

3. Application to selected electrochemical processes

3.1 The electrolytic reduction of acrylonitrile (ACN) to adiponitrile (ADN)

In one of the major electroorganic technologies the 82-90% selectivity of ADN production was ascribed primarily to efficient control of major byproducts: propionitrile and a $C_9H_{11}N_3$-trimer, as well as to the non-electrochemical formation of biscyanoethylether (Danly & Campbell, 1982). It is assumed here that in a hypothetical pilot - plant scale operation of a modified ADN process, batches produced within a set time period will be tested for ADN selectivity S immediately upon production. Denoting the random number of batches exhibiting S > 85% as N, and the single-event probability of S ≤ 85% as p, the probability that exactly N = n number of batches will exhibit S > 85% while k number of batches will exhibit the opposite in a sequential sampling is given by Eq.(1). Similarly, the probability that not more than N = n' batches will exhibit S > 85% is given by Eq.(3). The finding of a "substandard" batch is considered to be success in this context, since the elimination of such batches would be the ultimate goal. Table 3 contains selected numbers of exact failure-probabilities, and cumulative probabilities that no more than two failures will be observed at the indicated values of k and p.

K	p	N					P[N ≤ 2]
		0	1	2	5	10	
1 (*)	0.2	0.20	0.160	0.128	0.0655	0.0215	0.488
	0.5	0.50	0.250	0.125	0.0156	0.0005	0.875
	0.8	0.80	0.160	0.032	0.0003	1.9×10^{-8}	0.992
2	0.2	0.24	0.032	0.077	0.0786	0.0473	0.149
	0.5	0.25	0.125	0.188	0.0469	0.0027	0.563
	0.8	0.64	0.128	0.077	0.0012	7.2×10^{-7}	0.845
3	0.2	0.008	0.006	0.037	0.0551	0.0137	0.045
	0.5	0.125	0.063	0.187	0.0820	0.0019	0.038
	0.8	0.512	0.102	0.123	0.0034	2.0×10^{-7}	0.733
5	0.2	3.2×10^{-4}	0.003	0.003	0.0132	0.0344	0.035
	0.5	0.031	0.156	0.117	0.1231	0.0306	0.164
	0.8	0.328	0.066	0.197	0.0132	3.4×10^{-5}	0.590

Table 3. Failure/success probabilities in Section 3.1 at selected values of the single-event success probability p, success number k (an exact value of the random variable K of the number of substandard batches), and failure number n (an exact value of the random variable N of batches of acceptable quality). (*): geometric distribution

3.2 A nickel-iron alloy plating process

A novel NiFe alloy plating process is envisaged to have been carried out in several sequential experiments adhering to a tightly controlled Ni^{2+}/Fe^{2+} ionic ratio in the cell electrolyte kept within a narrow experimental temperature range. Defining alloy deposits of poor quality as a success (by the same reasoning as in Section 3.1), the experiments are assumed to indicate eight failures prior to three successes. In the absence of any knowledge of single-event success probabilities, the latter can be estimated to be p^* (UB) = (3-1)/(8+3-1) = 0.2 [Eq.(9)] and p^* (ML) = 3/(8+3) = 0.2727 [Eq.(10)]. Table 4 contains selected values of individual and cumulative probabilities for this process.

n	$P^*_{UB}[N = n]$	$P^*_{ML}[N = n]$	$\sum_r (P^*_{UB})$	$\sum_n (P^*_{ML})$
0	0.0080	0.0203	0.0080	0.0203
1	0.0192	0.0442	0.0272	0.0645
2	0.0307	0.0644	0.0579	0.1289
3	0.0410	0.0780	0.0989	0.2069
4	0.0492	0.0851	0.1480	0.2920
5	0.0551	0.0867	0.2031	0.3787
6	0.0587	0.0840	0.2618	0.4627
7	0.0604	0.0786	0.2922	0.5413
8	0.0604	0.0714	0.3526	0.6128
9	0.0590	0.0635	0.4116	0.6727
10	0.0567	0.0554	0.4683	0.7317
Mean	12		8	
Variance	60		29.34	
Mode	8		5	

Table 4. Failure/success probabilities in Section 3.2 at unbiased and maximum-likelihood estimator values of the single-event success probability based on earlier observations $n_0 = 8$; $k_0 = 3$. N is the random number of good quality alloy specimens in the presence of k = 3 alloy specimens of poor quality.
$P^*_{UB} = C(n+2;2)(0.2)^3(0.8)^n$; $P^*_{ML} = C(n+2;2)(0.2727)^3(0.7273)^n$

3.3 An electrolytic nanotechnological-size cadmium plating process with tagged Cd^{2+} ions

In a hypothetical study of its mechanism, a cadmium plating process is assumed to proceed until a monolayer of about 100 discharged Cd^{2+} ions (ionic radius = 0.097 nm; Dean, 1985) has fully been formed on an approximately 3 $(nm)^2$ deposition area by a 300 nA current pulse of 10 ms duration (corresponding to a current density of about 100 mA/cm^2 in conventional plating technology). Ten percent of the ions in the electrolyte are tagged (e.g., radioactively) for monitoring purposes. Their arrival to the surface with respect to untagged ions may be considered a sequence of Bernouilli trials with probability mass function

$$P[N = m\,;\, K = k] = C(n + k - 1;\, k - 1)0.1^k\, 0.9^n \quad (11)$$

and cumulative mass function

$$\begin{aligned} P[N \le n'\,;\, K = k] &= (0.1)^k[C(k\text{-}1;k\text{-}1) + C(k;k\text{-}1)(0.9) + C(k+1;k\text{-}1)(0.9)^2 + \ldots + C(n' + k - 1;k\text{-}1)(0.9)^{n'} \\ &= 1 - [\Gamma\,(n' + k + 1)/\{\,\Gamma(n' + 1)\,\Gamma(k)]\,\Psi_{0.9}\,(k;n') \\ &= 1 - (n' + k)!/[(n'!)(k!)]\,\Psi_{0.9}\,(k;n') \end{aligned} \quad (12)$$

in view of the fundamental relationship between the gamma function and factorials: $\Gamma\,(x + 1) = x!$

Selected individual and cumulative probabilities pertaining to the arrival of the first five tagged ions to the electrode surface are shown in Table 5.

N	P[N=n] [Eq.(11)]	P[N ≤ n] [Eq.(12)]
0	10^{-5}	10^{-5}
1	4.5×10^{-5}	5.5×10^{-5}
5	7.4×10^{-4}	0.0016
10	0.0035	0.0127
20	0.0129	0.0980
30	0.0196	0.2693
Mode = 36	0.0206	0.3916
40	0.0201	0.4729
Mean = 45	0.0185	0.5688
50	0.0163	0.6548
60	0.0114	0.7909
95	0.0017	0.9998
Variance	450	

Table 5. Failure/success probabilities and principal distribution parameters in Section 3.3, concerning the arrival of the first five tagged Cd^{2+} ions to the approximately 3 $(nm)^2$ deposition area. N is the random variable denoting the number of untagged ions that have arrived at the surface along with the arrival of the K = 5 tagged ions.

3.4 An aluminum anodizing process

In a typical conventional anodizing process (Pletcher & Walsh, 1990) using a 50 g/dm^3 chromium acid electrolyte, high corrosion-resistance opaque grayish white (possibly enamel) films are produced at an electrolyte temperature of 50 ^{0}C. To maintain a deposition rate of 13 - 20 μm/h the cell potential is gradually raised from zero to 30 V while the current density reaches an asymptote within the 10 - 15 mA/cm^2 range.

In a hypothetical research project with the objective of maintaining higher deposition rates, modified chromic acid concentration, and lower temperatures (i.e. lower energy input), a tolerance domain consisting of a combined range of acceptable operating variables in each bath (cell) is to be established, in order to meet new acceptable performance criteria. Considering bath - to - bath performance levels as Bernouilli trials, q is defined as the fraction of acceptable and p = 1 - q as the fraction of unacceptable quality (i.e. "success" in NBD parlance). Consequently, if the random variable N is the number of acceptable baths obtained prior to finding K unacceptable baths, the probability of finding exactly N = n acceptable baths on the k-th unacceptable bath is given by Eq.(1) rewritten as

$$P[N = n; K = k] = (n + k - 1)\,!/\ [(k - 1)!\ n!]p^k q^n \quad (13)$$

and specifically, when k = 1,

$$P[N = n; K = 1] = n!/[0!(n-0)!]pq^n = pq^n \quad (14)$$

Selected probability values are displayed in Table 6.

N	K	p/q		
		0.2/0.8	0.5/0.5	0.8/0.2
5	1 3	0.0656 0.0551	0.0156 0.0820	2.6×10^{-4} 3.4×10^{-3}
10	1 3	0.0214 0.0567	4.9×10^{-4} 0.0081	8.2×10^{-8} 3.5×10^{-6}
15	1 3	0.0070 0.0383	1.5×10^{-5} 5.2×10^{-4}	2.6×10^{-11} 2.3×10^{-9}
Mean	1 3	4 12	1 3	1/4 3/4
Variance	1 3	20 60	1/5 3/5	0.3125 0.9375
Mode	1 3	- 8	- 2	- 1/2

Table 6. Failure/success probabilities and principal distribution parameters in Section 3.4, related to the arrival of anodizing baths of unacceptable quality. N is the number of acceptable baths found at the k-th arrival.

3.5 Non-conducting oxide layer formation on Ti-MnO_2 and Ti-RuO_2 anodes

The possibility of non-conducting oxide layers forming on untreated titanium - manganese oxide and titanium –ruthenium oxide substrate has been known to be a source of failure for dimensionally stable anodes (DSA) over several decades (Smyth, 1966). In this illustration a recently developed experimental DSA is supposed to carry a prohibiting additive embedded in the conducting oxide matrix in order to reduce the presence of nonconductors. The reliability of the additive is estimated to be 98%, i.e. 2% of the DSA are believed to be susceptible to failure. In two independent tests, complying with Bernouilli trial conditions, twelve anode samples were taken randomly for each test from a large ensemble of anodes, finding four (Test 1) and five (Test 2) defective specimens. The defective ratios 4/12 = 0.333 and 5/12 = 0.417 might imply a prima facie rejection of the 2%-defectives-at-most claim by the anode manufacturer, as a questionable ("primitive") means of judgment. A careful NBD - based approach employing cumulative distributions (to account for all possible, not just the actually observed outcome) yields probabilities (via Eq.(2) and Eq.(4), respectively)

$$\text{Test 1: } P[\,N \le 8; K = 4] = 0.02^4\{C(3;3) + C(4;3)(0.98) + C(5;3)(0.98^2) + \ldots + C(11;3)(0.98^8)\}$$
$$= 1 - (1980)(5.05\text{x}10^{-4}) = 6.963 \times 10^{-5} \qquad (15)$$

and

$$\text{Test 2: } P[\,N \le 7; K = 5] = 0.02^3\{C(4;4) + C(5;4)(0.98) + C(6;4)(0.98^2) + \ldots + C(11;4)(0.98^7)\}$$
$$= 1 - (3960)(2.525 \times 10^{-4}) = 2.250 \times 10^{-6} \qquad (16)$$

which demonstrate, at a negligible numerical difference, an extremely low likelihood of finding up to eight and up to seven anode specimens, respectively, due to random effects. Hence, the 2% claim appears to be highly questionable.

4. Analysis and discussion

4.1 Computation of intermediate probabilities

The probability of a negative binomial variable N occurring (due to random causes) from a value $N = n_1$ to value $N = n_2$ can be expressed in the simplest form as

$$P[\, n_1 \le N \le n_2;\, K = k\,] = I_q(n_1\,;\, k) - I_q(n_2 + 1;k) \tag{17}$$

Under specific conditions, tabulations of the incomplete beta function (e.g. Beyer, 1968) may be employed for a quick estimation of probabilities (Appendix C).

4.2 Analysis of the contents of Sections 3.1 – 3.5

The primary role of Tables 2 - 6 resides in reaching a decision whether or not experimental observations indicate the presence of non - random effects causing the observed results. For the ADN - process in Section 3.1, the entries in Table 3 indicate that if, for instance, ten batches with $S > 85\%$ selectivity were found along with 1 - 5 batches with $S \le 85\%$, this finding would suggest a rather strong promise for the new process, inasmuch as there would be at most an about 0.5% chance for such a result arising from random reasons (at least in the $1 \le k \le 5$; $0.2 \le p \le 0.8$ ranges). Conversely, if e.g. both selectivity ranges were equally probable ($p = q = 0.5$), the finding of exactly two batches with $S > 85\%$ selectivity would be at most slightly promising for the new process, since an about 12 - 19% probability exists for random causes. However, if there were, for instance, at least five $S > 85\%$ observations, the process would be judged highly promising.

In a somewhat different manner, the modified process could also be deemed to be acceptable, if the coefficient of variation CV related to the GD describing the appearance of the first $S \le 85\%$ batch, one of the absolute measures of dispersion, and defined as the ratio of the standard deviation to the mean of the distribution, differed from unity only within a pre-specified (small) fraction. From Table 2, $CV = 1/\sqrt{q}$ when $k = 1$, and it follows that as p becomes progressively smaller, both q and CV approximate unity, the latter from values above. Stipulation of a not more than 10% downward difference from unity (i.e. $CV \le 1.1$) would specify $p \approx 0.2$, as the highest acceptable single - event probability of $S \le 85\%$ selectivity; put otherwise, the appearance of the first $S \le 85\%$ batch after four $S > 85\%$ batches would imply promise for the new process. If the stipulation were a more stringent 1% downward difference from unity, $CV \le 0.1$ would prescribe $p \approx 0.02$, i.e. an only 2% single event probability of $S \le 85\%$ selectivity.

As shown in Section 3.2, when single-event probabilities are not available from prior sources, their numerical values are highly sensitive to the method of parameter estimation. The largest individual probability: $p^*_{ML} = 0.0867$ occurring at mode = 5 signals at most an approximately 9% maximum chance for randomness-related observation of good quality deposits along with the observation of three low - quality deposits. It follows that the novel process can be considered effective. Cumulative probabilities yield the same qualitative result. If $p^*_{UB} = 0.2$ is accepted for the single-event success (i.e. poor deposit quality), experimental observations of failures (i.e. good deposit quality) with three successes would suggest process reliability in face of an about 6% chance of random occurrence. By contrast, randomness-related probability of up to ten failures along with three successes being almost 50%, the effectiveness of the new process would appear to be rather dubious, if such occurrences were observed experimentally. If $p^*_{ML} \approx 0.27$ is accepted, the almost two-thirds probability of finding up to ten failures due to random effects essentially rules out the new process as a viable alternative to the existing one. Since these estimators come from a single experiment, the conclusions should be accepted only provisionally until further experiments will have established a wider data base.

It is instructive to consider situations with a relatively high number of experimentally observed successes. Assuming, for the sake of argument, $n_0 = 3$ and $k_0 = 10$, the single event success probabilities $p^*(UB) = 9/12 = 0.7500$ and $p^*(ML) = 10/13 = 0.7692$ are expected to deliver near - identical probability estimates. As shown in Table 7, practical discrepancy between UB - based and ML - based probabilities is, indeed, nugatory. Similar conclusions can be reached for first-time success cases with large values of N: if, e.g., K = 1 and N = 10 are set, $P[N = 10] = 7.153 \times 10^{-7}$; $P[N \leq 10] = 0.99999976$ (with UB estimators), and $P[N = 10] = 3.299 \times 10^{-7}$; $P[N \leq 10] = 0.999999901$ (with ML estimators).

Anterior k_0	n_0	Stipulation k	n	$P[N = n]$	Prediction $\Psi_q(k; n)$	$P[N \leq n]$
10	3	1	2	0.0468 0.0410	0.00521 0.00410	0.9843 0.9877
10	3	15	7	0.0948 0.0792	6.81×10^{-8} 4.42×10^{-8}	0.8385 0.8870
15	7	20	10	0.1029 0.1004	6.91×10^{-10} 5.77×10^{-10}	0.9988 0.9990
15	7	10	10	0.0271 0.0194	2.04×10^{-8} 1.45×10^{-8}	0.9_64 0.9_64

Table 7. Selected probabilities based on single event probability estimators in Section 3.2 . In columns 5 – 7 the first entry is UB – based, the second entry is ML – based. In column 7 "9_6" denotes six consecutive nines to indicate the extent of closeness to unity (i.e., certainty).

In the ion-tagging scenario of Section 3.3, and as shown in Table 5, the largest individual probability of untagged ion arrival, predicted by the mode (n = 36) is only about 2%, but there is a nearly 40% cumulative chance that up to 36 untagged ions are in fact at the surface on that particular occasion. At the mean value n = 45, the cumulative probability is about 57%. Given skewness $(1 + 0.9)/\sqrt{(0.9)(5)} \approx 0.90$, and kurtosis $3 = 6/5 + (0.1)^2/[(0.9)(5)] \approx 4.2$, the asymmetric distribution may be considered to be moderately skew (Bulmer, 1979a) and somewhat leptokurtic (Bulmer, 1979b), with respect to the skew-free normal (Gaussian) distribution, serving as reference, whose kurtosis is exactly 3.

Monitoring the arrival of untagged ions in Section 3.3 prior to the presence of the first tagged ion at preset values of their probability may also be an important objective of process analysis. Values obtained via Eq.(5) and Eq.(6) have been rounded to the nearest upward or downward integer in Table 8. The entries in the second column are given by the numerical form of Eq.(5) : $n = - 21.8543 - 9.949122 \ln(P)$, and in the third column by the numerical form of Eq.(6) : $n' = - 9.49122 \ln(P)$. No untagged ions can be expected to arrive prior to the first tagged ion at a probability higher than about 9%, whereas cumulative probabilities of their prior arrival increase with their number. While there is only a 20% probability that one (or no) untagged ion precedes the first tagged ion, it is almost certain for 65 untagged ions to do so. At a higher single - event probability of tagged ion arrival, the number of previously arrived untagged ions is, of course, smaller: when, e.g. $p = 0.2$, about twenty or thirty such ions may be expected to arrive at a 99% and 99.9% probability, respectively, in contrast with forty three and sixty five, when $p = 0.1$.

Probability, P	The rounded number of untagged ions[1]	Up to the number of untagged ions[2]
0.001	44	-
0.005	28	-
0.01	22	-
0.05	7	-
0.09	1	-
0.10	-	0
0.20	-	1
0.50	-	6
0.90	-	21
0.95	-	27
0.99	-	43
0.999	-	65

Table 8. Selected numbers of untagged ions that have arrived at the electrode surface in Section 3.3 prior to the arrival of the first tagged ion at (arbitrary) preset probabilities ($p = 0.1$; $q = 0.9$). [1] Eq.(5); [2] Eq.(6)

The effect of the single-event success probability is also illustrated in Table 9 for the aluminizing baths of Section 3.4 . If this probability is low ($p = 0.2$) for unacceptable baths, the cumulative probabilities show that a relatively large number of acceptable baths can be produced prior to the first unacceptable bath. Conversely, only a relatively small number of acceptable baths can be expected, if this probability is high, before the arrival of the first unacceptable bath. Individual probabilities are much smaller: if, e.g. $p = 0.1$, it is essentially certain that (at least) up to sixty five acceptable baths would be found before the first unacceptable bath, but the chances of finding *exactly* sixty five baths is only $(0.1)(0.9^{65}) \approx 10^{-4}$.

Cumulative probability	Single event probability, p			
	0.1	0.2	0.5	0.8
0.20	1	0	-	-
0.30	2	1	-	-
0.50	6	2	0	-
0.80	14	6	1	0
0.90	21	9	2	0
0.95	27	12	3	1
0.99	43	20	6	2
0.995	49	23	7	2
0.999	65	30	9	3

Table 9. Selected "not more than" numbers of acceptable baths in Section 3.4 prior to the appearance of the first bath of unacceptable quality at a small, a medium, and a high success probability. Bath numbers are rounded up or down to the nearest integer.

Single-event probabilities pertaining to the event of acceptable (or unacceptable) baths can be readily computed by solving the nonlinear algebraic equation

$$q^n - q^{n+1} = P[q]; \quad n \text{ fixed} \tag{18}$$

The number of observed acceptable baths determines the largest probability obtained from the dP/dq = 0 condition, resulting in

$$P_{max} = n^n/(n+1)^{n+1} \text{ at } q_{max} = n/(n+1) \tag{19}$$

and is illustrated in Table 10. Vacancy in the blocks associated with n = 4, 5 and 6 indicates that the P[q] = 0.10 requirement cannot be satisfied inasmuch as the P_{max} values: 0.0819; 0.0670 and 0.0567 are significantly below 0.10. As shown in the last row of Table 10, the P[q] = 0.1 stipulation can be barely met even at n = 3, but it can be comfortably satisfied at n = 2, (and at n = 1,not shown explicitly; Eq.(18) readily yields $q = 1/2 + \sqrt{0.15} = 0.8873$, hence p = 0.1187. Also, from Eq.(19), $P_{max} = 1/4 = 0.25$, hence $q_{max} = p_{max} = 0.5$).

P[q] ↓	Single event probability of success, p				
n→	2	3	4	5	6
0.01	0.0102	0.0103	0.0104	0.0105	0.0107
0.02	0.0209	0.0213	0.0219	0.0224	0.0229
0.03	0.0320	0.0332	0.0345	0.0360	0.0378
0.05	0.0561	0.0602	0.0656	0.0730	0.0853
0.07	0.0833	0.0942	0.1131	0.1667	-
0.10	0.1331	0.1842	-	-	-
q_{max}	0.6667	0.7500	0.8000	0.8333	0.8571
p_{max}	0.3337	0.2500	0.2000	0.1667	0.1429
P_{max}	0.1481	0.1055	0.0819	0.0670	0.0567

Table 10. Single event success probabilities at selected values of P[q] and n in Eq.(18), and the largest attainable P[q] = P_{max} via Eq.(19) in Section 3.4

A more sophisticated (and time consuming) determination of single-event probabilities from a set of experimental observations would require (nonlinear) regression techniques applied to Eq.(18) carrying measurement replicates of the number of acceptable baths that precede the first unacceptable bath.

The situation described in Section 3.5 invites several ramifications with it arising from the sequential sampling plan SSP (Blank, 1980) applied to the statistical procedure. It is instructive to examine the effect of the single-event probabilities on the decision - making process. In compliance with Eq.(4), the cumulative mass function is generalized from Eq.(4) in terms of the incomplete beta function - based method as

$$P[N \le 8; K = 4] = 1 - 1980\,(q^9/9 - 3q^{10}/10 + 3q^{11}/11 - q^{12}/12) \tag{20a}$$

and

$$P[N \le 7; K = 5] = 1 - 3960\,(q^8/8 - 4q^9/9 + 6q^{10}/10 - 4q^{11}/11 + q^{12}/12) \tag{20b}$$

upon algebraic integration, involving the fundamental identities: $(1 - x)^3 = 1 - 3x + 3x^2 - x^3$; $(1 - x)^4 = 1 - 4x + 6x^2 - 4x^3 + x^4$. The variation of these probabilities with the single event probability of failure is depicted in Table 11.

Single event probability q of finding an anode of unacceptable quality	Probability of finding up to eight acceptable anodes out of twelve (Test 1)	Probability of finding up to seven acceptable anodes out of twelve (Test 2)
0.50	0.9270	0.8061
0.80	0.2054	0.0726
0.83	0.1324	0.0393
0.84	0.1114	0.0310
0.87	0.0645	0.0148
0.875	0.0528	0.0113
0.88	0.0464	9.50×10^{-3}
0.90	0.0256	4.33×10^{-3}
0.93	7.53×10^{-3}	8.76×10^{-4}
0.95	2.24×10^{-3}	1.84×10^{-5}
0.965	5.92×10^{-4}	5.46×10^{-5}
0.98	6.96×10^{-5}	2.25×10^{-6}
0.99	4.64×10^{-6}	3.78×10^{-7}

Table 11. The cumulative probability that, in Section 3.5, finding up to eight (Test 1, K = 4), and up to seven (Test 2, K = 5) anodes of acceptable quality is due to random effects.

Using Table 11 as a guide, a process analyst setting a cumulative probability of about two percent as the acceptance threshold for the 2% - defectives claim, would be inclined to accept it when (unknown to the analyst) $q \geq 0.93$ in Test 1, and when (again unknown to the analyst) $q \geq 0.88$ in Test 2. A different (and more exacting) analyst setting the claim - acceptance threshold to about 0.05% would accept the claim when (again unknown to the analyst) $q \geq 0.965$.

The point probabilities $P[N = 8] = C(11;3)(0.02^4)(0.98^8) = 2.24 \times 10^{-5}$ (Test 1), and $P[N = 7] = C(11;4)(0.02^5)(0.98^7) = 9.17 \times 10^{-7}$ (Test 2) would favour claim rejection only somewhat more strongly at this low level of single event success probability. This bias toward rejection is much more pronounced at higher p - values, as attested by the p = 0.5 (Test 1: 0.0403; Test 2: 0.0806) and p = 0.2 (Test 1: 0.0443; Test 2: 0.0221) cases, vis - à - vis the top two entries in Table 11. Cumulative probabilities put claim rejections on a firmer ground than point probabilities by increasing their statistical reliability.

In the alternative fixed sample plan FSP approach (Blank, 1980) the size of the anode ensemble is fixed prior to testing. While it is, in principle, a matter of arbitrary choice, it should not be smaller than the mean of the failure occurrences plus the number of successes (i.e. the mean number of acceptable anodes plus the number of defective anodes). Consequently, the ensemble size should be at least $8(0.98)/0.2 + 4 = 396$ in Test 1, and $7(0.98)/0.02 + 5 = 348$ in Test 2, indicating that SSP would be less time (and material) consuming than FSP.

Another decision scheme might be based on the probability that, in face of the 2% - defectives claim the number of acceptable anodes is between two pre-specified values. If, for the sake of argument, these values are 22 and 25 inclusive, and assuming that when a sufficiently large number of anodes were tested three defectives were found, the probability calculated via Eq.(17):

$$[22 \leq N \leq 25; K = 3] = I_{0.98}(22;3) - I_{0.98}(26.3) = 0.0062 \qquad (21a)$$

or, equivalently

$$P[22 \leq N \leq 25; K = 3] = (0.02^3)[C(24;2)(0.98^{22}) + C(25;2)(0.98^{23}) + C(26;2)(0.98^{24}) + C(27;2)(0.98^{25})] = 0.0062 \qquad (21b)$$

would permit inference of effectiveness for the new process. Similarly, if the decision criterion is based on the incidence of the first defective anode upon the appearance of N anodes of acceptable quality, and it is stipulated that this event occur from the 21st to the 24th test inclusively, the probability

$$P[21 \leq N \leq 24; K = 1] = (0.02)(0.98^{21} + 0.98^{22} + 0.98^{23} + 0.98^{24}) = 0.98^{21} - 0.98^{25} = 0.0508 \qquad (22)$$

would lead to the same conclusion. If the somewhat more stringent condition for the appearance of the first defective anode between the 14th and the 17th test inclusive were set, Eq.(22) would be rewritten as

$$P[14 \leq N \leq 17; K = 1] = (0.02)(0.98^{14} + 0.98^{15} + 0.98^{16} + 0.98^{17}) = 0.98^{14} - 0.98^{18} = 0.0585 \qquad (23)$$

and a process analyst, inclined to accept the 2% - defectives claim only up to a 4% cumulative probability due to random effects would most likely question the effectiveness of the process. The level of acceptance chosen by the process analyst is an admittedly subjective element in the decision process, but not even quantitative sciences can be fully objective at all times. This statement is all the more valid for applied probability methods as probabilities are prone to be influenced by individual experiences.

5. Caveats related to negative binomial distributions

One inviting pitfall in dealing with NBD - related probability calculations would be the attempt to apply combination strings where they do *not* apply. A case in point can be the 2% - defectives claim in Section 3.5 if, in computing e.g., the probability of finding eight acceptable anodes in the presence of four defective ones, the erroneous path: C(u;8)C(v;4)/C(u + v;12) were chosen. The latter provides the probability of selecting eight items out of u identical items simultaneously with selecting four items out of v identical items, the u and v items being of a different kind, with replacement. Apart from the conceptual error in this scheme, the arbitrary choice of u and v predicts widely different numerical values(if, e.g. u = 60 and v = 20, C(60;8) C(20;4)/C(120; 12) = 0.00117 is considerably different from, e.g., C(45;8) C(6;4)/C(51;12) = 0.0204).

Equally important is the stipulation of mutual independence of the Bernouilli trials. In the context of Section 3.5, e.g., the anode - producing process should have no "memory" of the quality of any previously produced specimen. Otherwise, the NBD - based probability calculations would produce biased, i.e. statistically unreliable results. Similar considerations apply as well to the other illustrative examples. Event interdependence would necessitate working with conditional probabilities; if E_1 were the event of a success and E_2 the event of failure, $P(E_1/E_2)$ would be the probability of a success occurring when a failure has occurred. This topic is further explored in the sequel.

6. Interdependence effects: the role of conditional probabilities and Bayes' theorem

6.1 Fundamental concepts

In terms of conventional set theory, the intersection ($A \cap B$) denotes the simultaneous existence ("coexistence") of two events, and (A/B) the existence of event A on the condition that event B has happened (is known to have happened). The probability expression $P(A \cap B) = P(A/B)P(B) = P(B/A)P(A)$ serves as the underpinning of Bayes' theorem (also known as Bayes' rule) involving conditional probabilities, written in a general form as

$$P(A_k/B) = P(B/A_k)P(A_k)/[P(B/A_1)P(A_1) + P(B/A_2)P(A_2) + \ldots + P(B/A_k)P(A_k) + \ldots + P(B/A_n)P(A_n)] \quad (24)$$

provided that the A_k, $k = 1,\ldots,n$ events are exclusive (i.e., independent of one another) and exhaustive (i.e., at least one of the events occurs). In the simpler instance of dual event sets (A, and A': not A), (B, and B': not B) it follows from Eq.(24) that

$$P(A/B) = P(B/A)P(A)/[P(B/A)P(A) + P(B/A')P(A')] \quad (25)$$

The denominator in Eq.(24) and Eq.(25) yields the overall probability of event B occurring. Independence of events A_k and B may consequently be defined as $P(B/A_k) = P(B)$ and $P(B/A) = P(B)$, etc. Similarly, the probability of intersecting events is simply the product of their individual probabilities: $P(A \cap B) = P(A)P(B)$, or equivalently, $P(B \cap A) = P(B)P(A)$, in the case of independence.

Employing similar arguments, the rest of the posterior probabilities may be written as

$$P(A'/B) = P(B/A')P(A')/[P(B/A')P(A') + P(B/A)P(A)] \quad (26)$$

$$P(A/B') = P(B'/A)P(A)/[P(B'/A)P(A) + P(B'/A')P(A')] \quad (27)$$

$$P(A'/B') = P(B'/A')P(A')/[P(B'/A')P(A') + P(B'/A)P(A)] \quad (28)$$

with the understanding that

$$P(A/B) + P(A'/B) = 1; \; P(A/B') + P(A'/B') = 1.$$

6.2 Application to the aluminum anodizing process in Section 3.4

In order to illustrate the Bayes' theorem - based approach, the assumption is made that the single event success probability p, i.e., the probability of an anodizing bath performing in an unacceptable manner, does *not* have a strictly defined value; in fact, it varies with a-priori experienced conditional probabilities. In this context A is defined as the event of p being lower or equal to a "believed" or preset value p^*, and the complementary event A' is that $p > p^*$. B is the event of the range of acceptable operating variables falling within a specified tolerance domain, and B' is the event of the range being outside the domain. For the sake of numerical demonstration, the prior probabilities P(A) = 0.85; P(B/A) = 0.97 and P(B/A') = 0.04 are set. They state, on the basis of some prior experience, respectively, that (i) there is an 85% chance for $p \leq p^*$; (ii) a 97% chance that if $p \leq p^*$, the range of operating variables ROV falls within the specified tolerance domain STD, and (iii) a 4% chance that if $p \leq p^*$, the range is outside the domain. The calculations assembled in Table 12 reinforce the $p \leq p^*$ hypothesis inasmuch as P(A/B) is near 100%, whereas P(A'/B) is less than 1%. The probabilities P(A/B') ≈ 0.15 and P(A'/B') ≈ 0.85 strongly imply that if the ROV were outside the STD, the $p \leq p^*$ hypothesis would be essentially untenable.

Probability statement	Bayes' theorem	Probability value, %
$p \leq p^*$ when ROV is within STD	P(A/B) = 0.97x0.85/(0.97x0.85+0.04x0.15)	99.28
$p > p^*$ when ROV is within STD	P(A'/B) = 0.04x0.15/(0.04x0.15+0.97x0.85)	0.72
$p \leq p^*$ when ROV is outside STD	P(A/B') = 0.03x0.85/(0.03x0.85+0.96x0.15)	15.04
$p > p^*$ when ROV is outside STD	P(A'/B') = 0.96x0.15/(0.96x0.15+0.03x0.85)	84.96

Table 12. Application of Bayes' theorem to the anodizing bath in Section 3.4

It is instructive to examine the sensitivity of these results to variations in P(A), P(B/A) and P(B/A') while other pertinent probabilities remain constant. The relationships

$$P(A/B) = 0.97P(A)/[0.04 + 0.93P(A)] \quad (29)$$

$$P(A/B') = 0.03P(A)/[0.96 - 0.93P(A)] \quad (30)$$

$$P(A/B) = 0.85P(B/A)/[0.006 + 0.85P(B/A)] \quad (31)$$

$$P(A/B) = 0.8245/[0.8245 + 0.15P(B/A')] \quad (32)$$

lead to a wealth of useful inferences. An increase in prior probability P(A) produces a general tendency toward unity for posterior probabilities P(A/B) and P(A/B'), although the effect on the latter is less rapid. If P(A) is close to unity, Bayes' theorem predicts little difference with respect to event B or event B'.

Considering specifically the N = 5, K = 3 scenario, and given the finding above that $P[p \leq p^*]$ is very close to unity, it is possible to find the numerical value of p^* that will satisfy the cumulative probability P^* set by the process analyst for accepting the performance of the new anodizing process, expressed as

$$P^* = P[N \leq 5; K = 3] = C(7;2)(p^*)^3[1 + q^* + (q^*)^2 + (q^*)^3 + (q^*)^4 + (q^*)^5] \quad (33)$$

or alternatively,

$$P^* = P[N \leq 5; K = 3] = 1 - I_q(5;3)\ ;\ q = q^* \quad (34)$$

Since $q^* = 1 - p^*$, and $\Gamma(9)/[\Gamma(6)\Gamma(3)] = 8!/(5!2!) = 168$, it follows from Eq.(34) that

$$P^* = 1 - 168[(q^*)^6/6 - 2(q^*)^7/7 + (q^*)^8/8] \quad (35)$$

If the process analyst sets, for example, $P^* = 1.5\%$, and if the single-event success probability p^* is about 7%, the performance of the new anodizing process can be inferred with great confidence. If five acceptable baths are found experimentally along with three unacceptable baths (the third one being the last one tested), an analyst would likely question positive claims regarding the new anodizing process.

In a somewhat more complicated situation several ranges of single-event probabilities would exist with related conditional probabilities. This case is illustrated by assuming three adjacent success probability ranges: $0.02 \leq p < 0.05$; $0.05 \leq p < 0.1$; $0.1 \leq p < 0.2$, their events denoted as A_1; A_2; A_3; ,respectively. With prior probabilities $P(A_1) = 0.4$; $P(A_2) = 0.5$; $P(A_3) = 0.07$, they represent a firm "belief" in p falling between 0.02 and 0.1, but considerable uncertainty about where it can be expected between these bounds. Similarly, the conditional probabilities, assumed to be $P(B/A_1) = 0.3$; $P(B/A_2) = 0.4$; $P(B/A_3) = 0.15$, reflect this state of affairs. Then, in Eq.(24) the probability of the OVR falling within STD, namely $P(B) = 0.3x0.4 + 0.4x0.5 + 0.15x0.07 = 0.3305$ raises serious doubt about the properness of OVR position, and not surprisingly, the only firm conclusion that can be drawn from the posterior probabilities: $P(A_1/B) = 0.3x0.4/0.3305 = 0.3631$; $P(A_2/B) = 0.4x0.5/0.3305 = 0.6051$ and $P(A_3/B) = 0.15x0.07/0.3305 = 0.03118$ is the rather weak likelihood of p falling between 0.1 and 0.2 due to *random* effects. Let the anodizing process produce ten baths of acceptable, and three baths of unacceptable quality in a subsequent test. Table 13 indicates that inside the domain of single-event success probabilities p must be lower than 0.07 in order to judge the anodizing process acceptable if $P[N = 10;K = 3] \leq 0.05$ is stipulated as a reasonable acceptance criterion for the process. Under such circumstances, acceptance would at worst carry with it an about 13% possibility of random effects accompanying the process analyst's decision.

Inherent subjectivity in conditional probabilities arising from reliance on personal experience as well as documented (subjective and/or objective) evidence had been claimed (usually by traditional statisticians) in the past as a weakness of Bayesian methods, but their advantages over traditional statistical approaches have been well recognized (e.g. Arnold, 1990; Manoukian, 1986; Utts & Heckard, 2002). The assertion that "...rational degree of belief is the only valid concept of probability..." (Bulmer, 1979c), represents a perhaps exaggerated, but thought-provoking pro-Bayesian view (Jeffreys, 1983). Section 6.2 portrays (albeit modestly) the usefulness of this segment of modern probability theory with an electrochemical flavour.

P	$P[N=10;K=3] = 66(p^*)^3q^{10}$	$P[N\leq10;K=3] = 1- 858(q^{11}/11 - q^{12}/6 +q^{13}/13)$
0.020	$4.314x10^{-4}$	$1.968x10^{-3}$
0.035	$1.982x10^{-3}$	$9.419x10^{-3}$
0.050	$4.940x10^{-3}$	0.0245
0.060	$7.678x10^{-3}$	0.0392
0.070	0.0109	0.0577
0.080	0.0147	0.0799
0.090	0.0187	0.1054
0.100	0.0230	0.1339
0.200	0.0567	0.4983

Table 13. Individual and cumulative probabilities for the three-p-range scenario in Section 6.2

7. Normal distribution – based approximations to NBD with large success occurrences

Rephrasing Theorem 4 - 13 (Arnold, 1990b) concerning large success numbers K, the random variable $(N_k - \mu)/\sigma$ closely approximates the standard random normal variate Z as K increases, and in the limit,

$$(N_k - \mu)/\sigma \rightarrow N(0;1);\ K \rightarrow \infty \tag{36}$$

with mean $\mu \equiv k(1 - p)/p$ and variance $\sigma^2 \equiv k(1 - p)/p^2$. Taking into account the conventional continuity correction required when a discrete distribution is approximated by a continuous distribution, it follows that

$$P[N_k = n_k;\ K = k] \approx \Phi(z'') - \Phi(z') \tag{37}$$

where

$$z' = (n_k - ½ - \mu)/\sigma\ ;\ z'' = (n_k + ½ - \mu)/\sigma \tag{38}$$

and $\Phi(z)$ is the cumulative standard normal distribution function, tabulated extensively in the textbook literature and monographs on probability and statistics.

To illustrate the scope of the normal distribution it is supposed that in Section 3.4 a preliminary study of the new anodizing process had indicated p = 0.4. It is further stipulated that in a subsequent experimental test twenty acceptable as well as twenty unacceptable baths were found in the usual manner (i.e. the last bath was unacceptable). Given that $\mu = 20\text{x}0.6/0.4 = 30$ and $\sigma^2 = 30/0.4 = 75$, Eq.(37) and Eq.(38) yield $z' = (19.5 - 30)/\sqrt{75} = -1.2124$ and $z'' = (20.5 - 30)/\sqrt{75} = -1.0970$, respectively. Hence, $\Phi(z') = 0.1131$ and $\Phi(z'') = 0.1357$, resulting in $P[N_k = 20;\ K = 20] \approx 0.1357 - 0.1131 = 0.0226$, versus the rigorous value of $C(39;19)(0.4^{20})(0.60^{20}) = 0.0277$. Since $P[N_k \leq 20;\ K = 20]$ must be larger than $P[N_k = 20;\ K = 20]$, it is not necessary to compute the former, if a lower than 2% cumulative probability were judged sufficient to accept the claim of better performance by the novel process. However, for the sake of completeness, Eq.(4) is shown to corroborate the properness of this reasoning:

$$P[N \leq 20;\ K = 20] = 1 - [\Gamma(41)/\Gamma(21)\Gamma(20)] \int_0^{0.6} u^{20} (1 - u)^{19} du = 0.1298 \tag{39}$$

along with the normal approximation $\Phi(z'') = 0.1357$. These findings do not signal, of course, a better performance.

Table 14 demonstrates that even at a relatively small number of successes the normal approximation to NBD is well within the same order of magnitude, albeit not uniformly so; this is also seen in the instance of cumulative probabilities. As a case in point, from Table 14: $P[0 \leq N \leq 7] = 0.2946$ is obtained employing rigorous NBD theory, whereas the normal approximation via the sum (0.0076 + 0.0017 + 0.0178 + 0.0270 + 0.0358 + 0.0469 + 0.0615 + 0.0700) = 0.2783 differs from the rigorous value only by a relative error of about - 6%. The simpler "shortcut": $P[0 \leq N \leq 7] \approx \Phi\ [(7.5 - 10)/\sqrt{20}] = \Phi\ (-0.56) = 0.2877$ equally qualifies as a good approximant on account of a relative error of about - 2.3%. The poorer approximation via $\Phi\ [(7 - 10)/\sqrt{20}] \approx \Phi\ (-0.67) = 0.2514$ with a - 14.7% relative error is the price to pay if the continuity correction is neglected. The last column in Table 14 demonstrates the unevenness of the error magnitudes. Using the relative error, or the error magnitude as a measure of approximation quality is the analyst's decision.

8. Conclusions

The illustrative examples, albeit not exhaustive, demonstrate the potential of NBD theory for analyzing a wide variety of electrochemical scenarios from a probabilistic standpoint. In view of a still rather limited employment of probability - based and statistical methods in the electrochemical research literature, a major intent of the material presented here is a "whetting of appetite" by stimulating cross fertilization between two important disciplines. There remains much more work to be done in this respect.

Success probability, p	Number of failures, n	P[N = n; K = 10]		Magnitude of normal approximation error
		NBD	Standard normal	
0.1	30	8.98×10^{-4}	0.0022	0.0013
0.1	15	2.69×10^{-4}	5.2×10^{-4}	7.0×10^{-4}
0.5	0	9.80×10^{-4}	0.0076	0.0066
0.5	1	0.0049	0.0117	0.0068
0.5	2	0.0134	0.0178	0.0044
0.5	3	0.0269	0.0270	0.0001
0.5	4	0.0436	0.0358	0.0078
0.5	5	0.0611	0.0469	0.0142
0.5	6	0.0764	0.0615	0.0149
0.5	7	0.0673	0.0700	0.0027
0.5	20	0.0093	0.0119	0.0026

Table 14. Comparison of individual probabilities in Section 7 via NBD and standard normal distribution theory

9. Acknowledgment

The manuscript was written on equipment provided by the Natural Sciences and Engineering Research Council of Canada (NSERC), and the Department of Chemical Engineering of the University of Waterloo.

10. Appendix A

The Bernouilli trial is a statistical experiment involving a binomial event with success probability p and failure probability $q = 1 - p$ (success and failure are opposite but arbitrarily defined events). Each trial is independent of any previous trial and p and q do not change from trial to trial. The exponent x in the pmf can take only two values, namely $x = 0$ (failure with probability q), or $x = 1$ (success with probability p). This is the Bernouilli distribution with pmf

$$P[X = x] = p^x (1 - p)^{1-x}; \; x = 0, 1 \qquad \text{(A.1)}$$

If, for instance, there are four anodes in a batch of Section 1 and only one among them is of acceptable quality (bad batch!), 3/4 is the probability of any of the anodes being defective, and 1/4 is the probability of any of the anodes being acceptable. Hence, Eq.(A.1) becomes $P[X=x] = (3/4)^x (1/4)^{1-x}$ with p(0) = 1/4 and p(1) = 3/4. Using the argument (Milton & Arnold, 1990) that in the particular case where x number of trials are needed to obtain three successes, each trial must end with a success and the remaining (x – 1) trials must result in exactly two successes and (x – 3) failures in some order, the probability

$$P[X = x] = C(x - 1;2)\, p^3 (1 - p)^{x-3} \quad x = 3,4,5 \tag{A.2}$$

can be generalized in a straightforward manner to obtain the pmf of the NBD

$$P[X = x] = C(x - 1;k - 1)\, p^k (1 - p)^{x-k} \tag{A.3}$$

and since x = n + k, where n is the number of failures when k successes have been observed, Eq.(1) is finally established. The C(x-1;2) combination in Eq.(A.2) represents the partitioning of (x – 1) elements into 2, and (x – 3) elements in accordance with probability theory, i.e. $C(x - 1;2) = (x - 1)!/([2!(x - 3)!]\ (=[x^2 -3x + 2]/2)$.

11. Appendix B

When k is sufficiently small, the $\Psi_q(k; n')$ function can be conveniently expressed in terms of polynomials carrying integer powers of its upper limit q, as shown in Table 15 (In the context of the cumulative probabilities, m = n').

Number of successes, k	$\Psi_q(k; m)$
1	$F(q) \equiv q^{m+1}/(m+1)$
2	$F(q) - q^{m+2}/(m+2)$
3	$F(q) - 2q^{m+2}/(m+2) + q^{m+3}/(m+3)$
4	$F(q) - 3q^{m+2}/(m+2) + 3q^{m+3}/(m+3) - q^{m+4}/(m+4)$
5	$F(q) - 4q^{m+2}/(m+2) + 6q^{m+3}/(m+3) - 4q^{m+4}/(m+4)] + f(5)$

Table 15. Polynomial expressions for the Ψ – function at the first five numbers of success. $f(5) \equiv q^{m+5}/(m + 5)$

12. Appendix C

The utility of incomplete beta function tables is shown by computing the cumulative probability that up to four failures appear prior to the appearance of the eighth (and last) success when p = 0.3, i.e., $1- I_{0.7}(5;8) = I_{0..3}(8;5)$ in terms of the incomplete beta function $I_x(a,b)$ tables (Beyer, 1968). As seen in the excerpt below with entries obtained from the tables rounded to four decimals, $I_{0.3}(8;5) \approx 0.01$; numerical integration of Eq.(A.4) yields 1 – (3960) (2.501 x 10^{-4}) = 0.0095. Similarly, $I_{0.35}(8;5) \approx 0.025$, and 1 – (3960) (2.461 x 10^{-4}) = 0.0255, with x = 1 – 0.35 = 0.65 as upper limit of integration.

$I_x(8;5)$	0.10	0.05	0.025	0.01	0.005
X	0.4410	0.3910	0.3489	0.3024	0.2725

Intermediate values may be approximated by various methods of interpolation. If, for example, the upper limit of the integral is set to x = 0.67, the value $I_{0.33}(8;5) \approx 0.01 + (0.025 –$

0.01)/(0.3489 - 0.3024) = 0.019 is about 8% higher than 0.0176 obtained via numerical integration. Alternatively, the data may be correlated via a properly selected regression. In the case discussed here, inspection of data suggests the semi-linearized form

$$\ln[I_{0.33}(8;5)] = a + b(x) \tag{A.4}$$

or a fully linearized form

$$\ln[I_{0.33}(8;5)] = c + d\ln(x) \tag{A.5}$$

Conventional least - squares fitting (e.g., Neter et al, 1990) yields a = - 10.00894;b = 17.74349; c = 2.836044; d = 6.231738 with coefficients of determination of 0.991 and 0.999, respectively, and $I_{0.33}(8;5)$ estimates 0.017598 and 0.017597. Disagreement only in the sixth decimal is a fortuitous finding, inasmuch as such closeness is not guaranteed, in general. The fact, that the pre-integral coefficient - carrying factorials and the Ψ – function often involve the multiplication of very large and very small numbers, might be viewed by some analysts as an incentive for preferring interpolation or regression methods.

13. Appendix D

List of symbols

a, b	general variables or arguments
C(a, b)	binomial coefficient (or combination): a!/[b!(a-b)!]
cmf	cumulative mass function
E_i	i-th probabilistic event
GD	geometric distribution
$I_q(\ldots)$	incomplete beta function {Eq.(4)]
K	random variable, denoting the number of successes; k its numerical value
M, N	random variables, denoting the number of failures; m, n their numerical value, respectively
ML	maximum likelihood
N(0;1)	standardized random normal distribution with zero mean and unit variance
NBD	negative binomial distribution
P[…]	probability of a variable or an event
$P[E_2/E_1]$	conditional probability of event E_2 occurring upon the occurrence of event E_1
p	single - event success probability
pmf	probability mass function
q	single-event failure probability
S	selectivity
u	“dummy” integration variable
UB	unbiased
Z	standard normal (Gaussian) variate, z its numerical value
μ	mean of the normal distribution
σ^2	variance of the normal distribution
$\Gamma(\ldots)$	gamma function of its argument
Φ	standard normal probability distribution function of variate Z
$\Psi(\ldots)$	auxiliary function [Eq.(4)]

Special symbols

Subscript 0: earlier observations (earlier data)
Superscript *: set (threshold) value

14. References

Arnold, S. F. (1990a). *Mathematical Statistics*, 535 – 570, Prentice Hall, ISBN 0-13-561051-6, Englewood Cliffs, New Jersey,USA

Arnold, S. F. (1990b). loc. cit., 148 -152.

Beyer, W. H. (Ed.) (1968). *Handbook of tables for Probability and Statistics*, 251– 265, CRC Press, ISBN 0-8493-0692-2, Boca Raton, Florida, USA

Blank, L. (1980). *Statistical Procedures for Engineering, Management and Science*, 208-211, McGraw Hill, ISBN 0-07-005851-2 2, New York, USA

Bulmer, M. G. (1979a). *Principles of Statistics*, 63, Dover, ISBN 0-486-63760-3, New York, USA

Bulmer, M. G. (1979b). *Principles of Statistics*, 65, Dover, ISBN 0-486-63760-3, New York, USA

Bulmer, M. G. (1979c). *Principles of Statistics*, 173, Dover, ISBN 0-486-63760-3, New York, USA

Danly, D. E., & Campbell, C. R. (1982). Experience in the Scale-up of the Monsanto Adiponitrile Process, In: *Technique of Electroorganic Synthesis, Scale-up and Engineering Aspects*, Weinberg, N. L. & Tilak, B. V., 285, Wiley and Sons, ISBN 0-471-06359-2(r.3), New York, USA

Dean, J. A. (Ed.) (1985). *Lange's Handbook of Chemistry*, 13th edn., **3**-121, McGraw Hill, ISBN 0-07-016192-5, New York, USA

Dougherty, E. R. (1990). *Probability and Statistics for the Engineering, Computing and Physical Sciences*, 149-152, Prentice Hall, ISBN 0-13-711995-X, Englewood Cliffs, New J ersey, USA

Evans, M., Hastings, N. & Peacock, B. (2000). *Statistical Distributions*, 3rd edn. 140-144, Wiley and Sons, ISBN 0-471-37124-6(pbk.), New York, USA

Jeffreys, H. (1983). *Theory of Probability*, 3rd edn. Oxford[Oxfordshire]: Clarendon Press, ISBN/ISSN 0198531931(pbk.), UK

Manoukian, E. B. (1986). *Modern Concepts and Theorems of Mathematical Statistics*, 39 – 40, Springer-Verlag, ISBN 0-387-96186-0, New York, Berlin, Heidelberg, Tokyo; 3-540-96186-0, Berlin, Heidelberg, New York, Tokyo

Milton, J.S. & Arnold, J. C. (1990). *Introduction to Probability and Statistics*, 2nd edn., 61, McGraw Hill, ISBN 0-07-042353-9, New York, USA

Pletcher, D. & Walsh, F. C. (1990). *Industrial Electrochemistry*, 2nd edn., 439, Chapman and Hall, ISBN 0-412-30410-4, London, UK

Smyth, D. M. (1966). The Heat Treatment of Anodic Oxide Film on Tantalum, In: *Journal of the Electrochemical Society*, Vol. 113, No. 12 (December 1966), pp. 1271-1274, ISN 0013-4651

Utts, J. M., & Heckard, R. F. (2002). *Mind on Statistics*, 319-319, Duxbury, ISBN 0-534-35935-3, Pacific Grove, California, USA

Walpole, R. E., Myers, R. H., Myers, Sh. L., & Ye, K. (2002) *Probability and Statistics for Engineers and Scientists*, 7th edn., 132-135, Prentice Hall, ISBN 0-13-041529-4, Upper Saddle River, New Jersey, USA

Weisstein, E. W. (n.d.) Negative Binomial Distribution, In: http://mathworld.wolfram.com/NegativeBinomialDistribution. *html*, Available from MathWorld-A Wolfram Web Source

3

Mathematical Modeling of Electrode Processes – Potential Dependent Transfer Coefficient in Electrochemical Kinetics

Przemysław T. Sanecki and Piotr M. Skitał
Rzeszów University of Technology
Poland

This chapter is dedicated to professor Zbigniew Galus, who consistently applied mathematical approach to electrochemical kinetics and to memory of professor Bogdan Jakuszewski who was succeeded in both theoretical and experimental electrochemistry.

1. Introduction

The connection between experimental results and mathematical modeling of electrode processes may become an inspiration for new results and deeper understanding of the nature of electrochemical processes and its kinetic description. Sometimes experiment is preceded by a theory, sometimes it is the other way round. To avoid over discovering of phenomena, uncertainties and even mistakes, a responsible validation of model results is required. The analysis of complex, multi-electron electrode processes with chemical step(s) provides the respective examples.

2. Elementary and apparent kinetic parameters in modeling of electrode processes

In chemical kinetics, elementary (one step) and complex (multi-step) processes are described. Exactly the same situation is observed in electrochemical kinetics where electron transfer steps and chemical steps are often coupled in various sequences. Therefore, electrochemical kinetics uses two kinds of kinetic parameters: elementary describing each single step of kinetic sequence and general (apparent, observed) relating to or describing the complex mechanism as a whole.

Consequently, there is a need of showing similarities, dissimilarities and relations between the two approaches to avoid possible confusion. The following significant problems are to be discussed here:

- The distinction among apparent and elementary kinetic parameters (Sanecki & Skitał, 2002a; Skitał & Sanecki, 2009).
- The accuracy of electrochemical kinetic parameters determination by the estimation method (Sanecki et al., 2003, 2006b).

- The analysis of complex current responses (Sanecki & Lechowicz, 1997; Sanecki & Kaczmarski, 1999; Sanecki & Skitał, 2002a; Sanecki et al., 2003; 2006a, 2006b, 2010; Skitał et al., 2010).
- The relationship between apparent and elementary kinetic parameters (Sanecki & Skitał, 2002a; Skitał & Sanecki, 2009) and its consequences.

Most of electrochemical processes are complex and electrodics deal with multicharge transfer reactions as well as with multicomponent systems not considered here (Bard & Faulkner, 2001; Sanecki & Skitał, 2002b). The extraction of elementary kinetic parameters from experimental responses of multicharge reactions always requires an appropriate kinetic model, even if its application is not clearly given. Even a single formula used for calculation of kinetic parameters is a model. Therefore, a need of appraisal of assumptions basing every kind of model is evident. If the formulae available in electrochemistry textbooks and monographs are not adequate for considered mechanism, a respective rigorous mathematical model of kinetic case is required. If the applied model is appropriate to considered experimental data, the physical sense of obtained kinetic parameters is clear and the ones are reliable for further discussion. We will show further that the simple model is not appropriate for processes with chemical step unless the chemical step is extremely fast or slow and a simplification is justified.

The difference between elementary and apparent kinetic parameters denotes their physical sense: (1)- elementary ones describe single step *i* (e.g. k_i i α_i), general ones (*apparent, observed*) (e.g. k_{app} i α_{app}) describe a sequence of steps (mechanism) as a whole i.e. electrochemical and chemical steps together, as well as the way they were obtained: (2)- elementary ones are obtained by estimation with the use of complex current response, apparent usually with the use of simplified model (formula) in relation to the same response. The use of elementary parameters method is a consequence of the fact that mechanism is described as a sequence of elementary steps. General and apparent kinetic parameters are useful when application of an analytical or estimation method to complex systems is not possible. An example of apparent kinetic parameters approach is the conception of general transfer coefficient e.g. applied to interpretation of Tafel plots of complex multi-electron processes (Bockris & Reddy, 1970). Similar discussion of apparent and elementary kinetic parameters can be found in monograph by Brenet and Traore (Brenet & Traore, 1971).

Acquisition of rigorous mathematical model of considered process is fundamental for any discussion of quantitative type. The actual numerical possibilities and availability of respective software are quite enough to solve practically each of mechanisms met in practice. The respective numerical procedure of nonlinear curve fitting, called multi-parameter estimation (MPE) is in general followed with validation of obtained kinetic parameters. The criterion of optimization is a minimization the difference between experimental and theoretical response by least squares method. In electrodics MPE operates on a set of CV responses for various scan rates (and eventually on responses for various concentrations). It is a fundamental rule in electrochemistry simulations. The respective fits should be optimized for different scan rates even if there is a possibility to determine it on single curve only. The considered set of CV curves for different scan rates should be covered by the only one set of kinetic parameters (Speiser, 1996a, 1996b). The power of CV is determined by the possibility to study the electrochemical system by using different time scales by changing potential scan rate (Speiser, 1985).

The form of $i=f(E)$ and $c=f(E)$ responses corresponding to a sum of steps as well as for individual intermediate species can be predicted for almost each of mechanisms, which is the important advantage of modeling. Another one is a possibility of simple moving from elementary to general kinetic parameters. For now, however, it has not been possible to complete it inversely. The limitation of the number of estimated parameters, even if series of kinetic runs are used, is a disadvantage of MPE method. The credibility of MPE method decreases with the number of them and confidence intervals become wider simultaneously. It is however a general problem which denotes all complex systems and is a result of their complexity and not of a calculation method. MPE is a global method i.e. it operates on whole original, non processed responses (please note a comment given with equations (5)–(8)).

The electrochemical responses are in most cases complex and reflect the influence of various factors (e.g. chemical, electron transfer, adsorption steps) on its shape. Therefore, a model should include all of them. Respective electrochemical current response or concentration response is a function of many variables and it is easy to show the influence of a single kinetic parameter on its shape. In other words, an electrochemical response is directly a superposition of one-electron current responses and indirectly of chemical kinetic parameters. This chapter contains a number of respective examples, which visualize the part of complex relationships between relevant kinetic parameters and the resulting electrochemical response, namely Fig. 2–8, 9, 11–14, 15D.

3. The accuracy of electrochemical kinetic parameters determination by estimation method

For consecutive reaction steps there is a problem with the ability of a calculation procedure to extract kinetic parameters of the second kinetic step when the overall not naturally resolved current response is rate limited by the first step. In the early fifties of the last century Schwemer and Frost (Frost & Schwemer, 1952; Schwemer & Frost, 1951) and Frost and Pearson (Frost & Pearson, 1961) first solved the problem of two consecutive second order reactions in chemical kinetics. The attention was focused on the accuracy of k_1 and k_2 determination (e.g. error values k_1 ±2–4%, k_2 ±5–10% were reported (Kuritsyn et al., 1974). Analogous problems exist in electrochemical kinetics. In the first approach, in voltammetry for ECE process, the errors of kinetic parameters for the first and second one-electron steps are determined during estimation routine procedure applied to experimental response. The statistical treatment is incorporated into an algorithm. There are two other ways: (a) applying an estimation procedure to theoretical (simulated) CV curves to obtain the kinetic parameters with the best possible accuracy for the system; (b) applying an estimation procedure to the same simulated CV curves altered by the addition of noise, a non perfect zero base line, and a slight ohmic drop in order to imitate experimental errors of real voltammetric curves. The ways (a) and (b) serve only for the purpose of modeling studies. In practice, the kinetic parameters are determined entirely with the use of experimental voltammograms. The method of validation based on theoretical (synthetic) responses is seldom applied but is worth recommendation. In particular, the fact that accuracy of the evaluated parameters depends on the k_2/k_1 ratio will be shown. This means that not only the system with the rate limited by the first electron transfer is considered, i.e. the case when $k_2>k_1$, but also that the whole spectrum of k_2/k_1 ratios is taken into account for the ECE reaction scheme.

Three approaches are possible for calculating confidence intervals of the estimated parameters. Firstly, a numerical method used for the estimation and fitting procedure is applied to a single experimental curve and the statistical treatment is incorporated into an algorithm (not accessible for user). This is a non-linear regression problem solved by the least squares method (Bieniasz & Speiser, 1998a, 1998b; Lavagnini et al., 1989; Scharbert & Speiser, 1989; Seber & Wild, 1989; Speiser, 1985). For example, refs. (Bieniasz & Speiser, 1998b; Lavagnini et al., 1989) discuss in details the error space of kinetic parameters obtained during a single estimation procedure. In the second approach, the errors are determined on the basis of repeated experiments as reported by Jäger and Rudolph (Jäger & Rudolph, 1997). The population of kinetic parameters obtained by estimation provide the mean value, median, standard deviation, confidence intervals, etc. In the third approach, the estimations are repeated for theoretical responses with different starting points; it turns out that for some kinetic parameters, the calculated values (e.g. k_1) are identical or almost identical (with errors that have no physically important meaning), for others (e.g. k_2) they are not identical and for some cases they differ even by an order of magnitude. The confidence intervals for the populations of particular kinetic parameter (k_1, k_2, etc.) become available. The third approach and the reproducibility of kinetic parameters k_1, α_1, k_2, α_2 by comparing input parameters with output ones evaluated according to procedures (a) is presented here. An example of first and second step kinetic parameters validation is presented in Fig. 1, respectively.

To summarize: after numerical fitting procedure (estimation) applied to experimental response, it is worth to repeat the procedure with the use of obtained theoretical data which imitate the experimental response. After that, both (derived from experimental and theoretical responses) error spaces of kinetic parameters should be critically compared (Sanecki et al., 2003). Generally, a precise determination of second stage kinetic parameters in a consecutive process is impossible if their magnitude is much higher than that in the first stage even if the errors of the kinetic parameters are evaluated from ideal theoretical CV curves. It means that such a result is an inner feature of the investigated system alone and not a result of any calculation procedure which is not able to change it.

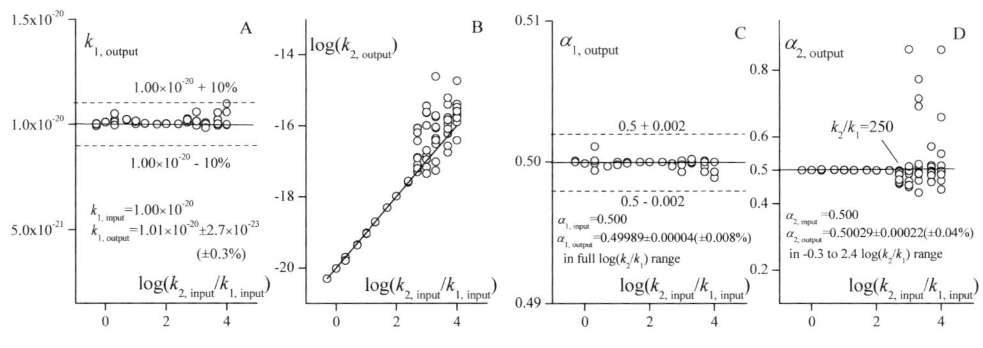

Fig. 1. Visualization of reproducibility of input kinetic parameters for theoretical CV curves of ECE reaction mechanism (1)-(3), point 5.1. Note that the rate constants are expressed on the saturated calomel electrode scale which results in unexpectedly small values of k_1 and k_2 (see also ref. (Sanecki et al., 2003)). [Reprinted from *J. Electroanal. Chem.*, Vol. 546, Sanecki, P., Amatore, C. & Skitał, P., The problem of the accuracy ..., 109-121, Copyright (2003), with permission from Elsevier.]

4. The list of applied kinetic models and description of numerical procedures

The problems formulated in this chapter were solved using *ESTYM_PDE* program. The *ESTYM_PDE* program has been designed to solve and estimate parameters of partial differential equations (PDE) describing one-dimensional mass and heat transfer coupled with chemical reaction. One of the program options enables solution of electrochemical reaction models. The numerical algorithm is based on an implementation of the method of orthogonal collocation on finite elements (OCFE) (Berninger et al., 1991; Gardini et al., 1985; Kaczmarski, 1996; Kaczmarski et al., 1997; Ma & Guiochon, 1991; Villadsen & Michelsen, 1978; Yu & Wang, 1989). After discretization of the space derivatives due to the method of OCFE, the obtained set of ordinary differential equations is solved with the backward differentiation formulae, implemented in the *VODE* procedure (Brown et al., n.d.). To estimate model parameters one of the best and fastest algorithms based on the least square fitting as proposed by Marquardt (Marquardt, 1963) in the version modified by Fletcher (Fletcher, n.d.) was used. The calculations of confidence level of estimated parameters were performed according to the method described in (Seber & Wild, 1989).

In the recovery process or estimation of model parameters, the set of differential equation must be solved many times as required by the least-squares algorithm. A single solution of the actual electrochemical reaction model takes 1-15 s (calculation of a single CV curve). The full estimation procedure, however, takes 0.5-2 hours. The time of estimation depends on the choice of initial values of estimated parameters and on the accuracy imposed for ODE solver. The highest applied accuracy of calculations was 1×10^{-15}. The accuracy is related to that of solving ordinary differential equations with VODE procedure and to absolute accuracy of the estimation of parameters. The last one was adjusted to be six orders of magnitude lower than expected value of an estimated parameter. The relative accuracy used in VODE was 1×10^{-13} and absolute accuracy was 1×10^{-15}. The Levenberg-Marquard procedure was applied as described in report (Fletcher, n.d.; Marquardt, 1963) without any changes. The same was for VODE procedure. All calculations were performed using the extended precision.

The algorithm applied for solving PDE's in *ESTYM_PDE* was also used in other programs for modelling adsorption chromatography processes (Kaczmarski et al., 1997, 2001; Kim et al., 2005, 2006a, 2006b). Other algorithms to estimate parameters of PDE are being used as well. Among them are *ELSIM* and *DigiSim®* programs in which a Simplex method and Levenberg-Marquard algorithm are applied, respectively. Simulation packages are described in papers (Bott et al., 1996; Bott , 2000; Feldberg, 1969) (*DigiSim®*) and (Bieniasz, 1997; Bieniasz & Britz, 2004) (*ELSIM*). The applied theory of electrochemical simulations is described in refs. (Bard & Faulkner, 2001; Bieniasz & Britz, 2004; Bieniasz & Rabitz, 2006; Britz, 2005; Feldberg, 1969; Gosser, 1993; Speiser, 1996a). The *ESTYM_PDE* program, similarly to other analogous programs available, provides the possibility of calculating concentration versus both space parameter and time for various geometries of the electrode.

To test a simulation software, Speiser (Speiser, 1996a) proposed to use ECE/DISP1 sequence (Amatore & Saveant, 1977). In paper (Sanecki et al., 2003), there is a comparison of available *Simulators* gathered by Speiser on the ground of the test (Table 1 and 2 and p.11 in (Speiser, 1996a)). We extended both of Speiser's tables by means of including our data for the example with the use of *ESTYM_PDE*. It is clear that all programs give practically the same results. The correctness of our calculations was also confirmed by the comparison of *ESTYM_PDE* with *DigiSim®* program. The results of calculations obtained for representative examples were exactly the same (Sanecki & Skitał, 2008).

Various examples of solving electrochemical problems by means of *ESTYM_PDE* software were described in our papers (Sanecki, 2001; Sanecki & Kaczmarski, 1999; Sanecki & Skitał, 2002a, 2007a, 2007b, 2008; Sanecki et al., 2003, 2006a, 2006b, 2010; Skitał & Sanecki, 2009; Skitał et al., 2010). The mathematical kinetic models of investigated mechanisms are presented in Table 1 with relevant references in which the models were applied with or without inclusion of the alpha variability parameter. Consecutively, the Scheme 1 illustrates the mutual interdependence of models.

Mechanism (model)	Reference
E_r	(Sanecki & Skitał, 2008)
E	(Sanecki & Skitał, 2008)
E_rC	(Sanecki & Skitał, 2007a, 2008)
E_rE_r	(Sanecki et al., 2006a)
$E_rE_r \| Hg(Me)$	(Sanecki et al., 2006a)
ECE	(Sanecki, 2001; Sanecki & Kaczmarski, 1999; Sanecki & Skitał, 2002a, 2007b, 2008; Sanecki et al., 2003; Skitał & Sanecki, 2009)
E_rCE_rC	(Sanecki & Skitał, 2008)
EC(C)E	(Sanecki & Skitał, 2007b)
ECE-ECE	(Sanecki, 2001; Sanecki & Kaczmarski, 1999; Sanecki & Skitał, 2007a, 2008)
ECE-EC(C)E	(Sanecki & Skitał, 2007b)
E_rCE_rC-E_rCE_rC	(Sanecki & Skitał, 2008)
E_rCE-ECE/E_rE-EE	(Sanecki et al., 2006b)
EE two-plate model with BET or Langmuir adsorption equation	(Sanecki et al., 2010; Skitał et al., 2010)

Table 1. Kinetic models prepared and/or applied to electrode processes.

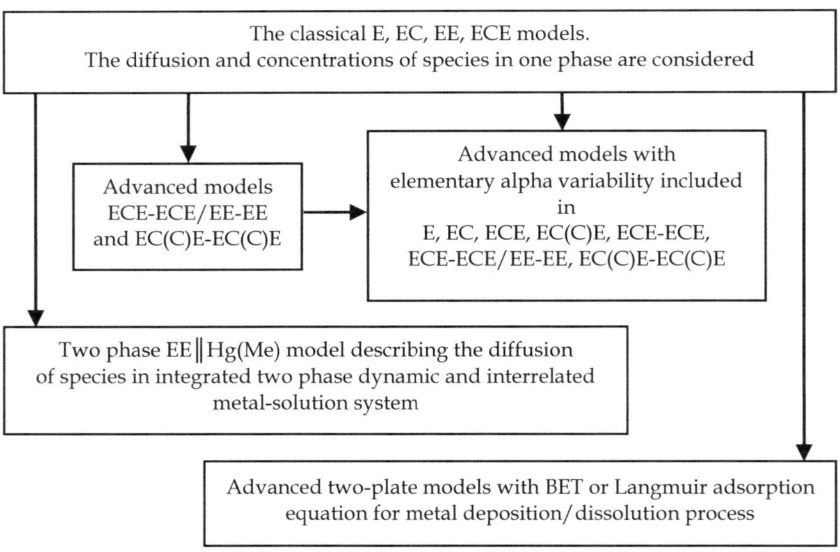

Scheme 1. Kinetic models prepared and/or applied to electrode processes and their mutual interdependence.

The MPE method is now widely applied in the determination of kinetic parameters of different dynamic processes e.g. for chromatography data with the use of respective mathematical model (Kaczmarski, 2007; Kim et al., 2005, 2006a).

5. The ECE process in cyclic voltammetry. The relationships between elementary and apparent kinetic parameters

Even textbook reaction mechanisms are not immune to changes.
Bernd Speiser

5.1 Stepwise - concerted mechanism competition and complex α_{app} plots

The reduction process going through two one-electron steps with chemical reaction between them is typical for a large number of processes in both organic and inorganic electrochemistry and is still of current interest e.g. (Andrieux et al., 1992; 1993, 1994, 1997; Antonello & Maran, 1997, 1998, 1999; Antonello et al., 2001, 2002a, 2002b; Costentin et al., 2009; Daasbjerg, 1999; Jaworski & Leszczyński, 1999; Najjar et al., 2007; Pause et al., 1999, 2001; Savéant, 1987, 1992, 1993; Severin et al., 1993; Speiser, 1996a; Workentin et al., 1995).

$$RX + e \overset{k_1, \alpha_1}{=} RX^{\bullet -} \tag{1}$$

$$RX^{\bullet -} \xrightarrow{k_f} R^{\bullet} + X^{-} \tag{2}$$

$$R^{\bullet} + e \overset{k_2, \alpha_2}{=} R^{-} \tag{3}$$

$$RX + e \overset{k_{concerted}}{=} R^{\bullet} + X^{-} \tag{4}$$

The first set (1)-(3) constitutes an ECE (stepwise, sequential) mechanism. The equation (4) describes the *concerted* mechanism that may or may not be followed by a second electron transfer (3).

Berzins and Delahay (Berzins & Delahay, 1953) were probably the first ones who successfully applied the modeling method to unravel contentious kinetic problem. Nicholson and Shain (Nicholson & Shain, 1964, 1965a) were the first ones to describe the full mathematical model of ECE process followed with its experimental verification (Nicholson & Shain, 1965b). The determination of elementary kinetic parameters from complex CV current responses and the analysis of the competition between stepwise and concerted mechanism presented in this chapter were realized by applying the ECE process mathematical model, identical with the one described by Nicholson and Shain.

During investigation on elementary transfer coefficient (ETC) variability (sub chapters 6 and 7), we came across a series of results in which α variability and its nonlinear complex plots together with stepwise-concerted mechanism transition were reported e.g. papers quoted here in place before eq. (1)-(4). We recognized the respective parameter as the apparent α since its value was generally calculated with the use of formulae (5) or (6) (Matsuda & Ayabe, 1955) as well as by the convolution method (7)-(8) (Bard & Faulkner, 2001; Galus,

1994; Speiser, 1996a). Both of the models were derived for processes without chemical step and therefore applying the dependences (5) and (6) for ECE process has no appropriate physical basis. Consequently, kinetic parameters, plots and conclusions, obtained for the applied model cannot have clear physical sense (unless the chemical step is extremely slow or extremely fast and the process becomes E or EE, respectively).

$$E_{p/2} - E_p = \Delta E_{p/2} = \frac{1.857 \cdot R \cdot T}{(\alpha \cdot F)} \tag{5}$$

(two selected characteristic points of CV curve are taken into consideration)

$$\frac{\partial E_p}{(\partial \log v)} = \frac{1.15 \cdot R \cdot T}{(\alpha \cdot F)} \tag{6}$$

(several selected characteristic points of CV curve are taken into consideration)

$$I(t) = \frac{1}{\pi^{1/2}} \int_0^t \frac{i(u)}{(t-u)^{1/2}} du \tag{7}$$

$$\ln k_{app}(E) = \ln D_0^{1/2} - \ln \frac{\left[I_{lim} - I(t)\right]}{i(t)} \tag{8}$$

where I_{lim} is the limiting current of convoluted (semiintegrated) curve; the CV curve as a whole is taken into consideration).

The stepwise-concerted competition problem was solved (Sanecki & Skitał, 2002a; Skitał & Sanecki, 2009) by respective simulations presented shortly here. To compare the stepwise and concerted mechanisms on the basis of both elementary and apparent kinetic parameters approach, a procedure reverse to that known from the literature was applied. The well-known procedure involves recording of CV experimental curves from which α_{app} values are calculated using peak width formula (5) or (6) or (7,8) method. The resulting $\alpha_{app}=f(E,v)$ dependences are used to conclude about possible mechanism transition.

In our approach, CV theoretical responses for the whole spectrum of well defined mechanisms from purely stepwise (low k_f value) to purely concerted (high k_f value), are generated. Then, just like for experimental curves, α_{app} values and $\alpha_{app}=f(E,v)$ patterns are determined as specific for the implemented mechanism. The procedure and its results are shown in Fig. 2 and Fig. 3, respectively.

Fig. 3A illustrates that the range of sharp peaks cannot be interpreted as a mechanism transition symptom since it remains in pure stepwise area. The order of magnitude of k_f necessary to reach concerted area, determined from $I_p v^{-0.5} = f(v)$ dependence (Fig. 3 B,C), (see also (Sanecki & Skitał, 2002a)), is $>10^9$ s^{-1}. The results confirm that α_{app} values obtained from peak width (5) or dE_p/dv dependence (6) for ECE process do not judge stepwise/concerted alternative hypothesis. Simultaneously, it turned out that non linear α_{app} patterns of type as presented in Fig. 3A are characteristic for ECE processes within very pure stepwise mechanism.

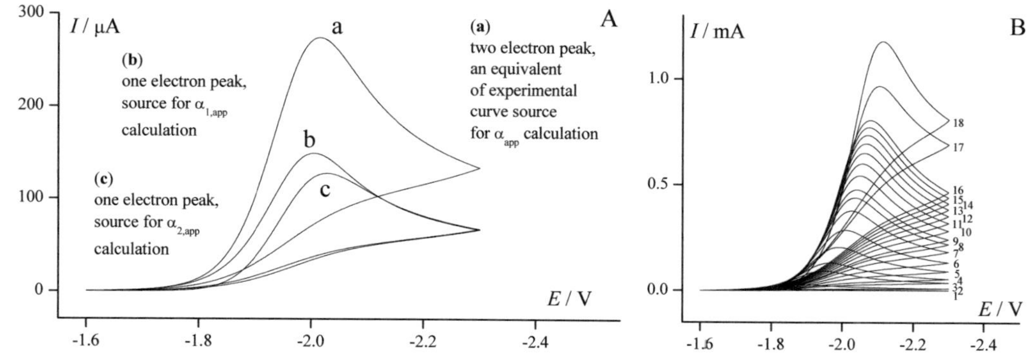

Fig. 2. (A) The theoretical CV responses obtained for the sequence of elementary steps given by eqs. (1) ÷ (3). (a): two-electron peak; (b): one-electron peak of step (1); (c): the same as (b) for step (3). (B) A series of two-electron peaks at various scan rates: 0.01, 0.1, 1, 2, 5, 10, 18, 25, 30, 40, 50, 60, 70, 80, 90, 100, 300, 500 V/s, increasing from the bottom to the top. [Reprinted from *Comput. Chem.*, Vol. 26, Sanecki, P. & Skitał P., The cyclic voltammetry simulation of ..., 297-311, Copyright (2002), with permission from Elsevier.]

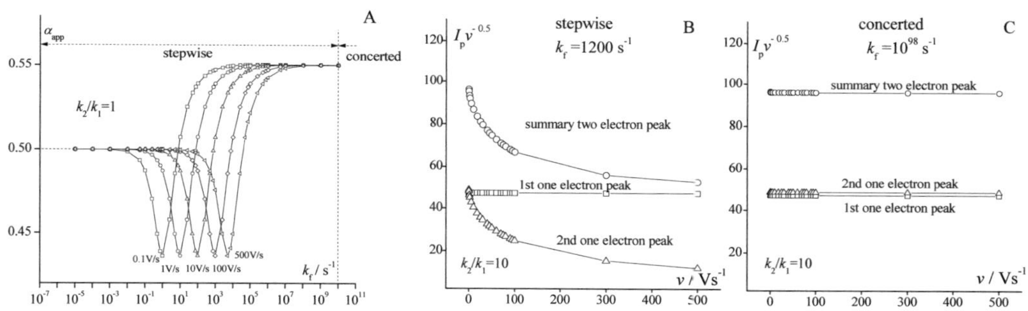

Fig. 3. The simulations of the sequence of (1) ÷ (3) steps. (A) The dependence of α_{app} on k_f at different scan rates. Input parameters: α_1=α_2=0.5. The output parameters were calculated from eq. (5). (B), (C). The relationship $i_p\, v^{-0.5}$ vs v as a diagnostic criterion. [Reprinted from *Comput. Chem.*, Vol. 26, Sanecki, P. & Skitał P., The cyclic voltammetry simulation of ..., 297-311, Copyright (2002), with permission from Elsevier.]

5.2 The analysis of complex current responses and α_{app} plots obtained as a result of convolution method

The result of application of convolution method to CV current responses is expected to be a linear type ln*k*=*f*(*E*) dependence. An α value can be determined from its slope (Bard & Faulkner, 2001). For ECE process, however, the line obtained from parent two-electron ECE peak is curved (the degree of its curvature depends on k_2/k_1 ratio and k_f value) and straight line approximation leads only to general, apparent parameters of no clear physical sense (Skitał & Sanecki, 2009). Moreover, the pseudo-linear plots clearly demonstrate involvement of two different steps of two-electron process (Fig. 4). Since linear regression is not sufficient here, the results of the convolution of two-electron ECE curves under discussion call for detailed mathematical procedure to evaluate kinetic parameters of individual steps. The simulation data (Fig. 4A) indicate that convolution method can be applied only to some boundary cases of ECE mechanisms as E or EE type with k_f from 10^9 value to $k_f=10^{98}$ as infinity, similarly as in Fig. 3 data.

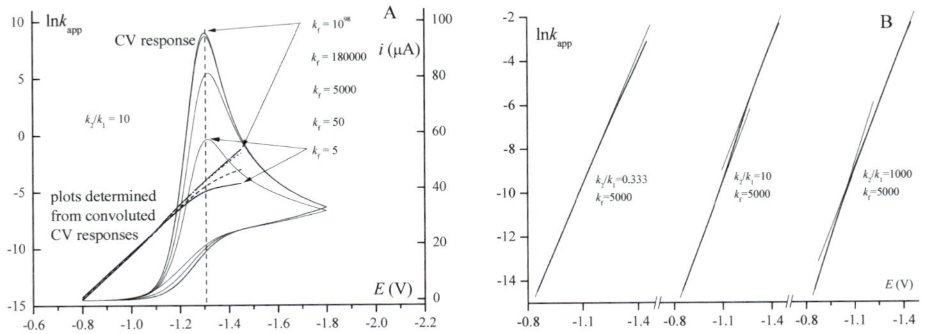

Fig. 4. (A) The influence of k_f on the shape of CV responses and $\ln k_{app}=f(E)$ dependences for simulated ECE mechanism. The interdependence between input CV curves and convolution results is shown. The length of non-linear curve fragment increases when k_f increases (from the bottom to the top). The rate constants k_{app} were calculated by convolution method (eq. (7) and (8)). (B) The influence of k_2/k_1 ratio on the shape of $\ln k_{app}=f(E)$ dependence obtained from convoluted CV responses for simulated ECE mechanism. Parameters of CV curves are as follows: k_1=0.03 cm s^{-1}, $\alpha_1=\alpha_2=0.5$, v= 1 Vs^{-1}. The k_2/k_1 and k_f values are presented on the plot. Note the increase of complexity of structure from the left to the right. For values of parameters and details see original paper (Skitał & Sanecki, 2009). [Reprinted from *Polish J. Chem.*, Vol. 83, Skitał P. & Sanecki P., The ECE Process in Cyclic Voltammetry. ..., 1127–1138, Copyright (2009), with permission from Polish Chemical Society.]

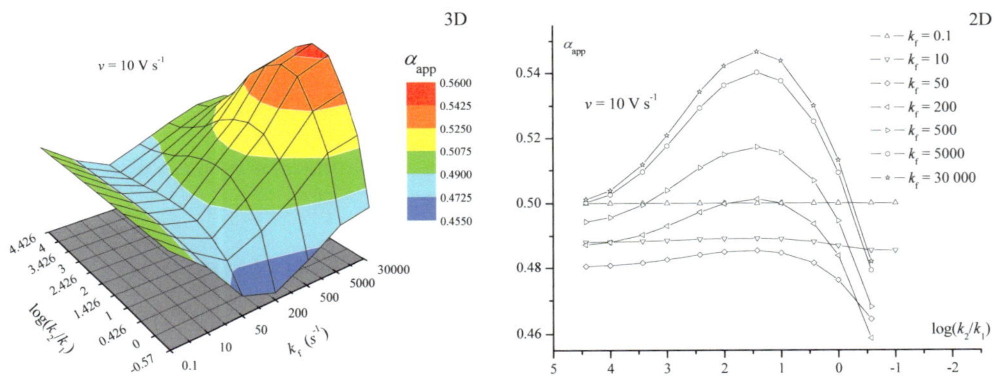

Fig. 5. The result of modeling of ECE process in cyclic voltammetry. An example of 3D plot: general parameter α_{app}, determined by convolution method, as a function of $\log(k_2/k_1)$ and k_f on 3D and 2D plots. Scan rate 10 V/s, k_1=0.03 cm s^{-1}; the values of kinetic parameters are in original paper (Skitał & Sanecki, 2009). For k_f value less than 0.1 s^{-1} (not included), where process can be simplified to E one, the respective 3D plot is flat. The 2D plot can be obtained by cutting the 3D one with a respective plane. [Reprinted from *Polish J. Chem.*, Vol. 83, Skitał P. & Sanecki P., The ECE Process in Cyclic Voltammetry. ..., 1127–1138, Copyright (2009), with permission from Polish Chemical Society.]

The fact of revealing of the complex structure of the considered ECE system in the form of broken linear or bent ln$k$=f(E) dependences (Fig. 4) makes the application of convolution method suitable only for coarse calculation of kinetic parameters or as precursor of estimation method. The results, in the form of α_{app} vs. k_f and k_2/k_1 plot presented in Fig. 5., indicate that the convolution method (7)-(8) applied leads to nonlinear α_{app} plots similar to these determined by means of eq. (5) and (6) and described in paper (Sanecki & Skitał, 2002a). The changes in α_{app} value and non-linear α_{app} patterns are present in stepwise mechanism zone and not in the range of its eventual change to concerted one.

Bent linear dependence ln$k$ vs. E from Fig. 4B is not surprising. It is well-known that such a processed electrochemical responses in form of bent straight lines reveal the complexity of the original response of consecutive process (Ružić, 1970, 1974). It means that system requires another kinetic model with elementary kinetic parameters. Similar situation can be found for some normal pulse polarography data (Sanecki & Lechowicz, 1997), some overlapped voltammetric data (Rusling, 1983) and in our simulation results (Fig. 4, 5).

The Fig. 5 data indicate, that linear, non-linear and various complex α_{app} plots of 2D type are only a special cases of 3D dependences. The dependences α_{app} on k_f as well as α_{app} on k_2/k_1 presented by us in (Sanecki & Skitał, 2002a) can be considered as the members of the same category.

In the light of presented facts, it is clear that variation of α_{app}, visualized by 2D and 3D plots is only an intriguing picture with no consequence for electrode kinetics since no change of mechanism was proven. As discussed earlier, the only cause of such plots origin is the fact, that paradigm of eqs. (5)-(6) as well as (7)-(8) (Bard & Faulkner, 2001; Greef et al., 1985) does not contain any chemical step.

In the process of modeling applied here (CV curves in Figs. 2, 4), no linear variability of ETC was either assumed or introduced into the model (α_i=0.5=const) i.e. normal Butler–Volmer's kinetics was applied. Therefore, the output α_{app} dependences (Figs. 3, 4 and 5) cannot contain α_{el} variability (Sanecki & Skitał, 2002a).

Conclusions

1. Stepwise and concerted processes are the limiting cases of ECE mechanisms with different chemical k_f constant (Sanecki & Skitał, 2002a; Sanecki & Skitał, 2007b).
2. For concerted mechanism it is possible to avoid the unclear defined $k_{concerted}$ constant (neither electrochemical nor chemical) in eq. (4) when kinetic description of ECE process with elementary kinetic parameters is applied.
3. The categories of nonlinear and complex α_{app} plots, as well as curved or bent lnk_{app}=$f(E)$ dependences as the result of application of EE or E or convolution kinetic model for ECE process, are a symptom of influence of chemical reaction on current response and not a change of mechanism. The observed non linear α_{app} variability and complex α_{app} plots are in accordance with Butler-Volmerian kinetics and are not a result of elementary α potential dependence.

6. The application of EC, ECE and ECE-ECE models with potential dependent transfer coefficient to selected electrode processes

The classical Butler-Volmer electrochemical kinetics with a priori assumed invariability of α ($\partial\alpha/\partial E$=0) provides approach which adequately describes the majority of observed electrode processes. The problem of α variability appears in some special cases of kinetic analysis e.g. during investigation of multi-stage electron transfer separated by a large potential interval, for series of substituted compounds with the same reactive group as well as for experimental long distance current-potential dependences of quasi-reversible processes (Corrigan & Evans, 1980; Matsuda & Tamamushi, 1979; Sanecki, 2001; Sanecki & Kaczmarski, 1999; Sanecki & Skitał, 2007b). The first case of the type was described by Pierce and Geiger as *alpha kinetic discrimination* (Pierce & Geiger, 1992). Currently, extended and still growing application of variety of solid electrodes (McDermott et al., 1992), including chemically modified and semiconductor electrodes, makes the problem of diversification and variability of transfer coefficient much more important than ever before.

The significance of the discussed problem is confirmed by simple calculations, based on relevant literature data (Angell & Dickinson, 1972; Garreau et al., 1979; Savéant & Tessier, 1982; Savéant & Tessier et al., 1977), which leads to the conclusion that ETC variability effect is usually 5-10 times stronger than the double layer correction effect (see also Sanecki & Skitał, 2007a).

Transfer coefficient is one of the fundamental concepts in electrode kinetics. The extensive theoretical support, including all modern theories of charge transfer, predicts the potential variability of ETC (Dogonadze, 1971; Hush, 1958; Levich, 1966; Marcus, 1956; Marcus, 1960; Marcus, 1977). The variation of ETC for reduction and oxidation processes of individual compounds and ions has been systematically investigated by Savéant's group, Corrigan and Evans and others (Angell & Dickinson, 1972; Bindra et al., 1975; Corrigan & Evans, 1980; Dogonadze, 1971; Frumkin, 1932; Garreau et al., 1979; Hush, 1958; Levich, 1966; Marcus, 1956; Marcus, 1960; Marcus, 1977; Matsuda & Tamamushi, 1979; McDermott et al., 1992; Nagy et al., 1988; Parsons & Passeron, 1966; Rifi & Covitz, 1974; Samec & Weber, 1973; Sanecki, 1986; Savéant & Tessier, 1975; Savéant & Tessier, 1982; Tyma & Weaver, 1980). The studies have been focused on the experimental detection of the ETC variability. The research goal has been achieved with minimal further continuation. In refs. (Angell & Dickinson, 1972; Bindra et al., 1975; Corrigan & Evans, 1980; Garreau et al., 1979; Matsuda & Tamamushi, 1979; McDermott et al., 1992; Nagy et al., 1988; Parsons & Passeron, 1966; Samec & Weber, 1973; Savéant & Tessier, 1975; Savéant & Tessier, 1982; Tyma & Weaver, 1980) one can easily find or calculate the $\partial\alpha/\partial E \neq 0$ values for the one electron processes and for resolved two electron reduction (Sanecki, 2001; Sanecki & Kaczmarski, 1999). The investigation of a problem of the potential dependent α has been continued by several authors (Chidsey, 1991; Finklea, 2001a, 2001b; Finklea & Haddox, 2001; Finklea et al., 2001; Haddox & Finklea, 2003; Miller, 1995; Smalley et al., 1995). What is more, well known analogy of α and β to α_B and β_B of the BrØnsted parameters (Albery, 1975; Frumkin, 1932; Rifi & Covitz, 1974), the ETC has also been interpreted as a Hammett-type reaction constant ρ (Sanecki, 1986). This interpretation, due to similarity of the homogenous and heterogeneous kinetics, allows the more flexible approach to the ETC value and is easy to understand by non electrochemists (Rifi & Covitz, 1974).

At present, the majority of electrochemists do not apply the ETC variability or consider it meaningless. Therefore, there is a need to indicate the special cases in which the effect may be of importance (Sanecki & Skitał, 2007a, 2007b). Our choice of respective experimental cases was focused on the following four examples (points 6.1-6.4). In turn, point 6.5 is devoted to comparison of IR_u drop and $\partial\alpha/\partial E$ parameter influence on CV responses.

6.1 Repetition of the important literature example of the observed elementary α variability case

Corrigan and Evans (Corrigan & Evans, 1980) showed that inclusion of α variability in form of $\partial\alpha/\partial E$ parameter into kinetic model of one-electron quasireversible reduction process of 2-nitro-2-methylpropane Eq. (9):

$$t\text{-BuNO}_2 + e^- \xrightarrow{k_1,\alpha_1} t\text{-BuNO}_2^{\bullet -} \xrightarrow{k_{ch}} t\text{-Bu}^{\bullet} + \text{NO}_2^- \quad (9)$$

leads to better fit between experimental and theoretical responses.

We obtained the result similar to that but for more extended scan rate range on GC and Hg electrodes in DMF solution, where the effect turned out to be more distinct. Our results are presented in Table 2 and in Fig. 6 ($\partial\alpha/\partial E = 0$ and $\partial\alpha/\partial E \neq 0$ for comparison) where the value $\partial\alpha/\partial E = 0.42\ V^{-1}$ was obtained as kinetic parameter by estimation for both electrodes.

The data shown in Table 2 indicate that our results, including the obtained $\partial\alpha/\partial E$ values, are in full agreement with the results obtained by Corrigan and Evans. In the light of presented facts the need of introducing α variability into kinetic model of the considered system is clear (Fig. 6). The better fit obtained for model with $\partial\alpha/\partial E$ included is confirmed by statistical data (Sanecki & Skitał, 2007a).

	(Corrigan & Evans, 1980)		(Sanecki & Skitał, 2007a)	
	Hg	Pt	GCE	Hg
$\partial\alpha/\partial E$, V^{-1}	0.37± 0.03	0.40	0.42±0.02	0.42±0.03
scan rate range	10-100 Vs^{-1}		0.1-100 Vs^{-1}	
medium	AN, TBAP	AN, TEAP	DMF, 0.3 M TBAP	
kinetic model	E		EC	

Table 2. CV reduction of 2-methyl-2-nitropropane. The comparison of our α variability results with analogical data published by Corrigan and Evans. [The Table reproduced by permission of The Electrochemical Society.]

6.2 Comparison of two ECE-ECE reduction steps with large potential difference (reduction of 1,3-benzenedisulfonyl difluoride)

A comparison of two identical reduction processes located at different potentials may be considered as a classical model for determination of α variability. The approach was first presented by Pearce and Geiger (Pierce & Geiger,1992). In turn, the α variability was determined by the comparison of two $-SO_2F$ group reduction on Hg electrode (Sanecki & Kaczmarski, 1999) ($\Delta\alpha$ method - the two different values of α were optimized by estimation). At present, the same system was investigated using the GCE electrode.

The results (Fig. 7) indicate that for electroreduction of 1,3-benzenedisulfonyl difluoride (1,3-BDF), the ECE kinetic model without α variability is not possible. There is a significant difference between the α's for both reduction stages i.e. between α_1 and α_3 as well as between α_2 and α_4 (note also the difference in a shape of the peaks). Fig. 7, the left column plots indicate that a model without $\Delta\alpha$ variability provides crossing experimental and theoretical plots. Both

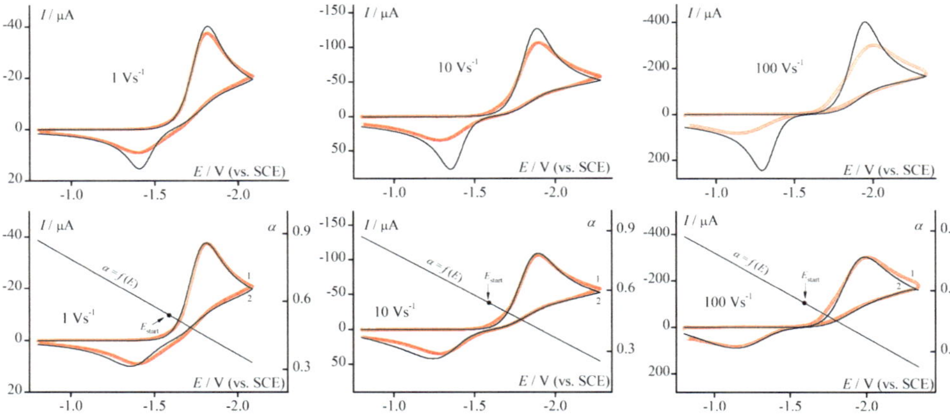

Fig. 6. CV reduction of 2-methyl-2-nitropropane. The comparison of two EC models: without (top row) and with (bottom row) α variability included. For parameters and details see original paper (Sanecki & Skitał, 2007a). [Reproduced by permission of The Electrochemical Society.]

of the approaches, i.e. introducing of different α values ($\Delta\alpha$ variability) and continuous α variability ($\partial\alpha/\partial E$), gave very similar results. The small potential difference and respective confidence intervals do not allow determination of the difference between α's within each of the two considered reduction stages (two-electron reduction of SO_2F). Moreover, the comparison of various electron transfers steps: anion radical creating (1) and radical reducing (2) is not appropriate for α variability determination.

The results and comparison of the two applied methods of α variability determination are presented in Table 3.

The method of α variability determination	**GC electrode** (Sanecki & Skitał, 2007a)	**Hg electrode** (Sanecki & Skitał, 2007a)	**Hg electrode** (Sanecki & Kaczmarski, 1999)
Estimation of the different α values	ΔE_p=0.705 V $\partial\alpha/\partial E$ = 0.20 ± 0.07	ΔE_p=0.749 V $\partial\alpha/\partial E$ = 0.51 ± 0.09	$\partial\alpha/\partial E$ = 0.44 ± 0.04
Estimation of the $\partial\alpha/\partial E$ parameter	$\partial\alpha/\partial E$ = 0.20 ± 0.03 V^{-1}	$\partial\alpha/\partial E$ = 0.41 ± 0.07	

Table 3. The comparison of α variability values for the electroreduction of 1,3-BDF. [The Table reproduced by permission of The Electrochemical Society.]

6.3 Comparison of two reduction steps within ECE process with a small potential difference. Reduction of *p*-toluenesulfonyl fluoride

The comparison of experimental CV responses for ECE reduction of *p*-toluenesulfonyl fluoride (TSF) (dissociative electron transfer process) with respective $\Delta\alpha$ variability model and variable α with $\partial\alpha/\partial E$ parameter included model is presented in paper (Sanecki & Skitał, 2007a). Due to small potential difference between the steps (56 mV at v=1 Vs^{-1}), where the steps are not naturally resolved, the difference between models with α variability included and not included is insignificant here. Moreover, the two ways of α variability introduction lead to almost the same dependences which suggests that the models are equivalent (Sanecki & Skitał, 2007a).

6.4 Comparison of the *p*-toluenesulfonyl fluoride and benzenesulfonyl fluoride reductions. The substituent effect

When the two consecutive steps of two-electron process are not naturally split and $k_2<k_1$ or $k_2\approx k_1$, it is possible to determine the elementary transfer coefficients with satisfactory accuracy. For such cases the difference of peak potentials is seen on calculated elementary steps parameters (Sanecki & Skitał, 2007a). When $k_2>k_1$, the respective confidence intervals for second electrochemical step are wider than those for the first one in case of 1,3-BDF electroreduction (Fig. 7).

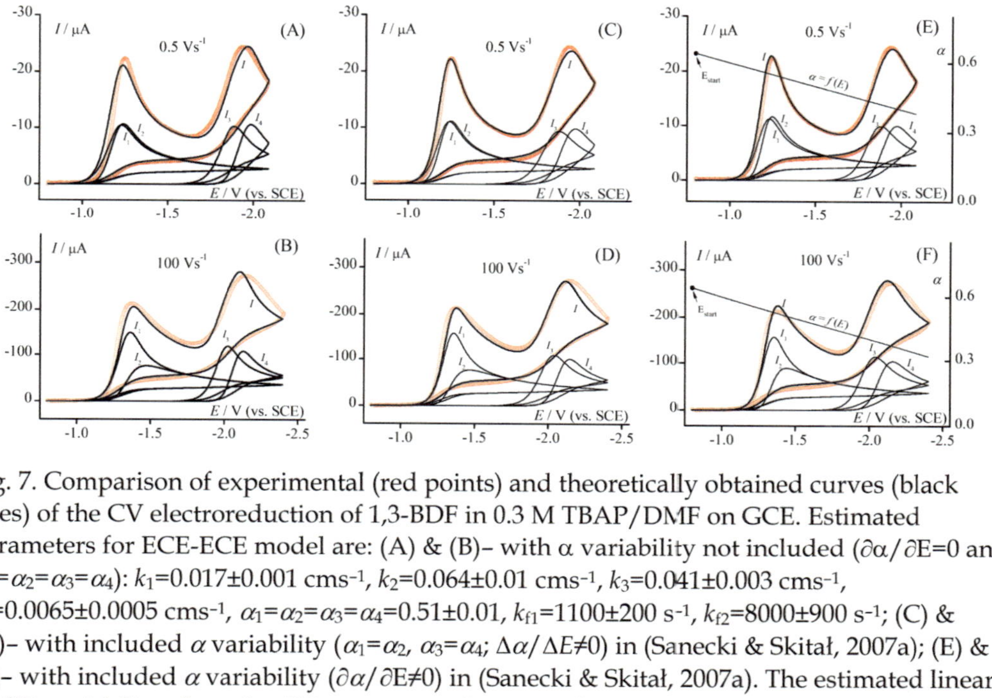

Fig. 7. Comparison of experimental (red points) and theoretically obtained curves (black lines) of the CV electroreduction of 1,3-BDF in 0.3 M TBAP/DMF on GCE. Estimated parameters for ECE-ECE model are: (A) & (B)- with α variability not included ($\partial\alpha/\partial E=0$ and $\alpha_1=\alpha_2=\alpha_3=\alpha_4$): $k_1=0.017\pm0.001$ cms^{-1}, $k_2=0.064\pm0.01$ cms^{-1}, $k_3=0.041\pm0.003$ cms^{-1}, $k_4=0.0065\pm0.0005$ cms^{-1}, $\alpha_1=\alpha_2=\alpha_3=\alpha_4=0.51\pm0.01$, $k_{f1}=1100\pm200$ s^{-1}, $k_{f2}=8000\pm900$ s^{-1}; (C) & (D)- with included α variability ($\alpha_1=\alpha_2$, $\alpha_3=\alpha_4$; $\Delta\alpha/\Delta E\neq0$) in (Sanecki & Skitał, 2007a); (E) & (F)- with included α variability ($\partial\alpha/\partial E\neq0$) in (Sanecki & Skitał, 2007a). The estimated linear $\alpha=f(E)$ variability along the CV response is shown on the plot. The scale for α is the right vertical axis. [Reproduced by permission of The Electrochemical Society.]

The effect of substituent on elementary α was considered in our previous papers (Sanecki, 2001; Sanecki & Kaczmarski, 1999). Now we compare the reduction of two compounds with $\Delta E_p\approx0.1$V to show the limit of possibility of α variability detection when two similar (with identical reducible group) compounds are compared (Fig. 8).

The two substituted compounds values of $\partial\alpha/\partial E$ obtained by comparison of respective kinetic parameters are quite reasonable but due to the small difference in α value must be charged with significant error. It suggests that for two substituted compounds, when difference in peak potential is under 0.1 V, the difference in α values is unrecognizable and variability is not a problem to consider. The similar picture as presented in Fig. 8 and the respective calculation results for two substituted compounds as well as for the series of iodobenzenes examples of α variability can be found in paper (Sanecki, 2001).

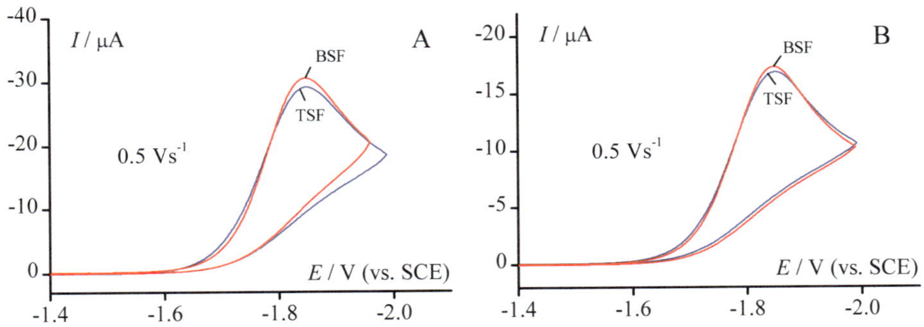

Fig. 8. The comparison of CV reduction of benzenesulfonyl fluoride (BSF) (red line) and TSF (blue line) in 0.3M TBAP/DMF on: (A) GC and (B) Hg electrodes. The curves on plots (A) and (B) were shifted to compare the slopes: plot (A) on GCE by ΔE= -0.122 V, plot (B) on Hg by ΔE= -0.098 V. Note the small difference of slopes which corresponds to a difference of α. The same effect is seen on estimated α values: α_1=0.67±0.02 for BSF and α_1=0.62±0.02 for TSF both on GCE. On Hg electrode for BSF α_1=0.67±0.02, for TSF α_1=0.62±0.02. It results in $\partial\alpha_1/\partial E$ = 0.41±V^{-1} for GCE and $\partial\alpha_1/\partial E$=0.61 V^{-1} for Hg electrode. For other parameters see original paper (Sanecki & Skitał, 2007a). [Reproduced by permission of The Electrochemical Society.]

6.5 Comparison of IR$_u$ drop and $\partial\alpha/\partial E$ parameter influence on CV responses

The influence of IR$_u$ effect and non-constant transfer coefficient effect on current response shape can be similar and there is a real risk of making a mistake of detecting a non existing $\partial\alpha/\partial E$ effect. Therefore, the comparison of both effects by simulations is presented in Fig. 9.

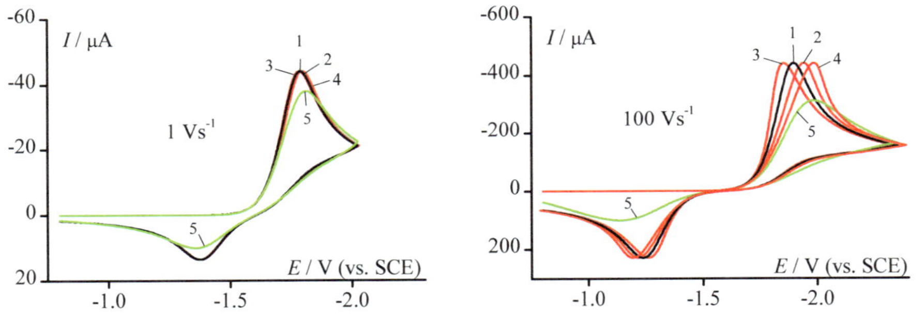

Fig. 9. Theoretical CV current responses for EC kinetic model. The influence of uncompensated resistance R$_u$ and included α variability for 1Vs^{-1} and 100Vs^{-1}. The scan rate value is the one difference between left and right plot. Curves (1),(2),(3),(4): variable α model with $\partial\alpha/\partial E$=0 and R$_u$=0Ω, 100Ω, -100Ω, 200Ω, respectively; curve (5): variable α model with $\partial\alpha/\partial E$=0.4 and R$_u$=0Ω. For other parameters see the paper (Sanecki & Skitał, 2007a). [Reproduced by permission of The Electrochemical Society.]

The IR_u compensation problems have been resolved by the method of estimation with the use of theoretical peaks (Sanecki et al., 2003). The series of data in 0.1–100 V/s scan rate range for EC, ECE, and ECE-ECE mechanisms, which correspond to the investigated compounds, were compared.

At first, series of theoretical EC responses were generated, for input set of kinetic parameters. The obtained responses were treated as pseudo-experimental ones and the output kinetic parameters were evaluated from them by a standard estimation procedure. The obtained results indicate that generally, for all kinetic models, IR_u attenuates the determination of exact values not only $\partial\alpha/\partial E$, but all input values of kinetic parameters. For EC model, the influence of IR_u effect is expressed on k_1 to a higher degree, and less on $\partial\alpha/\partial E$ and α_{start}.

The IR_u effect depends on a scan rate value. It is practically invisible at low v up to 1–2 Vs^{-1} (Fig. 9). The effect increases gradually with v and at 25, 50 and 100 V/s is clearly visible, for instance in the range 5–100 Vs^{-1} R_u=100 Ω produces non existing $\partial\alpha/\partial E$ = 0.01–0.04 V^{-1}. Hence, the low scan rate data are safe for $\partial\alpha/\partial E$ determination, even in case of lack of IR_u compensation (not recommended). The $\partial\alpha/\partial E \neq 0$ effect, however, is generally observed as a visible gap in fits for all scan rate peaks (examples: Fig. 9). The influence on the peak shape is different for both effects. The IR_u effect decreases the slope without change of the peak height and passes E_p towards negative potentials. The $\partial\alpha/\partial E$ effect decreases both the slope and the peak height with small influence on E_p.

An overcorrection done on IR_u affected responses is not able to provide a variable α effect, since after IR_u correction CV peaks are steeper which is exactly opposite to observed $\partial\alpha/\partial E$ effect on CV response (Fig. 9). On the contrary, the real risk is the CV response charged with uncompensated IR_u, which can be erroneously considered as comprising an $\partial\alpha/\partial E$ effect.

Other important results obtained from extensive simulations are as follows. For $IR_u>0$ (the case of undercompensation), the obtained $\partial\alpha/\partial E$ (and itself α) increases with v. For $IR_u<0$ (the case of overcompensation), the obtained $\partial\alpha/\partial E$ (and itself α) decreases with v and its negative values can be obtained. For $IR_u=0$ (the proper compensation), the obtained $\partial\alpha/\partial E$ are constant with v.

The recognized rules can be treated as diagnostic criteria. For the other ECE and ECE-ECE models under consideration, the above rules and conclusions hold, although they are less distinct for potential dependent α parameter due to small sensitivity of their kinetic systems.

7. Alkyl iodides electroreduction and α variability

The process of the electroreduction of organic iodides is an example of two-electron reductive cleavage reaction and a subject of extensive investigation in the organic electrochemistry (Andrieux et al., 1979; Caldwell & Hacobian, 1968a, 1968b; Colichman & Kung Liu, 1954; Hush, 1957; Hussey & Diefenderfer, 1967; Jaworski et al., 1992; Mairanovskii, 1969; Mairanovskii & Rubinskaya, 1972; Mairanovskii et al., 1975; Pause et al., 2000; Peters, 1991; Sanecki, 2001; Sawyer et al., 1995; Sease et al., 1968; Stackelberg, 1949;). The established mechanism is relevant for the cleavage of carbon-halogen bond and has been successfully applied also for the other categories of compounds (e.g. (Andrieux et al., 1996, 1997; Jakobsen et al., 1999)).

The influence of carbon chain length and the number of iodine atoms in the molecule was examined. The two-electron reductive cleavage with iodine elimination was numerically resolved into one-electron consecutive steps. The mechanism of the process was discussed in the frames of the two mathematical models: EC(C)E (with radical dimerisation) for one stage electroreduction of RI as well as ECE-EC(C)E kinetic model for two stage electroreduction of RI_2. In both models transfer coefficient variability was included. For electroreduction of diiodomethane the discrimination of elementary α between two reduction stages was determined (Sanecki & Skitał, 2007b). The α variability was determined using estimation procedure in which $\partial\alpha/\partial E$ was treated as other kinetic parameters and included into kinetic model (Corrigan & Evans,1980; McDermott et al.,1992).

The experimental facts and generally accepted relevant literature data suggested EC(E)E model with k_f of very high value (concerted process). Additionally, the estimation results denoted inclusion of ETC variability into the model. The obtained values of $\partial\alpha/\partial E$ parameter for monoalkyl iodides were in range between 0.33–0.36 V^{-1} with maximum confidence interval equal to ±0.09. The respective value for diodomethane was 0.20±0.04 V^{-1} (Fig. 10).

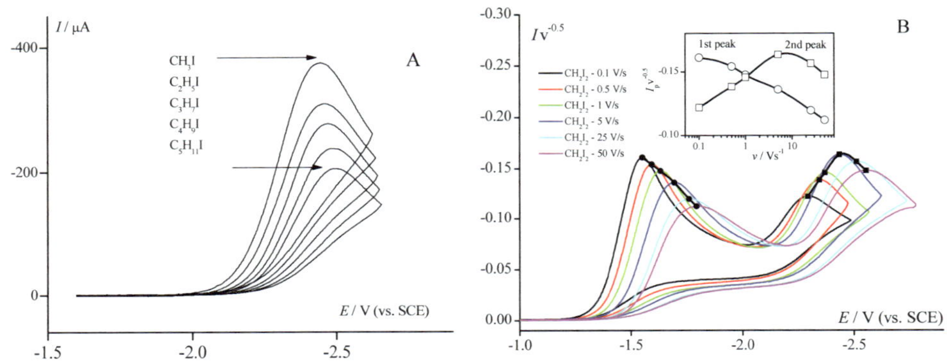

Fig. 10. (A) The CV electroreduction of five alkyl iodides in 0.3 M TBAP in DMF on GCE. The concentration of substrate was 4 mM. Scan rate $v = 5$ Vs^{-1}. Other details are specified in the original paper (Sanecki & Skitał, 2007b). (B) The CV electroreduction of CH_2I_2 in 0.3 M TBAP/DMF solution on GCE. The normalized CV current responses $Iv^{-0.5}$. The $I_p v^{-0.5}$ values for both reduction stages are shown as respective points on maximum current or, for clarity, separately in the sub-window. Note the different slope of kinetic $I_p v^{-0.5}$ vs v dependence in subwindow. [Reprinted from *Electrochim. Acta*, Vol. 52, Sanecki, P. & Skitał P., The electroreduction of alkyl iodides ..., 4675-4684, Copyright (2007), with permission from Elsevier.]

Conclusions

1. The electroreduction of mono alkyl iodides and diiodomethane on glassy carbon electrode can be described by EC(C)E and ECE-EC(C)E numerical model, respectively. The second electrochemical step of the EC(C)E process is generally slower then the first one.
2. The comparison of CV responses and the values of determined kinetic parameters for a sub-series of five mono iodides indicates that their slope and elementary transfer coefficient do not vary with a change of the alkyl substituent. The fact can be explained

by a small difference of reduction potentials caused by weak substituent effect of alkyl groups.

3. The determined $\partial\alpha/\partial E$ value for diiodomethane is equal to 0.20 V^{-1} and is about two times lower than that obtained in literature for aromatic compound electroreduction on Hg (Sanecki & Kaczmarski, 1999). For the other elementary processes the values in the range 0.174-0.4 V^{-1} were obtained (Savéant & Tessier, 1982). On the other hand, the respective values obtained for monoalkyl iodides are higher (Sanecki & Skitał, 2007b).
4. The respective comparison of elementary kinetic parameters for two-stage diiodomethane electroreduction on GCE indicates the presence of alpha kinetic discrimination with $\partial\alpha/\partial E$=0.15 V^{-1} according to Pearce and Geiger nomenclature (Pierce & Geiger, 1992).

8. Theoretical discovery of *isoalpha points*

8.1 Parameter $\partial\alpha/\partial E$ or non-continuous change of α ($\Delta\alpha$) as a source of revealing particular points on CV curve

The α variability was firstly used as a tool to reveal the particular properties of the investigated system in (Sanecki & Skitał, 2008). It was shown that respective simulations lead to discovery of the new meaningful phenomenon called *isoalpha* effect (Fig. 11,12). The inclusion of the $\partial\alpha/\partial E\neq0$ parameter or non-continuous variability of α ($\Delta\alpha$) into kinetic model of E, EC, ECE, ECE-ECE and of E_rCE_rC–E_rCE_rC mechanisms results in the change of the single CV curve into the set of CV curves comprising specific intersection point(s), called α independent current-potential point or *isoalpha point (iap)*. The place of *iap* appearance depends on v and/or k value (Fig. 12). The *isoalpha point* (E_{iap}, I_{iap}) can be, to some extend, treated as electrochemical analog of *isosbestic* point (Sanecki & Skitał, 2008).

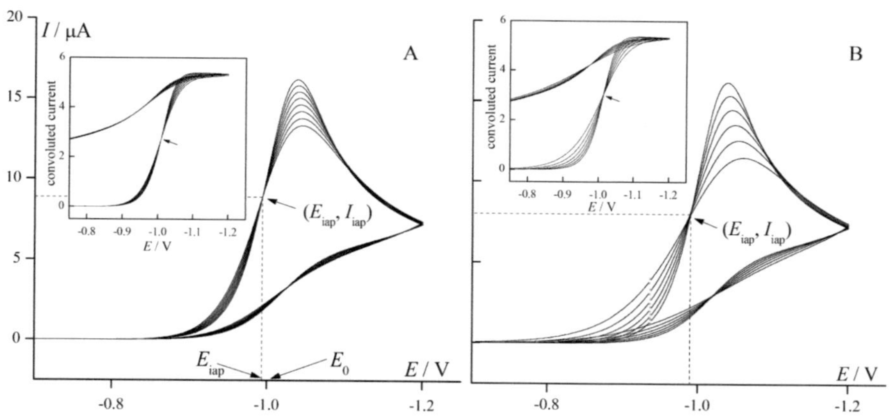

Fig. 11. The discovery of *isoalpha points* as a result of including of $\partial\alpha/\partial E\neq0$ parameter (A) or different α values (B) into kinetic model of one electron reduction step on the theoretical current response. The one-electron CV responses for E_{irr} process (A): $\partial\alpha/\partial E$ = 0; 0.2; 0.4; 0.6; 0.8; 1.0; 1.2; 1.4 increasing from the top to the bottom curves; (B): α = 1; 0.9; 0.8; 0.7; 0.6; 0.5 decreasing from the top to the bottom curves. The insets show the convoluted form of main plot with revealed *isoalpha point*. Other details are in original paper (Sanecki & Skitał, 2008). [Reprinted from *Electrochim. Acta*, Vol. 53, Sanecki, P. & Skitał P., The mathematical models of kinetics ..., 7711-7719, Copyright (2008), with permission from Elsevier.]

Parameter $\partial\alpha/\partial E$ into respective kinetic equation was previously applied by Corrigan and Evans (Corrigan & Evans, 1980) as well as McDermott and coauthors (McDermott et al., 1992). The application of the non-continuous ETC variability parameter ($\Delta\alpha$), in which each curve is generated for different α value was applied (Sanecki & Kaczmarski, 1999; Sanecki & Skitał, 2007a, 2007b; Sanecki et al., 2006b).

Theoretical curves visualized in Fig. 11A were obtained by the algorithm which includes $\partial\alpha/\partial E$ in the range 0 - 1.4 V^{-1}. Despite the fact, that experimental values of $\partial\alpha/\partial E$ do not exceed 0.5 V^{-1}, the application of wider range of $\partial\alpha/\partial E$ results in more extended set of curves. For the same reason α=1 as a starting value was chosen (Fig. 11B). When the variability starts at α=0.5, the intersection point (*iap*) remains the same but the plot is less convincing.

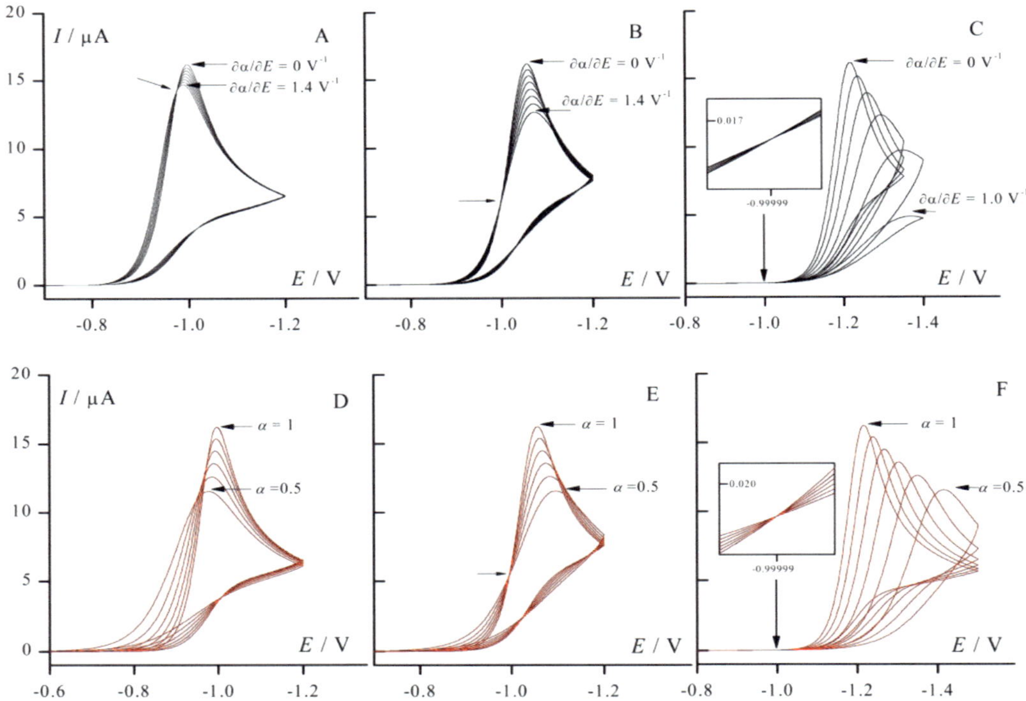

Fig. 12. The effect of including of $\partial\alpha/\partial E\neq0$ parameter (plots (A), (B), (C)) or different α values (plots (D), (E), (F)) into theoretical one-electron kinetic model E_{irr} as well as the influence of k value on isoalpha point current response. Plots (A), (B), (C) with $\partial\alpha/\partial E$ = 0; 0.2; 0.4; 0.6; 0.8; 1.0; 1.2; 1.4. (A) k=0.05, (B) k=0.005, (C) k=0.00001. Plots (D), (E), (F) with α = 1; 0.9; 0.8; 0.7; 0.6; 0.5; $\partial\alpha/\partial E$ = 0, (D) k=0.05, (E) k=0.005, (F) k=0.00001. The calculations were done with the use of both *ESTYM_PDE* (black lines, E_{irr}) and *DigiSim®* (red lines, E_rC with $k_f=1\times10^{10}$ s^{-1}). Notice, that the both programs provide identical current responses (plots (D), (E), (F)). The insets ((C) and (F)) show the enlarged form of background fragment. Other details are in original paper (Sanecki & Skitał, 2008). [Reprinted from *Electrochim. Acta*, Vol. 53, Sanecki, P. & Skitał P., The mathematical models of kinetics ..., 7711-7719, Copyright (2008), with permission from Elsevier.]

As a result of including of the ETC variation either through $\partial\alpha/\partial E$ parameter or by different values of α, a single curve becomes the set of curves which intersect in one characteristic point in which cathodic current is independent of α (Fig. 11 A and B, respectively). The result is similar to isosbestic point (Berlett et al., 2000; IUAPAC, 1997; Nakajima et al., 2004). The system resembling *iap* and isosbestic point, called *isopotential point* has been described in electrochemistry (Edens et al., 1991; Eichhorn & Speiser, 1994; Fitch & Edens, 1989). Its appearance is associated with the linked multi-sweep experiments when two distinguishable electroactive species interconvert.

The model for totally irreversible process applied with the use of *ESTYM_PDE* was described in paper by Nicholson and Shain (Nicholson & Shain, 1964) and is given in Bard and Faulkner's monograph (Bard & Faulkner, 2001). It is not included in *DigiSim*® software. In *DigiSim*®, however, the E_{irr} mechanism response was easily generated as E_rC with very high k_f value e.g. $k_f = 1\times10^{10}$ s^{-1} (Fig. 12 D,E,F).

A question arises why other authors dealing with the effect of potential-dependent α in voltammetry (Arun & Sangaranarayanan, 2004; Bieniasz & Speiser, 1998a; Bond & Mahon, 1997; Delahay, 1953; Nahir et al., 1994; Tender et al., 1994; Weber & Creager, 1994) did not observe so far such a phenomenon as *iap*, even if Marcus' λ parameter was discussed instead of α. Most probably there are three reasons. Firstly, the range of the phenomenon is not very wide e.g. it appears for $0.05 > k > 1\times10^{-4}$, at v =1Vs^{-1} (for k=0.05, Fig. 12 A,D the *iap* is diffused). At lower k values *iap* occurs but is invisible unless a magnification method is applied (Fig. 12 C,F). Secondly, is the phenomenon to be observed requires at least three (or more) superimposed curves and that was not fulfilled by the other authors. Even if one maintains all parameters constant except for α, no *iap* is observed and CV peak moves towards more negative potentials becoming flatter as α decreases. Examples of the lacking or not visible *iap* can be found in paper by Delahay (Fig. 2 in (Delahay, 1953)).

The third reason in form of examples was shown in Fig. 12 C,F. The data indicate that it is not possible to find *iap* without magnification of the plot and therefore some authors could not notice such a hidden effect. The observed *iap* phenomenon is not contradictory to the well established theory and provides its completion.

The similar, although less distinct *isoalpha points* are obtained for E_rC process when $\partial\alpha/\partial E\neq0$ parameter or different α values mode is introduced into the model (Fig. 12).The influence of scan rate and E_0 value on position *iap* is discussed in (Sanecki & Skitał, 2008). It is seen that *iap* moves down when scan rate increases and its value becomes closer to E_0 value.

8.2 ECE and ECEC mechanisms with included α variability

The kinetics should concern the steps and elementary parameters, which have clear physical meaning as it was performed earlier for E_{irr} process (Fig. 11,12). On the other hand, data applied for ECE process, allowed to generate dependences corresponding to those from Fig. 2 and may be important for ECE process kinetics. Examples of *isoalpha points* for both general (two-electron) peaks and elementary (one-electron) sub-peaks are presented in Fig. 13A,B.

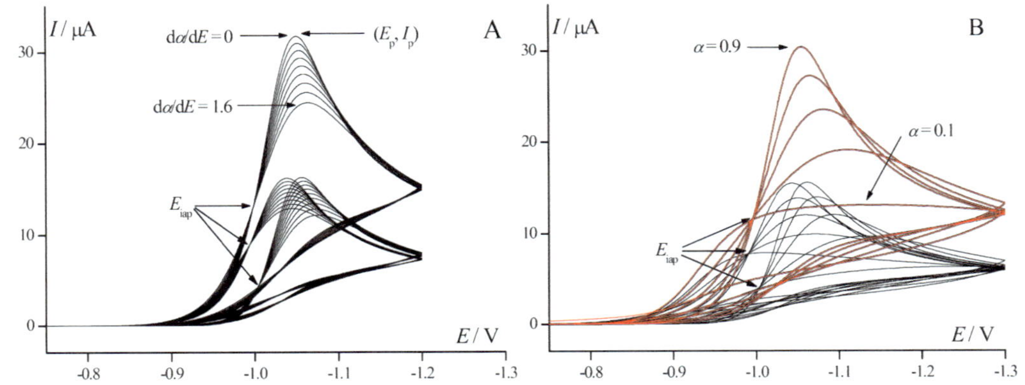

Fig. 13. (A) The *isoalpha* plots for $E_{irr}CE_{irr}$ mechanism. The $\partial\alpha/\partial E$ value is 0, 0.2, 0.4, 0.6, 0.8, 1.0, 1.2, 1.4, 1.6 from the top to the bottom. (B) The *isoalpha* plots for E_rCE_rC mechanism. The α value is 0.1, 0.3, 0.5, 0.7, 0.9 from the bottom to the top. The *isoalpha point* appears on both two-electron and one-electron CV responses. The calculations were done with the use of both *ESTYM_PDE* (black lines) and *DigiSim*® (red lines). Other details are in original paper (Sanecki & Skitał, 2008). [Reprinted from *Electrochim. Acta*, Vol. 53, Sanecki, P. & Skitał P., The mathematical models of kinetics ..., 7711-7719, Copyright (2008), with permission from Elsevier.]

Inclusion of $\partial\alpha/\partial E$ into ECE kinetic model, with other parameters being constant, results in three sets of curves: one for two-electron and two for one-electron stages (Fig. 13A, B). Each set has characteristic intersect point E_{iap} as it was observed earlier for one-electron process (Fig. 11,12).

8.3 E_rCE_rC–E_rCE_rC mechanisms with included α variability

The data presented in Fig. 13A for ECE and ECEC mechanism indicate one E_{iap} point on cathodic part of CV curve. For ECE-ECE and ECEC-ECEC sequences at least two E_{iap} points appear for both continuous (not presented) and non-continuous (Fig. 14A) variability of α, respectively. For the both mechanisms the position of *isoalpha points* is close to $E_{0,i}$ values.

It is worth to notice that apart from α potential dependent kinetic cases considered here, the, to some extent similar, intersection point appears for pure diffusion kinetics, non sensitive to the α value at all (Fig. 14B). The comparable diffusion kinetic case is presented in literature (Greef et al., 1985), Figure 6.5, p. 184) but without discussion in the context presented here.

9. Experimental confirmation of *isoalpha* points reality

This sub-chapter deals with α variability resulting from the change of electrode material properties caused by its modification. Under experimental conditions the change of electrode properties results in both: (1) a change of α i.e. the slope of log k vs. E dependence (electrochemical reaction constant $\alpha = -\rho$); (2) a change of electrochemical rate constant k.

Therefore, the experimental confirmation of predicted by theory *isoalpha point* requires a system with α change separately. For the most of experimental systems, the changes of α and k are coupled together and therefore a shift of CV curves along potential axis is possible and *isoalpha* point appearance may be disfigured. Despite of the above, experimental confirmation of existence of *isoalpha point* is possible by the use of the set of electrodes with slightly different electronic properties.

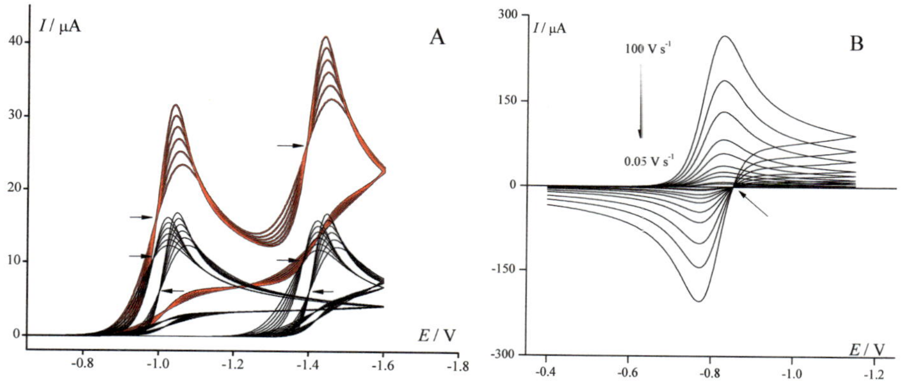

Fig. 14. (A) The theoretical CV current responses of E_rCE_rC–E_rCE_rC process with non-continuous α variability included. Parameter $\alpha_1=\alpha_2=\alpha_3=\alpha_4=1$, 0.9, 0.8, 0.7, 0.6, 0.5 decreases from the top curve to the bottom one. The calculations were done with the use of both *ESTYM_PDE* (black lines) and *DigiSim®* (red lines). Notice that both programs provide exactly the same current responses. Analogical picture is observed when $\partial\alpha/\partial E$ is the source of *isoalpha point.* (B) The appearance of another iso-current point for pure diffusion kinetic case. The one-electron CV responses for E_r process were generated. The v Vs^{-1} decreases from the top to the bottom. Other details are in original paper (Sanecki & Skitał, 2008). [Reprinted from *Electrochim. Acta*, Vol. 53, Sanecki, P. & Skitał P., The mathematical models of kinetics ..., 7711-7719, Copyright (2008), with permission from Elsevier.]

The Hg/Zn system, widely investigated in both aqueous and nonaqueous media, was chosen as a test ((Manzini & Lasia, 1994; Sanecki et al., 2006a) and literature therein). It turned out that Hg drop, after one or more CV cycles, behaves somehow differently even in the situation when metallic Zn was removed from it.

Fig. 15 presents the effect of Hg drop electrode pretreatment on CV response; for details reader is referred to the caption and original paper. The experimental results indicate (Fig. 15D) that an attempt for finding *iap* was successful. The experimental data, selected from Fig. 15 A,B,C, namely curves scanned for 0 and 1 minute intervals (Fig. 15A), provide the possibility of fitting experimental and simulated CV data which exploit *iap* property on rising part of the curve (Fig. 15D plot). The α variability parameter obtained by a comparison of α for both curves equals to $\partial\alpha/\partial E=0.13$. The observed variability of α is due to the change of electrode properties and not to solution parameters in Marcus' sense.

In our case each scan starts with the same concentration of both redox forms. Such conditions are not sufficient for the isopotential point formation of the type described in the literature (Edens et al., 1991; Eichhorn & Speiser, 1994; Fitch & Edens, 1989). In absence of changes of supporting electrolyte concentration, the only source of the *iap* appearance remains $\partial\alpha/\partial E$ or $\Delta\alpha$. Therefore, the occurrence of *iap* indicates that in the investigated system the effect of α variability takes place. Another experimental example of isoalpha

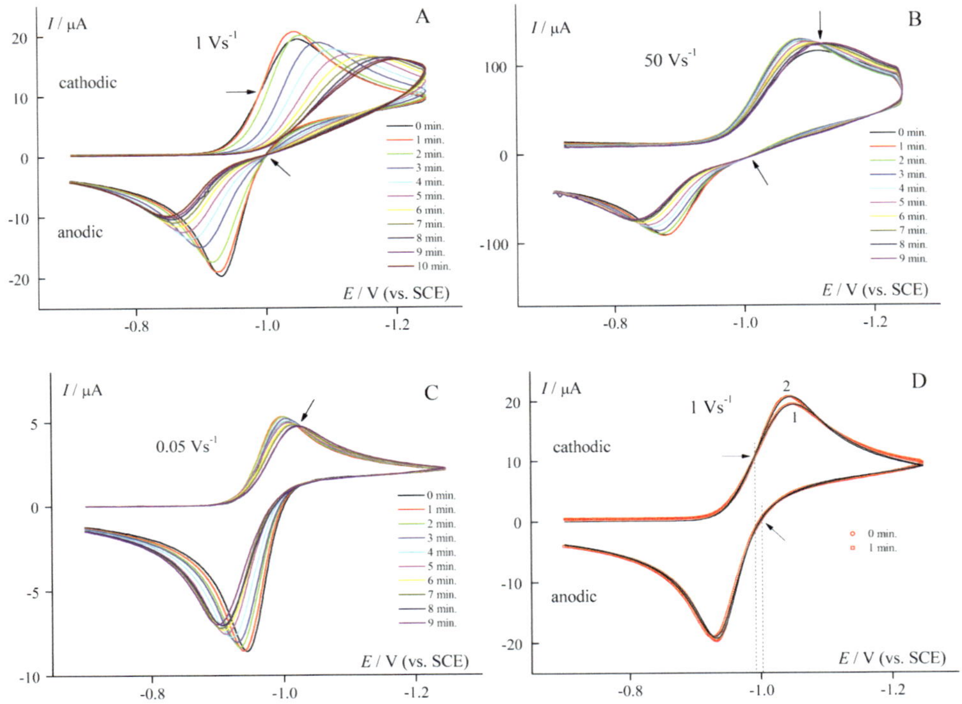

Fig. 15. The effect of non-linked multi scans CV experiment on the same mercury drop. The experimental system: 1mM Zn^{2+} in 1M $NaClO_4$. Plots (A) and (B) the scans were repeated in 1 minute intervals, at open circuit without polarization; $E_{0,exp}$= -0.970 V for Zn/Zn^{2+} system, E_{iap}= -1.000V. Plot (C) – the nine scans were repeated in 1 minute intervals, at starting potential -0.700 V. At the interval metallic Zn was removed from the drop by electrolysis. Plot (D) the comparison of Zn^{2+} electroreduction on pure Hg drop (curve 1, scan 0 min., red points) with the same process on the drop with minor trace of metallic Zn (curve 2, scan 1 min., red points) as well as the respective theoretical curves (black lines) according to extended EE‖Hg(Zn) kinetic model (Sanecki et al., 2006a). The curves 1 and 2 were selected from plot (A). The estimated alpha values: (scan 0) $\alpha_1=\alpha_2$=0.66±0.02, (scan 1) $\alpha_1=\alpha_2$=0.785±0.015. The observed *isopotential points* and scan rates are indicated on the plots. The other details are in original paper (Sanecki & Skitał, 2008). [Reprinted from *Electrochim. Acta*, Vol. 53, Sanecki, P. & Skitał P., The mathematical models of kinetics ..., 7711-7719, Copyright (2008), with permission from Elsevier.]

point occurrence is provided by hydrogen ions reduction (Fig. 16). Increasing concentration of atomic hydrogen with scan rate decrease is expected to be a source of modification of active electrode surface. In consequence, the *isoalpha point* appeares.

The presented experimental results (Fig. 15, 16) together with theoretical ones (Fig. 11-14) suggest that the change of electrode properties should be considered as an origin of occurrence of the *iap*. It can be supported by a quotation from Corrigan and Evans (Corrigan & Evans, 1980): "*For all of these cases, eq. $d\alpha/dE = F/2\,\lambda$ does not adequately describe the potential dependence of α because λ is a solution parameter which should not depend on surface conditions.*"). The results by Mattheiss and Warren (Matthesiss & Warren, 1977) showed that anymicroscopic inhomogeneity may significantly affect the Fermi level of electrons and, therefore, the kinetics of electrode process. The similar effect of the change of electrode properties caused by the electrodeposition process is well recognized for solid electrodes (Greef et al., 1985; Sanecki et al., 2006a; Sawyer et al., 1995).

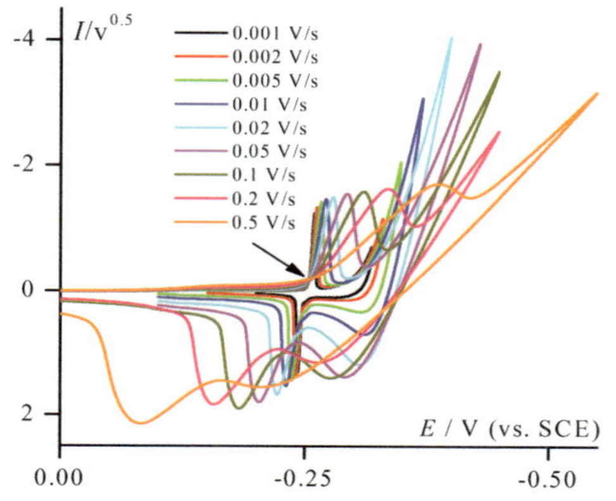

Fig. 16. The normalized complete CV responses $Iv^{-0.5}$ vs. potential E for hydrogen ions - reduction in 0.1M HCl solution. Electrode system: GCE/100 monomolecular Pd layers. The isopotential crossing point is indicated with the arrow (experimental data from ref. (Skitał et al., 2010)). Note the almost vertical position of rising part of CV curves which suggests high increasing of α values. Modification of active electrode surface can be explained by an increasing concentration of atomic hydrogen with scan rate decrease. [Reprinted from *Electrochim. Acta*, Vol. 55, Skitał P. & Sanecki P., The mathematical model of the stripping voltammetry ..., 5604–5609, Copyright (2010), with permission from Elsevier.]

Extended and still growing theory and application of solid electrodes (Horrocks et al., 1994; Kuznetsov & Ulstrup, 1999; Levich, 1966; Wantz et al., 2005) prove that presented results can be useful as an element of theory and practice concerning advanced solid electrodes and thin films dominating in both theoretical and applied electrochemistry (Swain, 2004).

We expect that another experimental *isoalpha points*, analogous to described above, will be found by other investigators.

10. Conclusions

1. Acquisition of rigorous mathematical model of considered mechanism comprising elementary kinetic parameters of electrochemical and chemical steps is fundamental for any discussion of quantitative type.
   MPE method with a proper mathematical model of the considered mechanism, applied to a series of electrochemical responses, is the simple, clear and rigorous way to determine the values of kinetic parameters with their confidence intervals. Additional error space for all estimated kinetic parameters, characteristic for considered mechanism like a fingerprint, can be easily determined basing on obtained theoretical responses. Every mechanism together with its mathematical model can generate electrochemical theoretical responses and its own error space, independently of error space obtained from experimental responses.

2. The common problems with extracting of kinetic parameters basing on selected characteristic points of complex electrochemical responses like half peak method or processed responses (e.g. convolution method) can be avoided by estimation of kinetic parameters on the whole original responses. The actual numerical possibilities and availability of respective software makes it possible to solve practically each of met in practice mechanisms by every researcher. The presented and quoted successful analysis of complex, multi-electron electrode processes with chemical step(s) provides the respective examples.
3. MPE method makes it possible to judge controversial or confused cases of mechanisms. A good example is here a stepwise/concerted mechanism. A uniform approach to understanding of the stepwise/concerted systems was confirmed by respective modeling for ECE mechanism where fluent passage from stepwise to concerted mechanism has been obtained. The only difference between them is the value of k_f.
4. The mathematical modeling of electrode processes linked with experimental results may become an inspiration for new results and deeper understanding of the nature of electrochemical processes and its description. A good example is here an including of alpha variability into kinetic models and discovery of *isoalpha points*.
5. The occurrence of *isoalpha point* indicates that, under constant composition of the solution, in investigated system $\partial a/\mathrm{d}E$ or Δa type variability takes place. The origin of experimentally found *isoalpha point* can be explained by the change of electrode surface conditions.

11. Acknowledgment

The authors are grateful to professor Christian Amatore for valuable explanations, discussions and cooperation.

12. Definitions, abbreviations and symbols

A mechanism can be defined as an system of chemical and electrochemical elementary steps with a set of respective elementary kinetic parameters.

A model is a mathematical description of the considered mechanism or system. Both the mechanism (system) and the model provide an electrochemical response, experimental and theoretical, respectively.

The mathematical model as well as respective electrochemical current or concentration response is a function of multiple variables. Among them there are system variables as electrochemical and chemical rate constants, equilibrium constants, transfer coefficients, standard potentials, etc. as well as variables of experimental type as scan rate, starting potential, concentration, temperature.

AN - acetonitryle
1,3-BDF - 1,3-benzenedisulfonyl difluoride
BSF - benzenesulfonyl fluoride
CV - cyclic voltammetry
DMF - dimethylformamide
$E_{0,i}$ –standard potential of step i (V)

E_{iap} – potential in the isoalpha point (V)
E_p – peak potential (V)
E_{start} – the value of potential where α variability starts
ETC (or α) – elementary transfer coefficient
GCE – glassy carbon electrode
I_{iap} – current in the isoalpha point (A)
iap – isoalpha point
k – electrochemical rate constant for one electron process (cm s^{-1})
k_i (i=1,2,3,...) – electrochemical rate constant for a step of multi electron process (cm s^{-1})
k_f – chemical first order rate constant (s^{-1})
k_{f1}, k_{f2} – chemical first order rate constant of multi stage process (s^{-1})
MPE – multi-parameter estimation
PDE – partial differential equation(s)
ESTYM_PDE - program for solving the models expressed by PDE and estimating their parameters
r; irr (in subscript) – reversible; irreversible
R_u – uncompensated resistance between the working and reference electrode
TBAP – tetrabutylammonium perchlorate
TEAP – tetraethylammonium perchlorate
TSF – *p*-toluenesulfonyl fluoride
v – scan rate
α – transfer coefficient of cathodic process (in electrode kinetics we use elementary kinetic parameters relating to, or describing the single step of kinetic sequence (k_1, k_2, α, α_1, α_2) as well as apparent (general) kinetic parameters i.e. relating to or describing the complex kinetic course or mechanism as a whole (e.g. k_{app}, α_{app}))
β – transfer coefficient of anodic process
α_{start} – the value of α to start with α variability
α_B , β_B – Brønsted acid-base catalysis coefficients
λ – solvent reorganisation factor in Marcus' theory

13. References

Albery, W. J. (1975). *Electrode Kinetics*, Clarendon Press, ISBN 0-19-855433-8, Oxford, pp. 100-101.

Amatore, C. & Saveant, J. M. (1977). ECE and disproportionation: Part V. Stationary state general solution application to linear sweep voltammetry. *J. Electroanal. Chem.*, Vol. 85, pp. 27-46, ISSN: 0022-0728.

Andrieux, C. P., Blocman, C., Dumas-Bouchiat, J.-M. & Savéant J. M. (1979). Heterogeneous and homogeneous electron transfers to aromatic halides. An electrochemical redox catalysis study in the halobenzene and halopyridine series. *J. Am. Chem. Soc.*, Vol. 101, pp. 3431-3441, ISSN: 0002-7863.

Andrieux, C. P., Le Gorande, A. & Savéant, J. M., (1992). Electron transfer and bond breaking. Examples of passage from a sequential to a concerted mechanism in the electrochemical reductive cleavage of arylmethyl halides. *J. Am. Chem. Soc.*, Vol. 114, pp. 6892-6904, ISSN: 0002-7863.

Andrieux, C. P., Differding, E., Robert, M. & Savéant, J.M. (1993). Controlling factors of stepwise versus concerted reductive cleavages. Illustrative examples in the electrochemical reductive breaking of nitrogen-halogen bonds in aromatic N-halosultams. *J. Am. Chem. Soc.,* Vol. 115, pp. 6592-6599, ISSN: 0002-7863.

Andrieux, C. P., Robert, M., Saeva, D. & Savéant, J. M., (1994). Passage from Concerted to Stepwise Dissociative Electron Transfer as a Function of the Molecular Structure and of the Energy of the Incoming Electron. Electrochemical Reduction of Aryldialkyl Sulfonium Cations. *J. Am. Chem. Soc.,* Vol. 116, pp. 7864-7871, ISSN: 0002-7863.

Andrieux, C. P., Savéant, J.-M., Tallec, A., Tardivel, R. & Tardy, C. (1996). Solvent Reorganization as a Governing Factor in the Kinetics of Intramolecular Dissociative Electron Transfers. Cleavage of Anion Radicals of α-Substituted Acetophenones. *J. Am. Chem. Soc.,* Vol. 118, pp. 9788-9789, ISSN: 0002-7863.

Andrieux, C. P., Savéant, J.-M., Tallec, A., Tardivel, R. & Tardy, C. (1997). Concerted and Stepwise Dissociative Electron Transfers. Oxidability of the Leaving Group and Strength of the Breaking Bond as Mechanism and Reactivity Governing Factors Illustrated by the Electrochemical Reduction of α-Substituted Acetophenones. *J. Am. Chem. Soc.,* Vol. 119, pp. 2420-2429, ISSN: 0002-7863.

Angell, D. H. & Dickinson, T. (1972). The kinetics of the ferrous/ferric and ferro/ferricyanide reactions at platinum and gold electrodes: Part I. Kinetics at bare-metal surfaces. *J. Electroanal. Chem.,* Vol. 35. pp. 55-72, ISSN: 0022-0728.

Antonello, S. & Maran, F. (1997). Evidence for the Transition between Concerted and Stepwise Heterogeneous Electron Transfer−Bond Fragmentation Mechanisms. *J. Am. Chem. Soc.,* Vol. 119, pp. 12595-12600, ISSN: 0002-7863.

Antonello, S. & Maran, F. (1998). Dependence of Intramolecular Dissociative Electron Transfer Rates on Driving Force in Donor−Spacer−Acceptor Systems. *J. Am. Chem. Soc.,* Vol. 120, pp. 5713-5722, ISSN: 0002-7863.

Antonello, S. & Maran, F. (1999). The Role and Relevance of the Transfer Coefficient α in the Study of Dissociative Electron Transfers: Concepts and Examples from the Electroreduction of Perbenzoates. *J. Am. Chem. Soc.,* Vol. 121, pp. 9668-9676, ISSN: 0002-7863.

Antonello, S., Formaggio, F., Moretto, A., Toniolo, C. & Maran, F. (2001). Intramolecular, Intermolecular, and Heterogeneous Nonadiabatic Dissociative Electron Transfer to Peresters. *J. Am. Chem. Soc.,* Vol. 123, pp. 9577-9584, ISSN: 0002-7863.

Antonello, S., Benassi, R., Gavioli, G., Taddei, F. & Maran, F. (2002a). Theoretical and Electrochemical Analysis of Dissociative Electron Transfers Proceeding through Formation of Loose Radical Anion Species: Reduction of Symmetrical and Unsymmetrical Disulfides. *J. Am. Chem. Soc.,* Vol. 124, pp. 7529-7538, ISSN: 0002-7863.

Antonello, S., Crisma, M., Formaggio, F., Moretto, A., Taddei, F., Toniolo, C. & Maran, F. (2002b). Insights into the Free-Energy Dependence of Intramolecular Dissociative Electron Transfers. *J. Am. Chem. Soc.,* Vol. 124, pp. 11503-11513, ISSN: 0002-7863.

Arun, P. M. & Sangaranarayanan, M. V. (2004). Current function for irreversible electron transfer processes in linear sweep voltammetry for the reactions obeying Marcus kinetics. *Chem. Phys. Lett.*, Vol. 387, pp. 317-321, ISSN: 0009-2614.

Bard, A. J. & Faulkner, L. R. (2001). *Electrochemical Methods, Fundamentals and Applications*, Wiley, ISBN: 0-471-04372-9, New York.

Berlett, B. S., Levine, R. L. & Stadtman, E. R. (2000). Use of Isosbestic Point Wavelength Shifts to Estimate the Fraction of a Precursor That Is Converted to a Given Product. *Anal. Biochem.*, Vol. 287, pp. 329-333, ISSN: 0003-2697.

Berninger, A. J., Whitley, R. D., Zhang, X. & Wang, N. H. L. (1991). A versatile model for simulation of reaction and nonequilibrium dynamics in multicomponent fixed-bed adsorption processes. *Comput. Chem. Eng.*, Vol. 15, pp. 749-768, ISSN: 0098-1354.

Berzins, T. & Delahay, P. (1953). Theory of Irreversible Polarographic Waves—Case of Two Consecutive Electrochemical Reactions. *J. Am. Chem. Soc.*, Vol. 75, pp. 5716-5720, ISSN: 0002-7863.

Bieniasz, L. K. (1997). ELSIM—a problem-solving environment for electrochemical kinetic simulations. Version 3.0-solution of governing equations associated with interfacial species, independent of spatial coordinates or in one-dimensional space geometry. *Comput. Chem.*, Vol. 21, p. 1-12, ISSN: 0097-8485.

Bieniasz, L. K. & Speiser, B. (1998a). Use of sensitivity analysis methods in the modelling of electrochemical transients Part 1. Gaining more insight into the behaviour of kinetic models. *J. Electroanal. Chem.*, Vol. 441, pp. 271-285, ISSN: 0022-0728.

Bieniasz, L. K. & Speiser, B. (1998b). Use of sensitivity analysis methods in the modelling of electrochemical transients: Part 3. Statistical error/uncertainty propagation in simulation and in nonlinear least-squares parameter estimation. *J. Electroanal. Chem.*, Vol. 458, pp. 209-229, ISSN: 0022-0728.

Bieniasz, L. K. & Britz, D. (2004). Recent Developments in Digital Simulation of Electroanalytical Experiments. *Polish J. Chem.*, Vol. 78, pp. 1195-1219, ISSN 0137-5083.

Bieniasz, L. K. & Rabitz, H. (2006). Extraction of Parameters and Their Error Distributions from Cyclic Voltammograms Using Bootstrap Resampling Enhanced by Solution Maps: Computational Study. *Anal. Chem.*, Vol. 78, pp. 8430-8437, ISSN 0003-2700.

Bindra, P., Brown, A. P., Fleischmann, M. & Pletcher, D. (1975). The determination of the kinetics of very fast electrode reactions by means of a quasi-steady state method: The mercurous ion/mercury system Part II: Experimental results. *J. Electroanal. Chem.*, Vol. 58, pp. 39-50, ISSN: 0022-0728.

Bockris J. O'M. & Reddy A. K. N. (1970). *Modern electrochemistry: An introduction to an interdisciplinary area*, Vol. 2, Macdonald, British SBN: 356 03262 0, London.

Bond, A. M. & Mahon, P. J. (1997). Linear and non-linear analysis using the Oldham–Zoski steady-state equation for determining heterogeneous electrode kinetics at microdisk electrodes and digital simulation of the microdisk geometry with the fast quasi-explicit finite difference method. *J. Electroanal. Chem.*, Vol. 439, pp. 37-53, ISSN: 0022-0728.

Bott, A. W., Feldberg, S. W. & Rudolph, M. (1996). Fitting Experimental Cyclic Voltammetry Data with Theoretical Simulations Using DigiSim® 2.1. *Curr. Seps.*, Vol. 15, pp. 67-71, ISSN: 0891-0006.

Bott, A. W. (2000). Simulation of Cyclic Voltammetry Using Finite Difference Methods. *Curr. Seps.*, Vol. 19, pp. 45-48, ISSN: 0891-0006.

Brenet J. P., Traore K., (1971). *Transfer Coeficients in Electrochemical Kinetics,* Akademic Press, ISBN: 0121309606, London and New York.

Britz, D. (2005). *Digital Simulation in Electrochemistry*, Springer, ISBN: 3-540-23979-0, Berlin.

Brown, P. N., Hindmarsh, A. C. & Byrne, G. D. (n.d.). *Variable Coefficient Ordinary Differential Equation Solver*, procedure available on http://www.netlib.org.

Caldwell, R. A. & Hacobian, S. (1968a). Polarographic and nuclear quadrupole resonance studies of some organic iodo compounds. I. Correlation of nuclear quadrupole resonance frequency and polarographic half-step potentia. *Austr. J. Chem.*, Vol. 21, pp. 1-8, ISSN: 0004-9425.

Caldwell, R. A. & Hacobian, S. (1968b). Polarographic and nuclear quadrupole resonance studies of some organic iodo compounds. II. The dependence of the half-step potential on the ionic character of the carbon-iodine bond. *Austr. J. Chem.*, Vol. 21, pp. 1403-1413, ISSN: 0004-9425.

Chidsey, C. E. D. (1991). Free Energy and Temperature Dependence of Electron Transfer at the Metal-Electrolyte Interface. *Science,* Vol. 251, pp. 919-922, ISSN: 0036-8075.

Colichman, E. L. & Kung Liu, S. (1954). Effect of Structure on the Polarographic Reduction of Iodo Compounds. *J. Am. Chem. Soc.,* Vol. 76, pp. 913-915, ISSN: 0002-7863.

Corrigan, D. A. & Evans, D. H. (1980). Cyclic voltammetric study of tert-nitrobutane reduction in acetonitrile at mercury and platinum electrodes: Observation of a potential dependent electrochemical transfer coefficient and the influence of the electrolyte cation on the rate constant. *J. Electroanal. Chem.,* Vol. 106, pp. 287-304, ISSN: 0022-0728.

Costentin, C., Donati, L. & Robert, M. (2009). Passage from stepwise to concerted dissociative electron transfer through modulation of electronic states coupling. *Chemistry*, Vol. 15, pp. 785-792, ISSN: 1521-3765.

Daasbjerg, K., Jensen, H., Benassi, R., Taddei, F., Antonello, S., Gennaro, A.& Maran, F. (1999). Evidence for Large Inner Reorganization Energies in the Reduction of Diaryl Disulfides: Toward a Mechanistic Link between Concerted and Stepwise Dissociative Electron Transfers? *J. Am. Chem. Soc.,* Vol. 121, pp. 1750-1751, ISSN: 0002-7863.

Delahay, P. (1953). Theory of Irreversible Waves in Oscillographic Polarography. *J. Am. Chem. Soc.,* Vol. 75, pp. 1190-1196, ISSN: 0002-7863.

Dogonadze, R. R. (1971). Theory of molecular electrode kinetic, In: *Reactions of Molecules at Electrodes*, Hush, N. S. (Ed.), pp. 135-227, Wiley-Interscience, ISBN: 0-471-42490-0, New York.

Edens, G. J., Fitch, A. & Lavy-Feder, A. (1991). Use of isopotential points to elucidate ion exchange reaction mechanisms: $Cr(bpy)_3^{3+}$ at montmorillonite clay-modified electrodes. *J. Electroanal. Chem.,* Vol. 307, pp. 139-154, ISSN: 0022-0728.

Eichhorn, E. & Speiser, B. (1994). Electrochemistry of oxygenation catalysts: Part 4. Solvent-composition-dependent isopotential points in cyclic voltammograms of Co(SALEN). *J. Electroanal. Chem.,* Vol. 365, pp. 207-212, ISSN: 0022-0728.

Feldberg, S. W. (1969). Digital simulation: A general method for solving electrochemicaldiffusion-kinetic problems. In: Electroanalytical Chemistry, Bard, A.J., (Ed.), Vol. 3, p. 199-296, Marcel Dekker, ISBN: 0824710371, New York.

Finklea, H. O. (2001a). Theory of Coupled Electron-Proton Transfer with Potential-Dependent Transfer Coefficients for Redox Couples Attached to Electrodes. *J. Phys. Chem. B*, Vol. 105, pp. 8685-8693, ISSN: 1520-6106.

Finklea, H. O. (2001b). Consequences of a potential-dependent transfer coefficient in ac voltammetry and in coupled electron–proton transfer for attached redox couples. *J. Electroanal. Chem.*, Vol. 495, pp. 79-86, ISSN: 0022-0728.

Finklea, H. O. & Haddox, R. M. (2001). Coupled electron/proton transfer of galvinol attached to SAMs on gold electrodes. *Phys. Chem. Chem. Phys.*, Vol. 3, pp. 3431-3436, ISSN: 14639076.

Finklea, H. O., Yoon, K., Chamberlain, E., Allen, J. & Haddox, R. (2001). Effect of the Metal on Electron Transfer across Self-Assembled Monolayers. *J. Phys. Chem. B*, Vol. 105, pp. 3088-3092, ISSN: 1520-6106.

Fitch, A. & Edens, G. J. (1989). Isopotential points as a function of an allowed cross reaction. *J. Electroanal. Chem.*, Vol. 267, pp. 1-13, ISSN: 0022-0728.

Fletcher, R. (n.d.). A modified Marquardt sub-routine for nonlinear least squares, AERE-R6799-Harwell.

Frost, A. A. & Schwemer, W. C. (1952). The Kinetics of Competitive Consecutive Second-order Reactions: The Saponification of Ethyl Adipate and of Ethyl Succinate. *J. Am. Chem. Soc.*, Vol. 74, pp. 1268-1273, ISSN: 0002-7863.

Frost, A. A. & Pearson, R. G. (1961). *Kinetics and Mechanism*, J. Wiley & Sons, ISBN: 047128355X, New York.

Frumkin, A. N. (1932). Bemerkung zur Theorie der Wasserstoffüberspannung. *Z. Phys. Chem.*, Vol. A160, pp. 116-118, ISSN: 0942-9352.

Galus Z. (1994). *Fundamentals of Electrochemical Analysis*, Ellis Horwood, ISBN: 83-01-11255-7, New York.

Gardini, L., Servida, A., Morbidelli, M. & Carra, S. (1985). Use of orthogonal collocation on finite elements with moving boundaries for fixed bed catalytic reactor simulation. *Comput. Chem. Eng.*, Vol. 9, pp. 1-17, ISSN: 0098-1354.

Garreau, D., Savéant, J. M. & Tessier, D. (1979). Potential dependence of the electrochemical transfer coefficient. An impedance study of the reduction of aromatic compounds. *J. Phys. Chem.*, Vol. 83, pp. 3003-3007, ISSN 0022-3654.

Gosser, D. K. J. (1993). *Cyclic Voltammetry: Simulation and Analysis of Reaction Mechanisms*, VCH Publishers, Inc., ISBN: 1-56081-026-2, New York.

Greef, R., Peat, R., Peter, L. M., Pletcher, D. & Robinson, J. (1985). Instrumental Methods in Electrochemistry, Ellis Horwood Limited, ISBN: 0-85312-875-8, Chichester.

Haddox, R. M. & Finklea, H. O. (2003). Proton coupled electron transfer of galvinol in self-assembled monolayers. *J. Electroanal. Chem.*, Vol. 550/551, pp. 351-358, ISSN: 0022-0728.

Horrocks, B. R., Mirkin, M. V. & Bard, A. J. (1994). Scanning Electrochemical Microscopy. 25. Application to Investigation of the Kinetics of Heterogeneous Electron Transfer at Semiconductor (WSe2 and Si) Electrodes. *J. Phys. Chem.*, Vol. 98, pp. 9106-9114, ISSN: 0022-3654.

Hush, N. S. (1957). Elektrodenreaktionen der Methylhalogenide. *Z. Elektrochem.*, Vol. 61, pp. 734-738, ISSN: 2590-1977.

Hush, N. S. (1958). Adiabatic Rate Processes at Electrodes. I. Energy-Charge Relationships. *J. Chem. Phys.*, Vol. 28, pp. 962-973, ISSN: 0021-9606.

Hussey, W. W. & Diefenderfer, A. J. (1967). Ortho-substituent effects in polarography. *J. Am. Chem. Soc.*, Vol. 89, pp. 5359-5362, ISSN: 0002-7863.

IUAPAC (1997). Compendium of Chemical Technology.

Jakobsen, S., Jensen, H., Pedersen, S. U. & Daasbjerg, K. (1999). Stepwise versus Concerted Electron Transfer-Bond Fragmentation in the Reduction of Phenyl Triphenylmethyl Sulfides. *J. Phys. Chem. A*, Vol. 103, pp. 4141-4143, ISSN: 1089-5639.

Jaworski, J. S., Kacperczyk, A. & Kalinowski, M. K. (1992). Hammett reaction constants for irreversible electroreduction of iodobenzenes in non-aqueous solvents. *J. Phys. Org. Chem.*, Vol. 5, pp. 119-122, ISSN: 1099-1395.

Jaworski, J. S. & Leszczyński, P. (1999). Solvent effect on kinetics of the chloride ion cleavage from anion radicals of 4-chlorobenzophenone. *J. Electroanal. Chem.*, Vol. 464, pp. 259-262, ISSN: 0022-0728.

Jäger, E.-G. & Rudolph, M. (1997). Cyclic voltammetric and impedance spectrometric investigations on addition/elimination reactions of Lewis bases accompanying the electrode reactions of a nickel chelate complex with a structural resemblance to the coenzyme F430. *J. Electroanal. Chem.*, Vol. 434, pp. 1-18, ISSN: 0022-0728.

Kaczmarski, K. (1996). Use of orthogonal collocation on finite elements with moving boundaries in the simulation of non-linear multicomponent chromatography. Influence of fluid velocity variation on retention time in LC and HPLC. *Comput. Chem. Eng.*, Vol. 20, pp. 49-64, ISSN: 0098-1354.

Kaczmarski, K., Mazzotti, M., Storti, G. & Morbidelli, M. (1997). Modeling fixed-bed adsorption columns through orthogonal collocations on moving finite elements. *Comput. Chem. Eng.*, Vol. 21, pp. 641-660, ISSN: 0098-1354.

Kaczmarski, K., Antos, D., Sajonz, H., Sajonz, P. & Guiochon, G. (2001). Comparative modeling of breakthrough curves of bovine serum albumin in anion-exchange chromatography. *J. Chromatogr A*, Vol. 925, pp. 1-17, ISSN: 0021-9673.

Kaczmarski, K. (2007). Estimation of adsorption isotherm parameters with inverse method – possible problems. *J. Chromatogr. A*, Vol. 1176, pp. 57-68, ISSN: 0021-9673.

Kim, H., Kaczmarski, K. & Guichon, G. (2005). Mass transfer kinetics on the heterogeneous binding sites of molecularly imprinted polymers. *Chem. Eng. Sci.*, Vol. 60, pp. 5425-5444, ISSN: 0009-2509.

Kim, H., Kaczmarski, K. & Guichon, G. (2006a). Optical and photoluminescent properties of sol-gel Al-doped ZnO thin films. *Chem. Eng. Sci.*, Vol. 61, pp. 1118-1121, ISSN: 0009-2509.

Kim, H., Kaczmarski, K. & Guichon, G. (2006b). Isotherm parameters and intraparticle mass transfer kinetics on molecularly imprinted polymers in acetonitrile/buffer mobile phases. *Chem. Eng. Sci.*, Vol. 61, pp. 5249-5267, ISSN: 0009-2509.

Kuritsyn, L. V., Sokolov, L. B., Savinov, V. M. & Ivanov, A. V. (1974). Otnositelnaya reaktsionnosposobnost aminogrupp i aromaticheskikh diaminov v reaktsii s khlorangidridnoi gruppoi. *Vysokomolekularnye Soedineniya*, Vol. 16B, pp. 532-535, UDK: 541.64:547.553.1/2.

Kuznetsov, A. M. & Ulstrup, J. (1999). *Electron Transfer in Chemistry and Biology*, J. Wiley & Sons, ISBN: 0-471-96749-1, Chichester.

Lavagnini, I., Pastore, P. & Magno, F. (1989). Comparison of the simplex, marquardt, and extended and iterated extended kalman filter procedures in the estimation of parameters from voltammetric curves. *Anal. Chim. Acta*, Vol. 223, pp. 193-204, ISSN: 0003-2670.

Levich, V. G. (1966). Present state of the Theory of Oxidation-Reduction in Solution (Bulk and Electrode Reactions). In: *Advances in Electrochemistry and Electrochemical Engineering*, Delahay, P. & Tobias, C. (Eds.), Vol. 4, pp. 249-371, Interscience, ISBN: 047020575X, New York.

Ma, Z. & Guiochon, G. (1991). Application of orthogonal collocation on finite elements in the simulation of non-linear chromatography. *Comput. Chem. Eng.*, Vol. 15, pp. 415-426, ISSN: 0098-1354.

Mairanovskii, S. G. (1969). O vliyanii strojenya organicheskogo depolarizatora na vielichinu spiecificheskogo effekta pri izmyenyenii pripoly kationa inliffyeryentnogo elyektrolita. *Elektrokhimiya*, Vol. 5, pp. 757-759, ISSN: 0424-8570.

Mairanovskii, S. G. & Rubinskaya, T. Ya. (1972). O spyetsifichyeskom vlivaniikatonov tyetraalkilammoniva pri elyektrovosstanovlyenii organivhyeskikh galoidproizvodnykh. *Elektrokhimiya*, Vol. 8, pp. 424-427, ISSN: 0424-8570.

Mairanovskii, S. G., Stradyn, J. P. & Bezuglyi, V. D. (1975). Polarografiya v organicheskoj khimii, *Izdatyelstvo Khimiya , UDK: 543:253:547*, Leningrad.

Manzini, M. & Lasia, A. (1994). Kinetics of electroreduction of Zn^{2+} at mercury in nonaqueous solutions. *Can. J. Chem.*, Vol. 72, pp. 1691-1698, ISSN: 0008-4042.

Marcus, R. A. (1956). On the Theory of Oxidation-Reduction Reactions Involving Electron Transfer. I. *J. Chem. Phys.*, Vol. 24, pp. 966-979, ISSN: 0021-9606.

Marcus, R. A. (1960). Exchange reactions and electron transfer reactions including isotopic exchange. Theory of oxidation-reduction reactions involving electron transfer. Part 4.—A statistical-mechanical basis for treating contributions from solvent, ligands, and inert salt. *Disc. Faraday Soc.*, Vol. 29, pp. 21-31, ISSN: 0014-7664.

Marcus, R. A. (1977). Theory and Applications of Electron Transfers at Electrodes and in Solution, In: *Special Topics in Electrochemistry*, Rock, P. A. (Ed.), pp. 161-179, Elsevier, ISBN: 0444416277, Amsterdam.

Marquardt, D. W. (1963). An Algorithm for Least-Squares Estimation of Nonlinear Parameters. *SIAM J. Appl. Math.*, Vol. 11, pp. 431-441, ISSN: 0036-1399.

Matsuda, H. & Ayabe, Y. Zur Theorie der Randles-Sevčikschen Kathodenstrahl-Polarographie. (1955). *Z. Elektrochem.*, Vol. 59, pp. 494-503, ISSN: 2590-1977.

Matsuda, K. & Tamamushi, R. (1979). Potential-dependent transfer coefficients of the Zn(II)/Zn(Hg) electrode reaction in aqueous solutions. *J. Electroanal. Chem.*, Vol. 100, pp. 831-839, ISSN: 0022-0728.

Matthesiss, L. F. & Warren, W. W. (1977). Band model for the electronic structure of expanded liquid mercury. *Phys. Rev. B*, Vol. 16, pp. 624-638, ISSN: 1098-0121.

McDermott, M. T., Kneten, K. & McCreery, R. L. (1992). Anthraquinonedisulfonate adsorption, electron-transfer kinetics, and capacitance on ordered graphite electrodes: the important role of surface defects. *J. Phys. Chem.*, Vol. 96, pp. 3124-3130, ISSN 0022-3654.

Miller, C. J. (1995). Heterogeneous Electron Transfer Kinetics at Metallic Electrodes. In: *Physical electrochemistry : principles, methods, and applications*, Rubinstein, I. (Ed.), pp. 27-80, Marcel Dekker, ISBN: 0824794524, New York.

Nagy, Z., Leaf, G. K., Minkoff, M. & Land, R. H. (1988). Extension of dc transient techniques to reactions with a potential dependent charge transfer coefficient. *Electrochim. Acta*, Vol. 33. pp. 1589-1593, ISSN: 0013-4686.

Nahir, T. M., Clark, R. A. & Bowden, E. F. (1994). Linear-Sweep Voltammetry of Irreversible Electron Transfer in Surface-Confined Species Using the Marcus Theory. *Anal. Chem.*, Vol. 66, pp. 2595-2598, ISSN: 0003-2700.

Nakajima, R., Tsuruta, M., Higuchi, M. & Yamamoto, K. (2004). Fine Control of the Release and Encapsulation of Fe Ions in Dendrimers through Ferritin-like Redox Switching. *J. Am. Chem. Soc.*, Vol. 126, pp. 1630-1631, ISSN: 0002-7863.

Nicholson, R. S. & Shain, I. (1964). Theory of Stationary Electrode Polarography. Single Scan and Cyclic Methods Applied to Reversible, Irreversible, and Kinetic Systems. *Anal. Chem.*, Vol. 36, pp. 706-723, ISSN: 0003-2700.

Nicholson, R. S. & Shain, I. (1965a). Theory of Stationary Electrode Polarography for a Chemical Reaction Coupled Between Two Charge Transfers. *Anal. Chem.*, Vol. 37, pp. 178-190, ISSN: 0003-2700.

Nicholson, R. S. & Shain, I. (1965b). Experimental Verification of an ECE Mechanism for the Reduction of p-Nitrosophenol, Using Stationary Electrode Polarography. *Anal. Chem.*, Vol. 37, pp. 190-195, ISSN: 0003-2700.

Najjar, F., André-Barrès, C., Baltas, M., Lacaze-Dufaure, C., Magri, D.C., Workentin, M.S., Tzédakis, T (2007). Electrochemical reduction of G3-factor endoperoxide and its methyl ether: evidence for a competition between concerted and stepwise dissociative electron transfer. *Chemistry*, Vol. 13, 1174-1179, ISSN: 1521-3765.

Parsons, R. & Passeron, E. (1966). The potential-dependence of the transfer coefficient in the cr(II)/cr(III) reaction. *J. Electroanal. Chem.*, Vol. 12, pp. 524-529, ISSN: 0022-0728.

Pause, L., Robert, M. & Savéant, J. M. (1999). Can Single-Electron Transfer Break an Aromatic Carbon−Heteroatom Bond in One Step? A Novel Example of Transition between Stepwise and Concerted Mechanisms in the Reduction of Aromatic Iodides. *J. Am. Chem. Soc.*, Vol. 121, pp. 7158-7159, ISSN: 0002-7863.

Pause, L., Robert, M. & Savéant, J. M. (2001). Stepwise and Concerted Pathways in Photoinduced and Thermal Electron-Transfer/Bond-Breaking Reactions. Experimental Illustration of Similarities and Contrasts. *J. Am. Chem. Soc.*, Vol. 123, pp. 4886-4895, ISSN: 0002-7863.

Peters, D. G. (1991). Halogenated Organic Compounds. In: *Organic Electrochemistry: An introduction and guide*, Lund, H. & Baizer, M. M. (Ed.), chapter 8, pp. 362-395, Marcel Dekker, Inc., ISBN: 0-8247-8154-6, New York.

Pierce, D. T. & Geiger, W. E. (1992). Electrochemical kinetic discrimination of the single-electron-transfer events of a two-electron-transfer reaction: cyclic voltammetry of the reduction of the bis(hexamethylbenzene)ruthenium dication. *J. Am. Chem. Soc.*, Vol. 114, pp. 6063-6073, ISSN: 0002-7863.

Rifi, M. R. & Covitz, F. H. (1974). *Introduction to Organic Electrochemistry*, Marcel Dekker, Inc., ISBN: 0824760638, New York.

Rusling, J. F. (1983). Determination of rate constrants of pseudo-first-order electrocatalytic reactions from overlapped voltammetric data. *Anal. Chem.*, Vol. 55, pp. 776-781, ISSN: 0003-2700.

Ružić, I. (1970). Logarithmic analysis of two overlapping d.c. polarographic waves II. Multistep electrode raction. *J. Electroanal. Chem.*, Vol. 25, pp. 144-147, ISSN: 0022-0728.

Ružić, I., (1974). On the theory of stepwise electrode processes. *J. Electroanal. Chem.*, Vol. 52, pp. 331-354, ISSN: 0022-0728.

Samec, Z. & Weber, J. (1973). The influence of chemisorbed sulphur on the kinetic parameters of the reduction of Fe^{3+} ions on a platinum electrode on the basis of the marcus theory of electron transfer. *J. Electroanal. Chem.*, Vol. 44, pp. 229-238, ISSN: 0022-0728.

Sanecki, P. (1986). An interpretation of the electrochemical transfer coefficient as a reduction constant. *Electrochim. Acta,* Vol. 31, pp. 1187-1191, ISSN: 0013-4686.

Sanecki, P. & Lechowicz, J. (1997). The Problem of Complex Curves in Normal Pulse Polarography, *Electroanalysis,* Vol. 9, pp. 1409-1415, ISSN: 1521-4109.

Sanecki, P. & Kaczmarski, K. (1999). The Voltammetric Reduction of Some Benzene-sulfonyl Fluorides, Simulation of its ECE Mechanism and Determination of the Potential Variation of Charge Transfer Coefficient by Using the Compounds with Two Reducible Groups. *J. Electroanal. Chem.,* Vol. 471, pp. 14-25, ISSN: 0022-0728. Erratum published in (2001). *J. Electroanal. Chem.*, Vol. 497, pp. 178-179.

Sanecki, P. (2001). A numerical modelling of voltammetric reduction of substituted iodobenzenes reaction series. A relationship between reductions in the consecutive-mode multistep system and a multicomponent system. Determination of the potential variation of the elementary charge transfer coefficient, *Comput. Chem.*, Vol. 25, pp. 521-539, ISSN: 0097-8485.

Sanecki, P. & Skitał P. (2002a) The cyclic voltammetry simulation of a competition between stepwise and concerted dissociative electron transfer. The modeling of alpha apparent variability. The relationship between apparent and elementary kinetic parameters, *Comput. Chem.*, Vol. 26, pp. 297-311, ISSN: 0097-8485.

Sanecki, P. & Skitał P. (2002b). A comparison of the multistep consecutive reduction mode with the multicomponent system reduction mode in cyclic voltammetry. *Comput. Chem.*, Vol. 26, pp. 333-340, ISSN: 0097-8485.

Sanecki, P., Amatore, C. & Skitał, P. (2003). The problem of the accuracy of electrochemical kinetic parameters determination for the ECE reaction mechanism. *J. Electroanal. Chem.*, Vol. 546, pp. 109-121, ISSN: 0022-0728.

Sanecki, P., Skitał, P. & Kaczmarski, K. (2006a). An integrated two phases approach to Zn^{2+} ions electroreduction on Hg. *Electroanalysis*, Vol. 18, pp. 595-604, ISSN: 1521-4109.

Sanecki, P., Skitał, P. & Kaczmarski, K. (2006b). Numerical modeling of ECE-ECE and parallel EE-EE mechanisms in cyclic voltammetry. Reduction of 1,4-benzenedisulfonyl difluoride and 1,4-naphthalenedisulfonyl difluoride. *Electroanalysis*, Vol. 18, pp. 981-991, ISSN: 1521-4109.

Sanecki, P. & Skitał, P. (2007a). The Application of EC, ECE and ECE-ECE Models with Potential Dependent Transfer Coefficient to Selected Electrode Processes. *J. Electrochem. Soc.*, Vol. 154, pp. F152-F158, ISSN 0013-4651.

Sanecki, P. & Skitał, P. (2007b). The electroreduction of alkyl iodides and polyiodides The kinetic model of EC(C)E and ECE-EC(C)E mechanisms with included transfer coefficient variability. *Electrochim. Acta*, Vol. 52, pp. 4675-4684, ISSN: 0013-4686.

Sanecki, P. & Skitał, P. (2008). The mathematical models of kinetics of the E, EC, ECE, ECEC, ECE–ECE and ECEC–ECEC processes with potential-dependent transfer coefficient as a rationale of isoalpha points. *Electrochim. Acta,* Vol. 53, pp. 7711-7719, ISSN: 0013-4686.

Sanecki, P., Skitał, P. & Kaczmarski, K. (2010). The mathematical models of the stripping voltammetry metal deposition/dissolution process. *Electrochim. Acta,* Vol. 55, pp. 1598-1604, ISSN: 0013-4686.

Sawyer, D. T., Sobkowiak, A. & Roberts, J. L. (1995). *Electrochemistry for Chemists,* J. Wiley, Inc., ISBN:0-471-59468-7 , New York.

Savéant, J. M. & Tessier, D. (1975). Convolution potential sweep voltammetry V. Determination of charge transfer kinetics deviating from the Butler-Volmer behaviour. *J. Electroanal. Chem.,* Vol. 65, pp. 57-66, ISSN: 0022-0728.

Savéant, J. M. & Tessier, D. (1977). Potential dependence of the electrochemical transfer coefficient. Reduction of some nitro compounds in aprotic media. *J. Phys. Chem.,* Vol. 81. pp. 2192-2197, ISSN: 0022-3654.

Savéant, J. M. & Tessier, D. (1982). Variation of the electrochemical transfer coefficient with potential. *Faraday Discuss. Chem. Soc.,* Vol. 74, pp. 57-72, ISSN: 0301-7249.

Savéant, J. M. (1987). A simple model for the kinetics of dissociative electron transfer in polar solvents. Application to the homogeneous and heterogeneous reduction of alkyl halides. *J. Am. Chem. Soc.,* Vol. 109, pp. 6788-6795, ISSN: 0002-7863.

Savéant, J. M. (1992). Dissociative electron transfer. New tests of the theory in the electrochemical and homogeneous reduction of alkyl halides. *J. Am. Chem. Soc.,* Vol. 114, pp. 10595-10602, ISSN: 0002-7863.

Savéant, J. M. (1993). Electron transfer, bond breaking, and bond formation. *Acc. Chem. Res.,* Vol. 26, pp. 455-461, ISSN: 0001-4842.

Schwemer, W. C. & Frost, A. A. (1951). A Numerical Method for the Kinetic Analysis of Two Consecutive Second Order Reactions. *J. Am. Chem. Soc.,* Vol. 73, pp. 4541-4542, ISSN: 0002-7863.

Scharbert, B. & Speiser, B. (1989). Chemical information from electroanalytical data. Part 1—Determination of system parameters for quasi-reversible electron transfer reactions from cyclic voltammetric test data and data for the reduction of cerium (IV) bis(octaethylporphyrinate). *J. Chemometr.,* Vol. 3, pp. 61-80, ISSN: 0886-9383.

Sease, J. W., Burton, F. G. & Nickol, S. L. (1968). Mechanism of electrolytic reduction of carbon-halogen bond. II. A rho sigma study. *J. Am. Chem. Soc.,* Vol. 90, pp. 2595-2598, ISSN: 0002-7863.

Seber, G. A. F. & Wild, C. J. (1989). *Nonlinear Regression,* John Wiley & Sons, Inc, ISBN: 0471617601, New York.

Severin, M. G., Arévalo, M. C., Maran, F. & Vianello, E. (1993). Electron-transfer bond-breaking processes: an example of nonlinear activation-driving force relationship in the reductive cleavage of the carbon-sulfur bond. *J. Phys. Chem.,* Vol. 97, pp. 150-157, ISSN 0022-3654.

Skitał, P. M. & Sanecki, P. T. (2009). The ECE Process in Cyclic Voltammetry. The Relationships Between Elementary and Apparent Kinetic Parameters Obtained by Convolution Method. *Polish Journal of Chemistry*, Vol. 83, pp. 1127–1138, ISSN: 0137-5083.

Skitał, P., Sanecki, P. & Kaczmarski, K. (2010). The mathematical model of the stripping voltammetry hydrogen evolution/dissolution process on Pd layer. *Electrochim. Acta*, Vol. 55, pp. 5604–5609, ISSN: 0013-4686.

Smalley, J. F., Feldberg, S. W., Chidsey, C. E. D., Linford, M. R., Newton, M. D. & Liu, Y.-P. (1995). The Kinetics of Electron Transfer Through Ferrocene-Terminated Alkanethiol Monolayers on Gold. *J. Phys. Chem.*, Vol. 99, pp. 13141-13149, ISSN: 0022-3654.

Speiser, B. (1985). Multiparameter estimation: extraction of information from cyclic voltammograms. *Anal. Chem.*, Vol. 57, pp. 1390-1397, ISSN: 0003-2700.

Speiser, B. (1996a). Numerical simulation of electroanalytical experiments: recent advance in methodology, In: *Electroanalytical Chemistry, A Series of Advances*, Bard, A. J. & Rubinstein, I. (Eds.), Vol. 19, pp. 1-106, Marcel Dekker, ISBN: 082479379X, New York.

Speiser, B. (1996b). Electron Transfer and Chemical Reactions—Stepwise or Concerted? On the Competition between Nucleophilic Substitution and Electron Transfer. *Angew. Chem. Int. Ed. Engl.*, Vol. 35, pp. 2471-2474, 1521-3773.

Stackelberg, M. & Stracke, W. (1949). The polarographic behavior of unsaturated and halogenated hydrocarbons. *Z. Elektrochem.*, Vol. 53, pp. 118-125, ISSN: 2590-1977.

Swain, G. M. (2004). Electrically Conducting Diamond Thin Films: Advanced Electrode Materials for Electrochemical Technologies. In: *Electroanalytical Chemistry, A Series of Advances*, Bard, A. J. & Rubinstein, I. (Eds.), Vol. 22, pp. 182-278, Marcell Dekker, Inc., ISBN: 0824747194 , New York.

Tender, L., Carter, M. T. & Murray, R. W. (1994). Cyclic Voltammetric Analysis of Ferrocene Alkanethiol Monolayer Electrode Kinetics Based on Marcus Theory. *Anal. Chem.*, Vol. 66, pp. 3173-3181, ISSN: 0003-2700.

Tyma, P. D. & Weaver, M. J. (1980). Further observations on the dependence of the electrochemical transfer coefficient upon the electrode potential. *J. Electroanal. Chem.*, Vol. 111, pp. 195-210, ISSN: 0022-0728.

Wantz, F., Banks, C. E. & Compton, R. G. (2005). Edge Plane Pyrolytic Graphite Electrodes for Stripping Voltammetry: a Comparison with Other Carbon Based Electrodes. *Electroanalysis*, Vol. 17, pp. 655-661, ISSN: 1521-4109.

Weber, K. & Creager, S. E. (1994). Voltammetry of Redox-Active Groups Irreversibly Adsorbed onto Electrodes. Treatment Using the Marcus Relation between Rate and Overpotential. *Anal. Chem.*, Vol. 66, pp. 3164-3172, ISSN: 0003-2700.

Workentin, M., Maran, F. & Wayner, D. D. M., (1995). Reduction of Di-tert-Butyl Peroxide: Evidence for Nonadiabatic Dissociative Electron Transfer *J. Am. Chem. Soc.*, Vol. 117, pp. 2120-2121, ISSN: 0002-7863.

Villadsen, J. & Michelsen, M. L. (1978). *Solutions of Differential Equation Models by Polynomial Approximation*, Prentice-Hall, ISBN: 0138222053, New York.

Yu, Q. & Wang, N. H. L. (1989). Computer simulations of the dynamics of multicomponent ion exchange and adsorption in fixed beds—gradient-directed moving finite element method. *Comput. Chem. Eng.*, Vol. 13, pp. 915-926, ISSN: 0098-1354.

4

Electron-Transfer-Induced Intermolecular [2 + 2] Cycloaddition Reactions Assisted by Aromatic "Redox Tag"

Kazuhiro Chiba* and Yohei Okada
Tokyo University of Agriculture and Technology
Japan

1. Introduction

Electron transfer is a fundamental and ubiquitous process in chemical and biological reactions and thus has been extensively studied. A wide variety of unique molecular systems have been developed that connect electron donor and electron acceptor using different types of bridges to investigate electron-transfer reactions. Typically, laser flash photolysis techniques are used in combination with photo-excitable electron donors to provide insights into the chemical and physical properties of electron-transfer reactions. Electron donors are photochemically activated to induce electron transfer toward the electron acceptor, which is then detected spectroscopically. One recent achievement in this field is the discovery of long-range photo-induced electron transfer through a DNA helix, which offers interesting applications as novel conductive materials.

In organic synthesis, electron-transfer-induced reactions have been used extensively to achieve various chemical transformations and construct a wide variety of organic compounds, including natural products, pharmaceutical products, and functional materials. To trigger these electron-transfer-induced reactions, photochemical processes are widely employed. One-electron oxidants and reductants also can be used to initiate electron-transfer-induced reactions. In this context, electrochemical approaches have been utilized to trigger either one- or two-electron transfers through electrode processes that afford electron-transfer-induced reactions. Based on electrochemical processes, various functional group transformations and a wide variety of carbon-carbon bond formation reactions can be accomplished in a controlled manner. For example, Kolbe electrolysis is a well-established process for forming carbon-carbon bonds (Fig. 1). In this reaction, decarboxylation is anodically induced to generate carbon free radicals, which are then homocoupled to make a new carbon-carbon bond.

In addition to free radicals, electrochemical processes also are efficient for the generation of several reactive organic species, including ions and radical ions that can be introduced into organic syntheses as intermediates. In this chapter, we describe electron-transfer-induced carbon-carbon bond formation reactions based on the generation of carbon radical cations as reactive intermediates through electrode processes. Electrochemical studies of the reaction mechanisms have led to the development of new intermolecular [2 + 2] cycloaddition reactions.

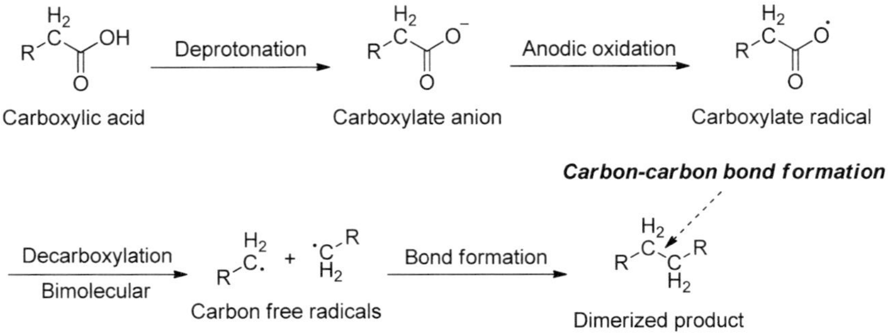

Fig. 1. Reaction mechanism of Kolbe electrolysis

2. Electron-transfer-induced cycloaddition reactions

Cycloaddition reactions play important roles in organic synthesis, allowing complicated ring systems to be synthesized in one step. Numerous synthetic strategies based on cycloaddition reactions have been established to construct various frameworks. As an example, Diels-Alder reactions have been studied from not only synthetic but also mechanistic aspects. Recently, several types of enzymatic Diels-Alder-like reactions also have been reported, leading to the recognition that they are critical in biological systems. In this field, electron-transfer-induced cycloaddition reactions are one of the most intriguing research subjects, both practically and theoretically. To generate radical ions as reactive intermediates for such electron-transfer-induced cycloaddition reactions, photochemical processes are commonly used with photosensitizers. In addition, one-electron oxidants and reductants also are effective for triggering electron-transfer-induced cycloaddition reactions.

In this context, electrochemical approaches can afford reactive organic species, including radicals, ions, and radical ions, at electrodes, which can accomplish various types of electron-transfer-induced cycloaddition reactions without oxidants or reductants. Oxidation or reduction potentials can be controlled easily and the reaction conditions can be simply designed using a combination of supporting electrolytes and typical polar organic solvents, initiating both electron-transfer-induced intra- and intermolecular cycloaddition reactions. We have been developing a series of electron-transfer-induced cycloaddition reactions initiated by anodic oxidation using lithium perchlorate (LPC)/nitromethane (NM) electrolyte solution. An LPC/NM electrolyte solution can stabilize anodically generated carbocation intermediates, facilitating carbon-carbon bond formation reactions. Previously, we have reported electron-transfer-induced intermolecular [4 + 2] and [3 + 2] cycloaddition reactions using an LPC/NM electrolyte solution (Fig. 2). These reactions were conducted under constant potential conditions using carbon felt (CF) working electrodes, platinum counter electrodes, and Ag/AgCl reference electrodes. Hydroquinones and phenols were anodically oxidized to generate corresponding quinones and phenoxonium cations as reactive intermediates, which were then trapped by olefin nucleophiles to construct various substituted [4 + 2] and [3 + 2] cycloadducts.

3. Electron-transfer-induced intermolecular [2 + 2] cycloaddition reactions

Among reported electron-transfer-induced cycloaddition reactions, the [2 + 2] reactions have received attention because they appear to be involved in DNA lesions and repair.

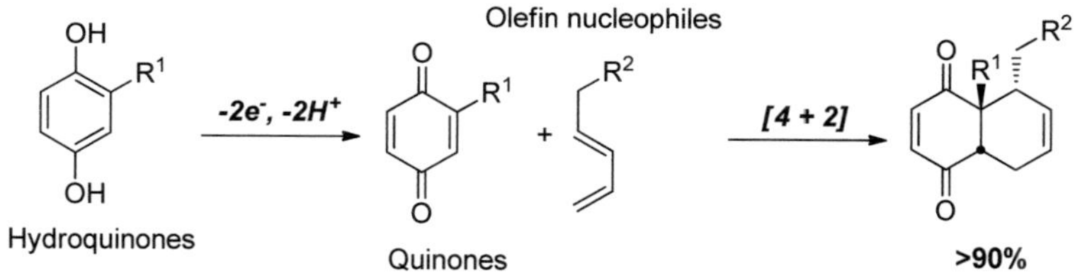

Chiba, K.; Jinno, M.; Kuramoto, R.; Tada, M. *Tetrahedron Lett.* **1998**, *39*, 5527-5530.

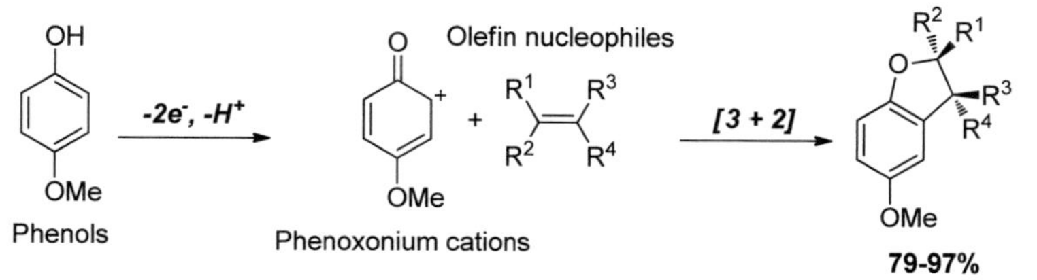

Chiba, K.; Fukuda, M.; Kim, S.; Kitano, Y.; Tada, M. *J. Org. Chem.* **1999**, *64*, 7654-7656.

Fig. 2. Electron-transfer-induced intermolecular [4 + 2] and [3 + 2] cycloaddition reactions

Electron-transfer-induced cycloreversion reactions of four-membered rings also have been represented through radical ion intermediates. Many mechanistic studies on these electron-transfer-induced [2 + 2] cycloaddition reactions and cycloreversion reactions of four-membered ring have been reported.

Enol ethers were then introduced into a LPC/NM electrolyte solution. Enol ethers can generate corresponding radical cations through anodic oxidation, which are then employed as reactive intermediates for cycloaddition reactions. We found that intermolecular [2 + 2] cycloaddition reactions proceeded to construct cyclobutane rings when anodic oxidation of enol ethers were conducted in the presence of olefin nucleophiles in LPC/NM electrolyte solution using CF electrodes as both working and counter, and Ag/AgCl electrodes as reference under constant potential conditions (Fig. 3). These reactions were completed with a catalytic amount of electricity, and starting materials were recovered quantitatively when no potentials were applied. In addition, no cyclobutane ring formation was observed through the anodic oxidation of enol ethers in the absence of olefin nucleophiles followed by their addition, even under radiation conditions. Thus, the intermolecular [2 + 2] cycloaddition reactions clearly responded to the application of electricity, and the corresponding radical cations of enol ethers did not accumulate, indicating that their immediate trapping by olefin nucleophiles was required for the reactions.

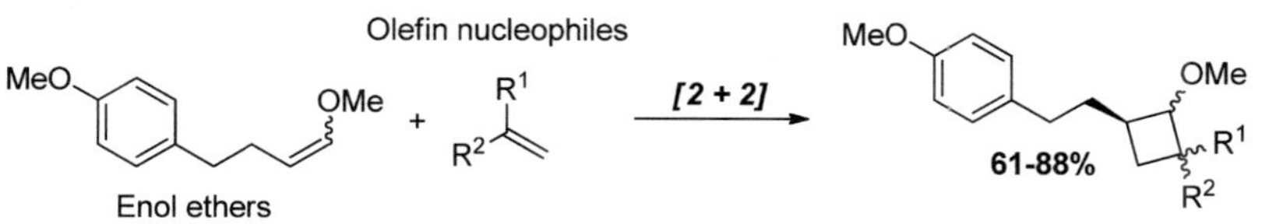

Chiba, K.; Miura, T.; Kim, S.; Kitano, Y.; Tada, M. *J. Am. Chem. Soc.* **2001**, *123*, 11314-11315.

Fig. 3. Electron-transfer-induced intermolecular [2 + 2] cycloaddition reactions

As described above, examples of electron-transfer-induced intermolecular [2 + 2] cycloaddition reactions utilized the enol ether that possessed an alkoxyphenyl group. When similar enol ethers that did not possess an alkoxyphenyl group were prepared and used, [2 + 2] cycloaddition reactions did not occur; instead, olefin cross-metathesis reactions were induced through anodic oxidation in the presence of olefin nucleophiles (Fig. 4).

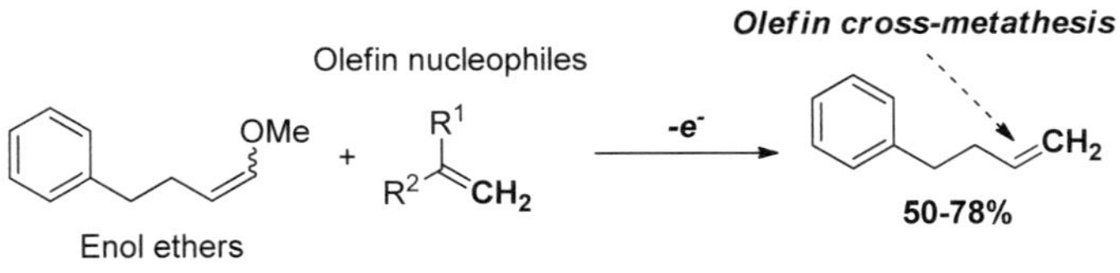

Miura, T.; Kim, S.; Kitano, Y.; Tada, M. Chiba, K. *Angew. Chem. Int. Ed.* **2006**, *45*, 1461-1463.

Fig. 4. Electron-transfer-induced olefin cross-metathesis reactions

These results indicate that the alkoxyphenyl group was essential for the formation of the cyclobutane ring. This can be explained by electron-transfer-induced [2 + 2] cycloaddition reactions beginning with the anodic oxidation of enol ethers to generate their radical cations, which were then trapped by olefin nucleophiles, resulting in the corresponding cyclobutyl radical cations as electron acceptor. The electron-rich alkoxyphenyl group was expected to function as an effective electron donor to complete the formation of the cyclobutane ring through intramolecular electron transfer from the alkoxyphenyl group to the cyclobutyl moiety. In contrast, the phenyl group was not an effective electron donor for the reactions, leading not to the [2 + 2] cycloaddition reaction but to the olefin cross-metathesis reaction (Fig. 5).

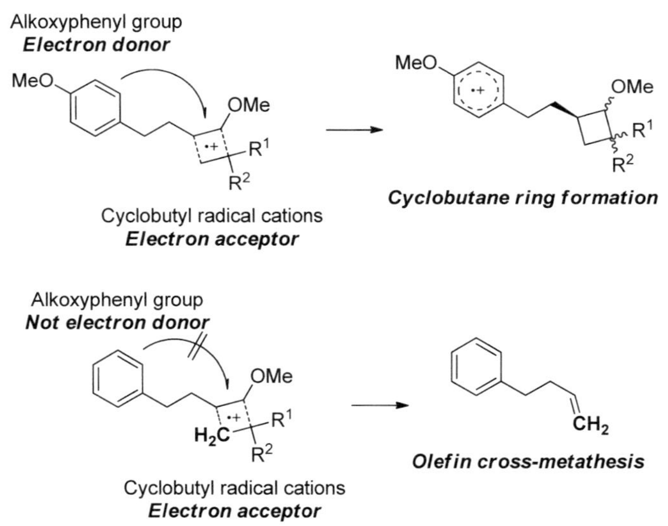

Fig. 5. Plausible reaction mechanisms of electron-transfer-induced [2 + 2] cycloaddition reactions and olefin cross-metathesis.

On the basis of these plausible reaction mechanisms, we envisioned that the alkoxyphenyl group of radical cation intermediates derived from olefin nucleophiles also could function as an effective electron donor for completion of the formation of the cyclobutane ring through similar intramolecular electron transfer. Therefore, 4-allylanisole (**1**) and 1-ethoxyprop-1-ene (**2**) were chosen as olefin nucleophiles that possessed an alkoxyphenyl group and aliphatic enol ether. The anodic oxidation of 1-ethoxyprop-1-ene (**2**) in the presence of an excess of 4-allylanisole (**1**) gave the corresponding cyclobutane ring-containing product (**3**) in high yield (Fig. 6). An excess of olefin nucleophile was essential for effective trapping of transient enol ether radical cations. When similar olefin nucleophiles not possessing an alkoxyphenyl group, such as allylbenzene (**4**), were used in place of 4-allylanisole (**1**), no [2 + 2] cycloaddition reaction occurred, even in the presence of the electron-rich alkoxyphenyl compound, anisole (**5**). These results support the plausible reaction mechanisms. Thus, the anodic oxidation of 1-ethoxyprop-1-ene (**2**) triggered the reaction to generate a radical cation, which then reacted with 4-allylanisole (**1**) to form the corresponding cyclobutyl radical cation as the electron acceptor. Electron transfer from the intramolecular electron donor, the alkoxyphenyl group of the radical cation intermediate, to the cyclobutyl moiety, completed the formation of the cyclobutane ring. In contrast, the intermolecular electron donor, anisole (**5**), was not effective in the reaction even when the corresponding cyclobutyl radical cation formed between the radical cation of 1-ethoxyprop-1-ene (**2**) and allylbenzene (**4**). Thus, since the radical cation remained on the alkoxyphenyl group, and the reaction completed with a catalytic amount of electricity, the alkoxyphenyl radical cation could act as an electron acceptor to oxidize the starting enol ether, completing the overall reaction (Fig. 7).

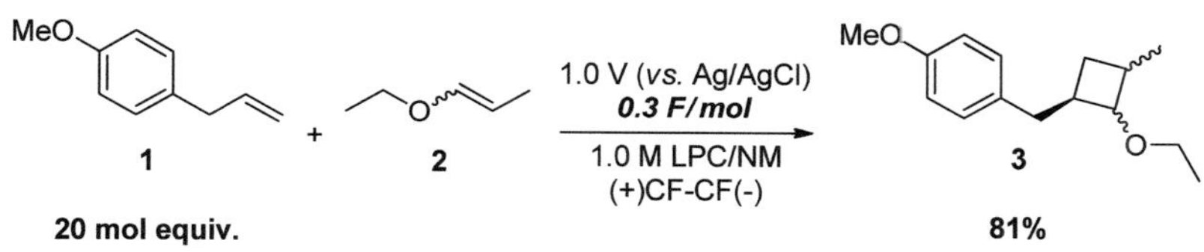

Okada, Y.; Akaba, R.; Chiba, K. *Org. Lett.* **2009**, *11*, 1033-1035.

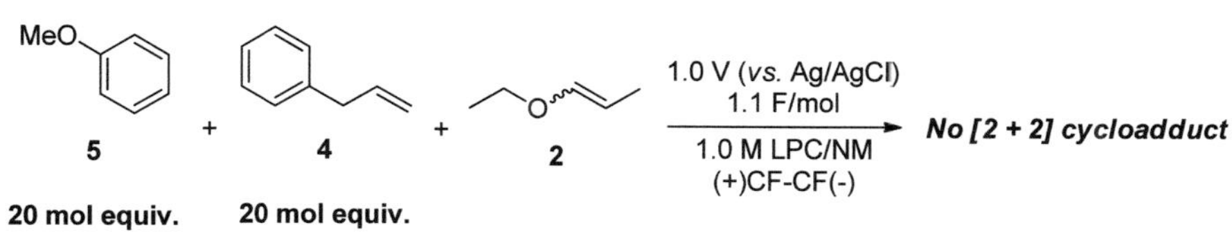

Fig. 6. Electron-transfer-induced intermolecular [2 + 2] cycloaddition reaction between 4-allylanisole (**1**) and 1-ethoxyprop-1-ene (**2**).

Moreover, 3,4-duhydro-2*H*-pyran (**6**), the cyclic version of 1-ethoxyprop-1-ene (**2**), can be introduced into the reaction to afford the corresponding bicyclic cyclobutane ring-containing product (**7**) in excellent yield through intermolecular reaction with 4-allylanisole (**1**) (Fig. 8). As described, the bicyclic framework can be efficiently prepared in one step.

4. Cyclic voltammetric studies on electron-transfer-induced intermolecular [2 + 2] cycloaddition reactions

These synthetic results prompted cyclic voltammetric studies on electron-transfer-induced intermolecular [2 + 2] cycloaddition reactions with the goal of understanding the details of

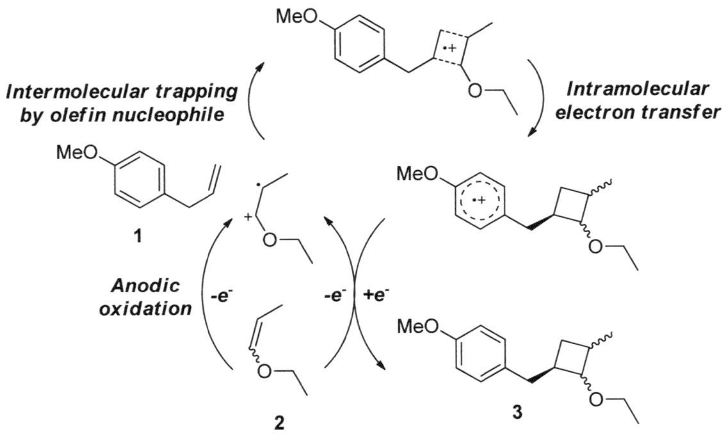

Fig. 7. Plausible reaction mechanism of the electron-transfer-induced intermolecular [2 + 2] cycloaddition reaction between 4-allylanisole (**1**) and 1-ethoxyprop-1-ene (**2**).

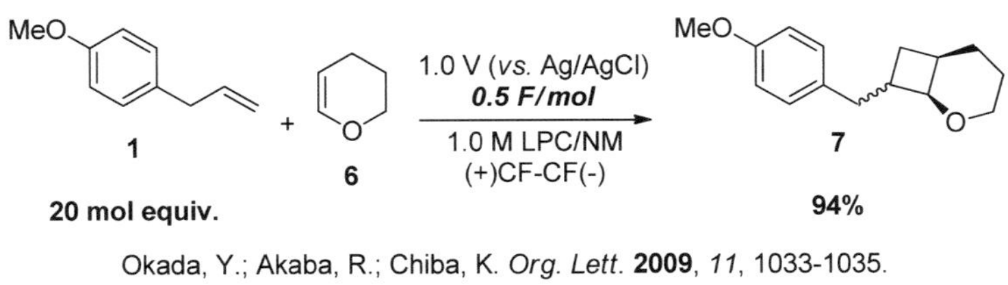

Fig. 8. Electron-transfer-induced intermolecular [2 + 2] cycloaddition reaction between 4-allylanisole (**1**) and 3,4-dihydro-2*H*-pyran (**6**).

their reaction mechanisms. For this purpose, electron-transfer-induced intermolecular [2 + 2] cycloaddition reaction between 4-allylanisole (**1**) and 1-ethoxyprop-1-ene (**2**) was chosen as a model, because the electrocatalytic nature is prominent. Cyclic voltammograms were recorded using a glassy carbon working electrode, platinum counter electrode, and Ag/AgCl reference electrode in an LPC/NM electrolyte solution. Peak oxidation potentials of 4-allylanisole (**1**), 1-ethoxyprop-1-ene (**2**), and the cyclobutane ring-containing product (**3**) were shown at 1.51 V, 1.18 V, and 1.50 V, respectively. The oxidation potential of 1-ethoxyprop-1-ene (**2**) was significantly lower than that of 4-allylanisole (**1**), enabling the selective anodic oxidation of 1-ethoxyprop-1-ene (**2**), even in the presence of an excess of 4-allylanisole (**1**). In addition, the oxidation potential of the cyclobutane ring-containing product (**3**) was similar to that of 4-allylanisole (**1**), indicating that anodic oxidations of both 4-allylanisole (**1**) and the cyclobutane ring-containing product (**3**) could occur on their electron-rich alkoxyphenyl groups to give the corresponding alkoxyphenyl radical cations. Thus, based on their oxidation potentials, the alkoxyphenyl radical cation of the cyclobutane ring-containing product (**3**) could oxidize 1-ethoxyprop-1-ene (**1**) to generate the neutral cyclobutane ring-containing product (**3**) and the radical cation of 1-ethoxyprop-1-ene (**1**). Furthermore, in these sequential reactions, anodic backward discharge also was possible; thus, the alkoxyphenyl radical cation of the cyclobutane ring-containing product (**3**) also might be reduced at the anode to complete the overall electrocatalytic pathway of the reactions, when lower constant potential conditions were employed for the reaction (Fig. 9).

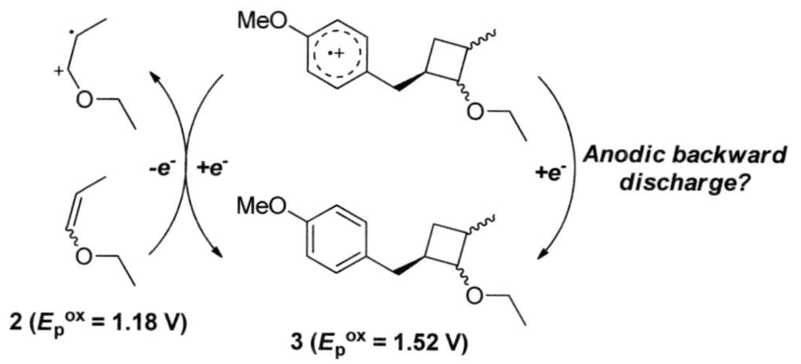

Fig. 9. Plausible reduction mechanism of the alkoxyphenyl radical cation of the cyclobutane ring-containing product (**3**).

Such anodic backward discharge would be a key process in the EC-backward-E electrochemistry. EC-backward-E electrochemistry can be defined as sequential reactions involving interactive electron transfers through a certain chemical transformation. Initial electron transfer (E) occurred between the starting substrate and the electrode to trigger a certain chemical transformation (C), and the resulting product induced subsequent backward electron transfer (bE) at the electrode. To address the reduction mechanism of the alkoxyphenyl radical cation of the cyclobutane ring-containing product (**3**), the reaction was monitored by cyclic voltammetry. The cyclic voltammogram of 1-ethoxyprop-1-ene (**2**) showed an oxidation peak clearly *ca.* 1.18 V (Fig. 10). However, when the cyclic voltammogram of 1-ethoxyprop-1-ene (**2**) was recorded in the presence of an excess of 4-allylanisole (**1**), its oxidation peak was hardly visible. This observation indicated that the alkoxyphenyl radical cation of the cyclobutane ring-containing product (**3**) significantly decreased the oxidation current of 1-ethoxyprop-1-ene (**2**) because of the anodic backward discharge. In contrast, a clear oxidation peak was observed, even when the cyclic voltammogram of 1-ethoxyprop-1-ene (**2**) was recorded in the presence of an excess of allylbenzene (**4**) (Fig. 11). In this case,

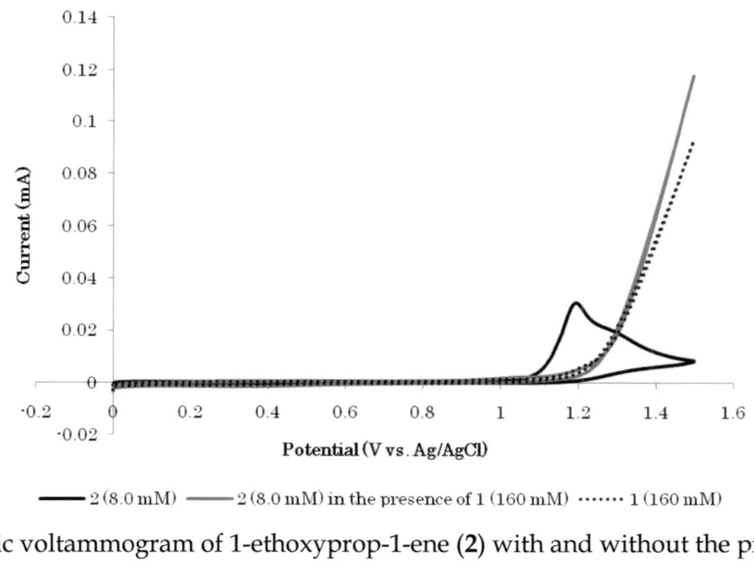

Fig. 10. Cyclic voltammogram of 1-ethoxyprop-1-ene (**2**) with and without the presence of 4-allylanisole (**1**).

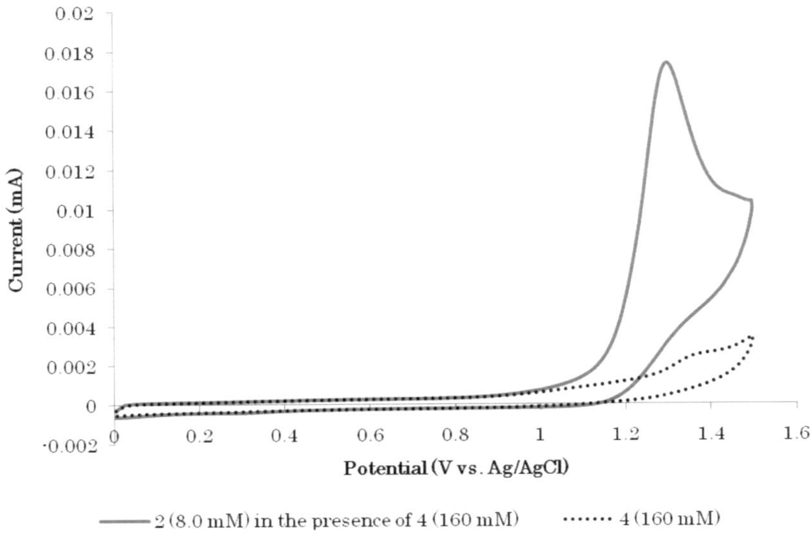

Fig. 11. Cyclic voltammogram of 1-ethoxyprop-1-ene (**2**) in the presence of allylbenzene (**4**).

allylbenzene (**4**) was not able to form the corresponding cyclobutane ring-containing product through reaction with the radical cation of 1-ethoxyprop-1-ene (**2**). Therefore, EC-backward-E electrochemistry at the anode was conjugated to the formation of the cyclobutane ring between the anodically generated radical cation of 1-ethoxyprop-1-ene (**2**) and 4-allylanisole (**1**).

To observe this EC-backward-E electrochemistry, the chemical transformation (C) needed to occur relatively faster than diffusion from the electrode, indicating that intermolecular trapping of the anodically generated radical cation of 1-ethoxyprop-1-ene (**2**) by 4-allylanisole (**1**) was a rapid process (Fig. 12). Moreover, the conversion of the cyclobutyl

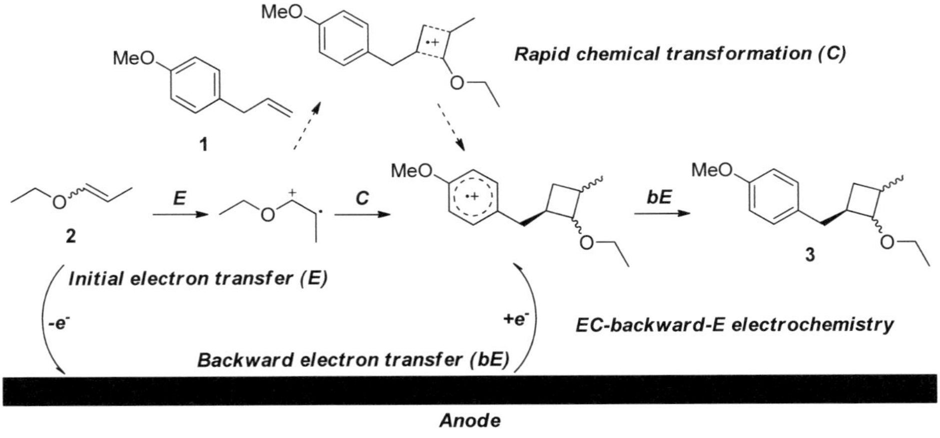

Okada, Y.; Akaba, R.; Chiba, K. *Tetrahedron Lett.* **2009**, *50*, 5413-5416.

Fig. 12. EC-backward-E electrochemistry observed through the electron-transfer-induced intermolecular [2 + 2] cycloaddition reaction between 4-allylanisole (**1**) and 1-ethoxyprop-1-ene (**2**).

radical cation intermediate to the alkoxyphenyl radical cation intermediate could be rationalized as extremely rapid intramolecular electron transfer, since the lifetime of cyclobutyl radical cations previously reported was very short and could not be detected even by nanosecond time-resolved laser flash photolysis studies.

The alkoxyphenyl group was confirmed to be crucial for the reactions and its role could be defined as a "redox tag" (Fig. 13). During sequential electron-transfer processes, the alkoxyphenyl "redox tag" initially functioned as an electron donor that induced intramolecular electron transfer to form the cyclobutane ring. As described above, the intramolecular electron transfer from the alkoxyphenyl group to the cyclobutyl radical cation has been demonstrated as a key step for the formation of cyclobutane ring and the alkoxyphenyl radical cation intermediate generated. The subsequent role of the alkoxyphenyl "redox tag" was as an electron acceptor to complete the overall reactions. As the oxidation potential of the alkoxyphenyl group was relatively high, it must be reduced either through the oxidation of 1-ethoxyprop-1-ene (**2**) or through anodic backward discharge.

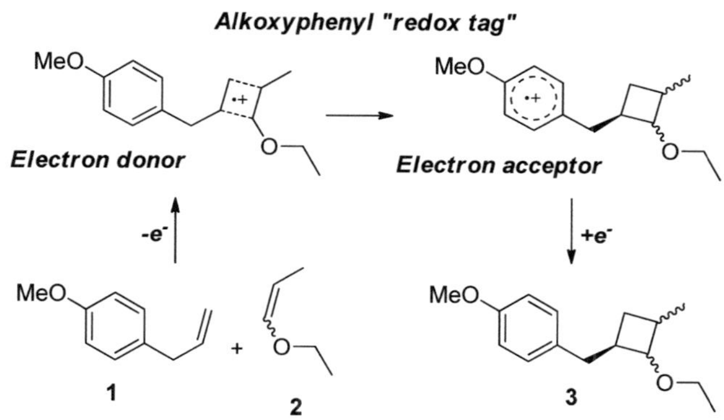

Fig. 13. Plausible function of alkoxyphenyl "redox tag."

5. Electron-transfer-induced intermolecular [2 + 2] cycloaddition reactions assisted by aromatic "redox tag"

These mechanistic results prompted a search for new electron-transfer-induced intermolecular [2 + 2] cycloaddition reactions based on the aromatic "redox tag" strategy. For this purpose, electron-transfer-induced intermolecular [2 + 2] cycloaddition reaction of 3,4-dihydro-2*H*-pyran (**6**) was chosen as a model, because of its relatively simple stereochemistry. Initially, both 3-allylanisole (**8**) and 2-allylanisole (**9**) could effectively trap the anodically generated radical cation of 3,4-dihydro-2H-pyran (**6**) to construct the corresponding bicyclic cyclobutane ring-containing products (**10,11**) in excellent yields (Fig. 14). These results indicate minimal positional effects of the substituent on the aromatic ring, and that the "redox tag" function might be dependent on the electron-density of the aromatic ring. Several functional groups were introduced into the aromatic ring to control its electron density, which led to new electron-transfer-induced intermolecular [2 + 2] cycloaddition reactions (Fig. 15).

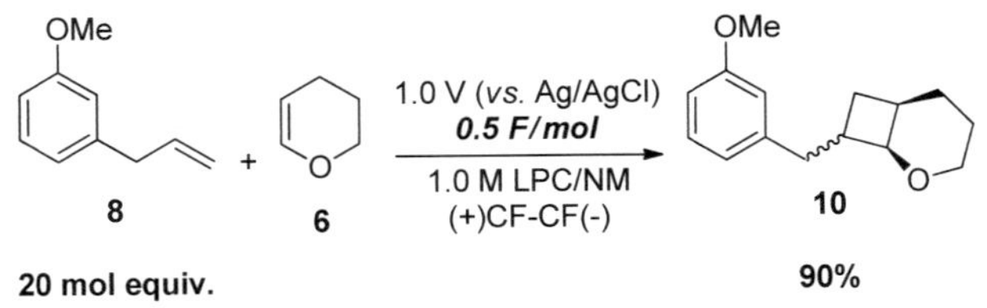

Okada, Y.; Nishimoto, A.; Akaba, R.; Chiba, K. *J. Org. Chem.* **2011**, *76*, 3476-3476.

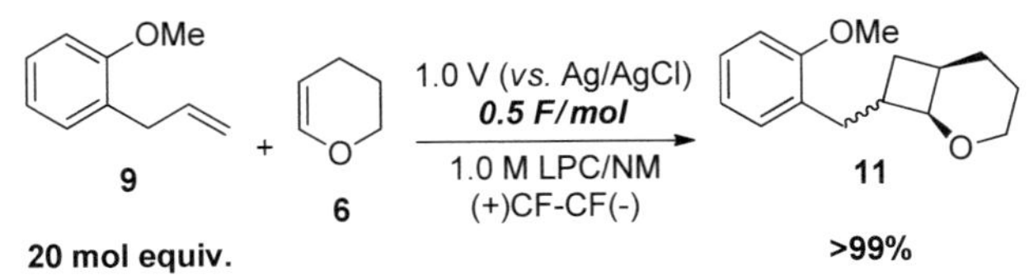

Okada, Y.; Akaba, R.; Chiba, K. *Org. Lett.* **2009**, *11*, 1033-1035.

Fig. 14. Electron-transfer-induced intermolecular [2 + 2] cycloaddition reaction between 3-allylanisole (**8**) or 2-allylanisole (**9**) and 3,4-dihydro-2*H*-pyran (**6**).

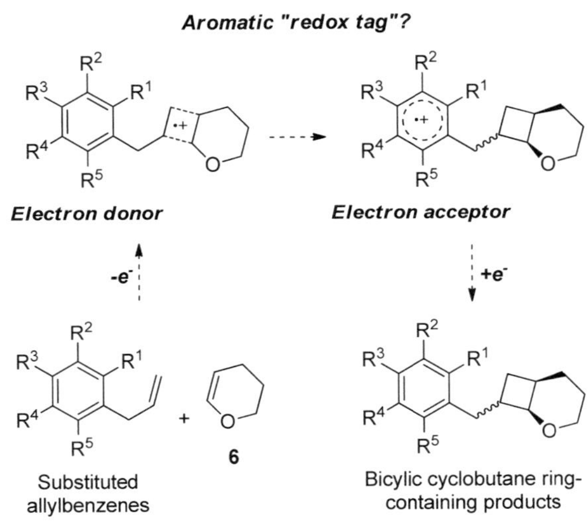

Fig. 15. New electron-transfer-induced intermolecular [2 + 2] cycloaddition reactions based on aromatic "redox tag" strategy.

For this purpose, various substituted allylbenzenes as olefin nucleophiles were prepared to test their function as an aromatic "redox tag." The anodic oxidations of 3,4-dihydro-2*H*-pyran (**6**) were attempted in the presence of an excess of these substituted allylbenzenes.

Results indicated that both 4-allyl-2-methylanisole (**12**) and 1-allyl-4-phenoxybenzene (**13**) functioned as effective olefin nucleophiles in the reaction to afford the corresponding bicyclic cyclobutane ring-containing products (**14, 15**) in excellent yields (Fig. 16); however, 4-allyl-1,2-dimethoxybenzene (**16**) and 5-allyl-1,2,3-trimethoxybenzene (**17**) were less efficient for the reaction and resulted in production of only small amounts of the corresponding bicyclic cyclobutane ring-containing products (**18,19**), even with stoichiometric amounts of electricity

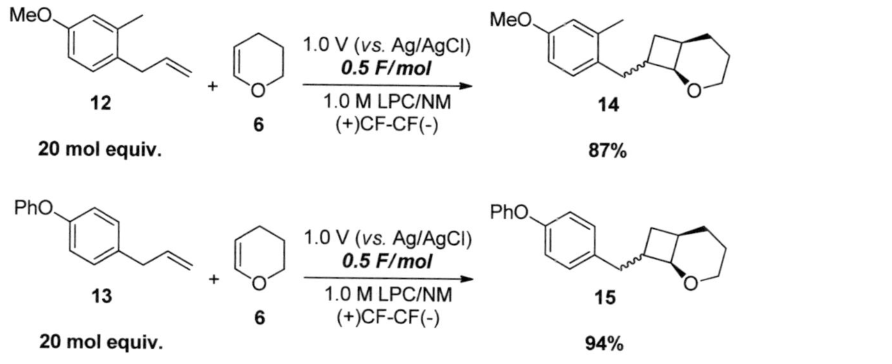

Okada, Y.; Nishimoto, A.; Akaba, R.; Chiba, K. *J. Org. Chem.* **2011**, *76*, 3476-3476.

Fig. 16. Electron-transfer-induced intermolecular [2 + 2] cycloaddition reaction between 4-allyl-2-methylanisole (**12**) or 1-allyl-4-phenoxybenzene (**13**) and 3,4-dihydro-2*H*-pyran (**6**).

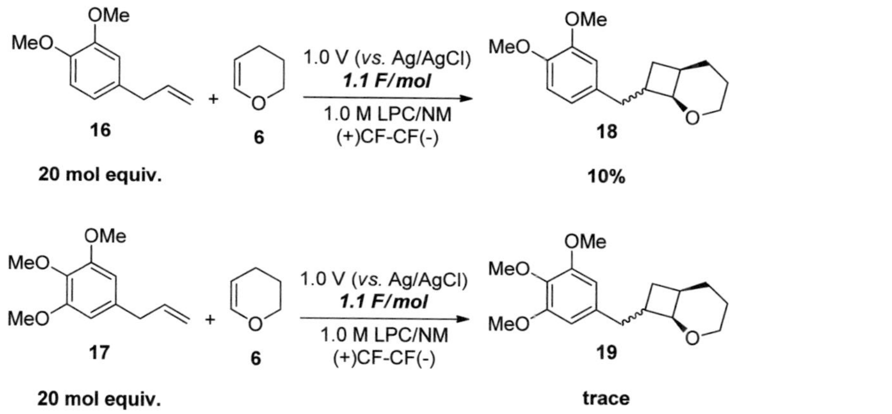

Okada, Y.; Akaba, R.; Chiba, K. *Org. Lett.* **2009**, *11*, 1033-1035.

Fig. 17. Electron-transfer-induced intermolecular [2 + 2] cycloaddition reaction between 4-allyl-1,2-dimethoxybenzene (**16**) or 5-allyl-1,2,3-trimethoxybenzene (**17**) and 3,4-dihydro-2*H*-pyran (**6**).

(Fig. 17). Moreover, 1-allyl-4-methylbenzene (**20**) also functioned as an aromatic "redox tag" to induce the corresponding [2 + 2] cycloaddition reaction, which was in sharp contrast with non-substituted allylbenzene (**4**) (as described above), and the efficiencies of the alkylphenyl "redox tag" relied significantly on the number of alkyl substituent (Fig. 18). Thus, 1-allyl-2,4-dimethylbenzene (**22**) and 2-allyl-1,3,5-trimethylbenzene (**23**) could react with 3,4-dihydro-2*H*-pyran (**6**) effectively to give the corresponding bicyclic cyclobutane ring-containing products (**24, 25**). These observations clearly indicated that the aromatic "redox tag" function was closely related to electron density, which could be quantified as an oxidation potential. Then, the oxidation potentials of several substituted allylbenzenes were measured to represent an appropriate value of oxidation potential for the aromatic ring to function as a "redox tag;" namely, the lower oxidation potentials were favorable, as long as they were greater than that of 3,4-dihydro-2*H*-pyran (**6**) (Table 1). The electron densities of both 4-allyl-1,2-dimethoxybenzene (**16**) and 5-allyl-1,2,3-trimethoxybenzene (**17**) were significantly increased because of the strong electron-donating nature of the alkoxy group, precluding selective anodic oxidation of 3,4-dihydro-2*H*-pyran (**6**) in the presence of 4-allyl-1,2-dimethoxybenzene (**16**) or 5-allyl-1,2,3-trimethoxybenzene (**17**). In contrast, the electron densities of the aromatic rings also could be adjusted to an appropriate value even with relatively low electron-donating alkyl groups (Fig. 19).

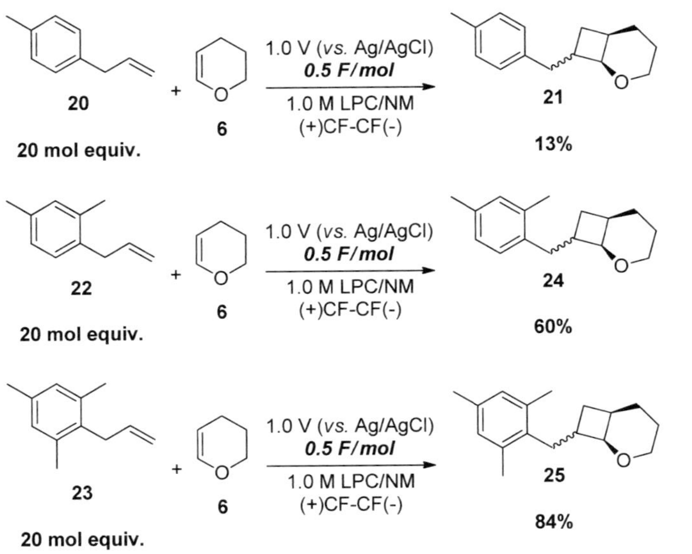

Okada, Y.; Nishimoto, A.; Akaba, R.; Chiba, K. *J. Org. Chem.* **2011**, *76*, 3476-3476.

Fig. 18. Electron-transfer-induced intermolecular [2 + 2] cycloaddition reaction between 1-allyl-4-methylbenzene (**20**), 1-allyl-2,4-dimethylbenzene (**22**), or 2-allyl-1,3,5-trimethylbenzene (**23**) and 3,4-dihydro-2*H*-pyran (**6**).

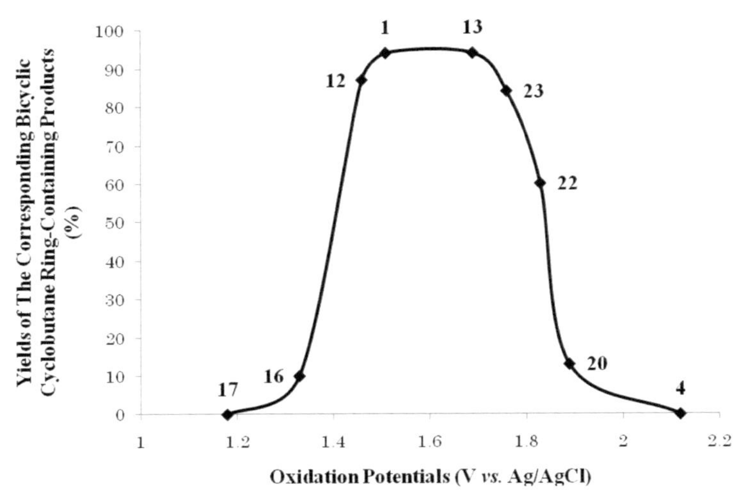

Fig. 19. Relation between oxidation potentials of various substituted allylbenzenes and yields of the corresponding bicyclic cyclobutane ring-containing products.

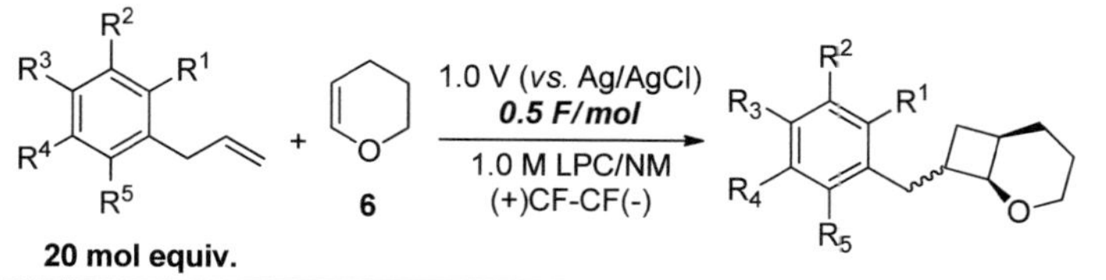

Substrates	Oxidation Potentials (*vs.* Ag/AgCl)	Yields[a] (%)
17 (R^1 = H, R^2 = MeO, R^3 = MeO, R^4 = MeO, R^5 = H)	E_p^{ox} = 1.18 V	trace
16 (R^1 = H, R^2 = MeO, R^3 = MeO, R^4 = H, R^5 = H)	E_p^{ox} = 1.33 V	10
6	E_p^{ox} = 1.41 V	-
12 (R^1 = Me, R^2 = H, R^3 = MeO, R^4 = H, R^5 = H)	E_p^{ox} = 1.46 V	87
1 (R^1 = H, R^2 = H, R^3 = MeO, R^4 = H, R^5 = H)	E_p^{ox} = 1.51 V	94
13 (R^1 = H, R^2 = H, R^3 = PhO, R^4 = H, R^5 = H)	E_p^{ox} = 1.59 V	94
23 (R^1 = Me, R^2 = H, R^3 = Me, R^4 = H, R^5 = Me)	E_p^{ox} = 1.76 V	84
22 (R^1 = Me, R^2 = H, R^3 = Me, R^4 = H, R^5 = H)	E_p^{ox} = 1.83 V	60
20 (R^1 = Me, R^2 = H, R^3 = H, R^4 = H, R^5 = H)	E_p^{ox} = 1.89 V	13
4 (R^1 = H, R^2 = H, R^3 = H, R^4 = H, R^5 = H)	E_p^{ox} = 2.12 V	n.d.

[a]Yields of the corresponding bicyclic cyclobutane ring-containing products determined by NMR

Table 1. Oxidation potentials of various substituted allylbenzenes.

6. Electron-transfer-induced cycloreversion reactions of cyclobutane rings assisted by aromatic "redox tags"

The electron impact mass spectrum of the *trans*-bicyclic cyclobutane ring-containing product (**7**) possessed a fragmentation pattern that contained a base peak at m/z 148, which was assigned to the radical cation of 4-allylanisole (**1**). This result suggests that the high-energy radical cation of the *trans*-bicyclic cyclobutane ring-containing product (**7**) produced in the mass spectrometer participated in the cycloreversion reaction of the cyclobutane ring. Based on this observation, the electron-transfer-induced cycloreversion reactions could be initiated through electrode processes.

The cyclic voltammogram of the *trans*-bicyclic cyclobutane ring-containing product (**7**) was then recorded under same conditions as described above to measure its peak oxidation potential at 1.54 V, which was similar to that of 4-allylanisole (**1**). Based on these values, the anodic oxidation of the *trans*-bicyclic cyclobutane ring-containing product (**7**) also was expected to occur on the electron-rich alkoxyphenyl group to give the corresponding alkoxyphenyl radical cation, which might induce cycloreversion reaction of the cyclobutane ring. Indeed, anodic oxidation of the *trans*-bicyclic cyclobutane ring-containing product (**7**) was attempted to give the cycloreversion product, 4-allylanisole (**1**), in moderate yield (Fig. 20). Analysis of the reaction mixture revealed that this reaction was accompanied by formation of a small amount of *cis*-bicyclic cyclobutane ring-containing product (**7**), verifying that the radical cation generated on the alkoxyphenyl group contributed to cleavage of the carbon-carbon bond that constitutes the cyclobutane ring. Intramolecular electron transfer between the alkoxyphenyl group and the cyclobutyl moiety was reversible; therefore, the cycloreversion reaction required an excess amount of electricity (Fig. 21). In this case, the amount of 4-allylanisole (**1**), derived from the cycloreversion reaction, was not sufficient to trap the radical cation of 3,4-dihydro-2*H*-pyran (**6**); thus, the reaction was irreversible. Although the

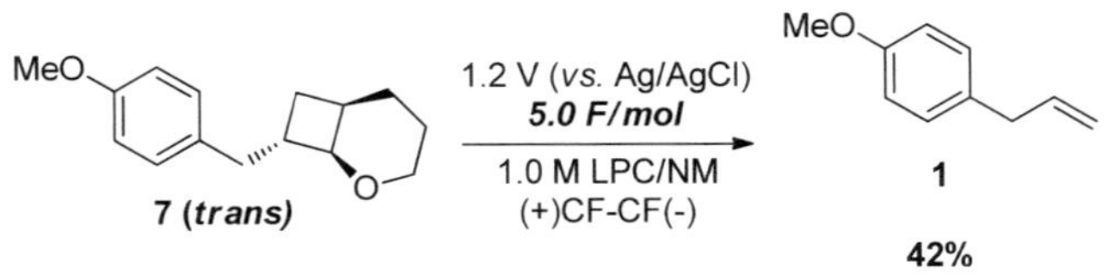

Fig. 20. Electron-transfer-induced cycloreversion reaction of the *trans*-bicyclic cyclobutane ring-containing product (7).

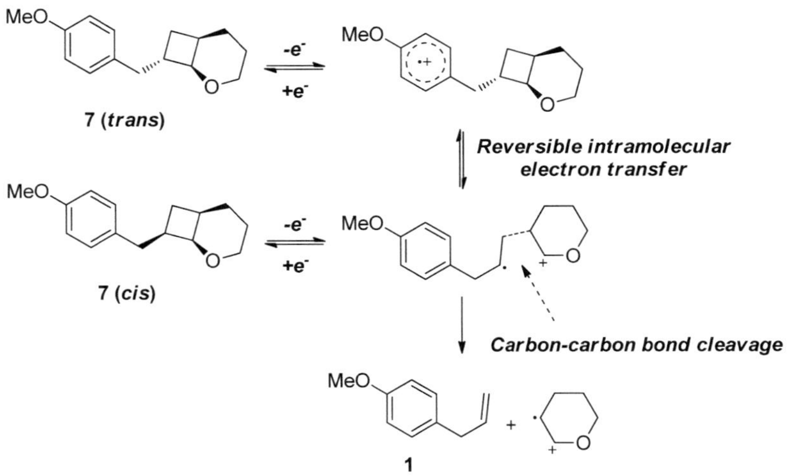

Fig. 21. Reaction mechanism of the electron-transfer-induced cycloreversion reaction of the *trans*-bicyclic cyclobutane ring-containing product (7).

cycloreversion reaction was highly chemoselective, the isolated yield was moderate because the cycloreversion product, 4-allylanisole (**1**), also was anodically oxidized under the reaction conditions, leading to its decomposition. Moreover, anodic oxidation of the *trans*-bicyclic cyclobutane ring-containing product (**25**) was attempted to give the cycloreversion product, 2-allyl-1,3,5-trimethylbenzene (**23**), in moderate yield. These results suggest that the aromatic "redox tag" also could facilitate the cycloreversion reaction of cyclobutane rings through the production of the corresponding aromatic radical cations (Fig. 22).

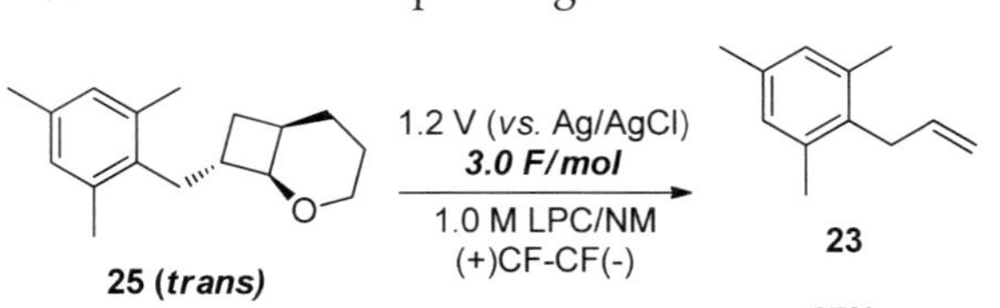

Fig. 22. Electron-transfer-induced cycloreversion reaction of the *trans*-bicyclic cyclobutane ring-containing product (**25**).

In contrast with the electron-transfer-induced cycloreversion of the *trans*-bicyclic cyclobutane ring-containing product (**7**), that of the all *trans*-cyclobutane ring-containing product (**3**) was less efficient; its peak oxidation potential was measured at 1.52 V (Fig. 23). Based on this value, the anodic oxidation of the all *trans*-cyclobutane ring-containing product (**3**) also was expected to occur on its electron-rich alkoxyphenyl group, producing the corresponding radical cation. However, it could not facilitate the cycloreversion reaction of the cyclobutane ring efficiently. Apparently, the ring strain of the bicyclic structure was responsible for driving the cycloreversion reactions. Thus, the following mechanisms were proposed for the electron-transfer-induced cycloreversion reactions assisted by the aromatic "redox tag" (Fig. 24). The initial oxidation of the bicyclic cyclobutane ring-containing products occurred on the aromatic ring. Through reversible intramolecular electron transfer between the aromatic ring and cyclobutyl moiety, the corresponding cyclobutyl radical cations formed, leading to their cycloreversion reactions.

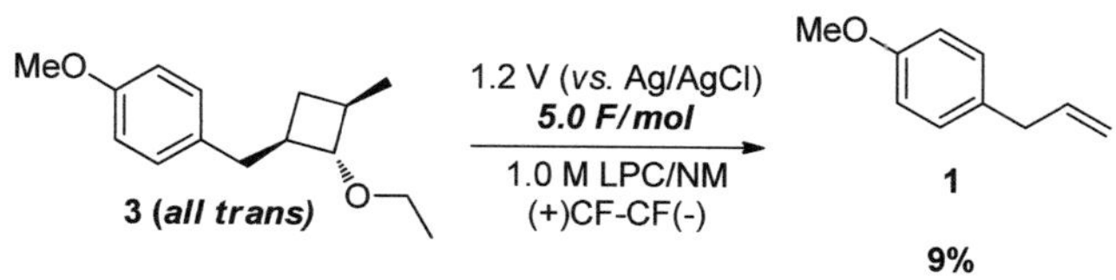

Fig. 23. Electron-transfer-induced cycloreversion reaction of the all *trans*-cyclobutane ring-containing product (**3**).

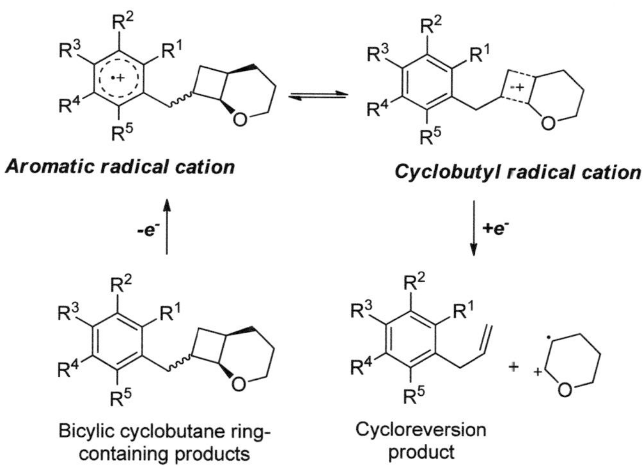

Fig. 24. Plausible reaction mechanism of electron-transfer-induced cycloreversion reaction of bicyclic cyclobutane ring-containing products.

7. Conclusion

The electrochemistry of electron transfer at the electrodes was key for generating reactive intermediates, leading to both carbon-carbon bond formation reactions and cleavage reactions. In particular, new electron-transfer-induced intermolecular [2 + 2] cycloaddition reactions between anodically generated enol ether radical cations and olefin nucleophiles were discovered to produce cyclobutane rings. Through mechanistic studies based on an electrochemical approach, these carbon-carbon bond formation reactions were found to be assisted by the aromatic "redox tag," which could also facilitate several carbon-carbon bond cleavage reactions leading to the cycloreversion of cyclobutane rings. The aromatic "redox tag" functioned as both electron-donor and electron-acceptor in the sequential electron-transfer processes.

Electron-transfer-induced reactions play an important role in organic synthesis. In particular, electrochemical approaches that involve electron transfer at the electrode can regulate organic transformations in a highly controlled manner, which should lead to additional applications in both academic and industrial fields.

8. Acknowledgments

This work was partially supported by a Grant-in-Aid for Scientific Research from the Ministry of Education, Culture, Sports, Science, and Technology. In addition, we thank Professor Dr. Ryoichi Akaba at Gunma National College of Technology for his valuable suggestions and comments regarding the reaction mechanisms.

9. References

Arata, M.; Miura, T. & Chiba, K. (2007). Electrocatalytic Formal [2+2] Cycloaddition Reactions between Anodically Activated Enyloxy Benzene and Alkenes. *Organic Letters*, Vol. 9, No. 21, pp. 4347-4350, ISSN 1523-7052

Chiba, K.; Jinno, M.; Kuramoto, R. & Tada, M. (1998). Stereoselective Diels-Alder reaction of electrogenerated quinones on a PTFE-fiber coated electrode in lithium perchlorate / nitromethane. *Tetrahedron Letters*, Vol. 39, No. 31, pp. 5527-5530, ISSN 0040-4039

Chiba, K.; Fukuda, M.; Kim, S.; Kitano, Y. & Tada, M. (1999). Dihydrobenzofuran Synthesis by an Anodic [3 + 2] Cycloaddition of Phenols and Unactivated Alkenes. *The Journal of Organic Chemistry*, Vol. 64, No. 20, pp. 7654-7656, ISSN 0022-3263

Chiba, K.; Miura, T.; Kim, S.; Kitano, Y. & Tada, M. (2001). Electrocatalytic Intermolecular Olefin Cross-Coupling by Anodically Induced Formal [2+2] Cycloaddition between Enol Ethers and Alkenes. *Journal of the American Chemical Society*, Vol. 123, No. 45, pp. 11314-11315, ISSN 0002-7863

Fry, A. J. & Britton, W. E. (Eds.). (April 30, 1986). *Topics in Organic Electrochemistry*, Springer, ISBN 978-0306420580, New York, New York, USA

Little, R. D. & Weinberg, N. L. (Eds.). (August 30, 1991). *Electroorganic Synthesis*, CRC Press, ISBN 978-0824785840, New York, New York, USA

Lund, H. & Hammerich, O. (Eds.). (December 14, 2000). *Organic Electrochemistry* (Fourth Edition), CRC Press, ISBN 978-0824704308, New York, New York, USA

Miura, T.; Kim, S.; Kitano, Y.; Tada, M. & Chiba, K. (2006). Electrochemical Enol Ether/Olefin Cross-Metathesis in a Lithium Perchlorate/Nitromethane Electrolyte Solution. *Angewandte Chemie International Edition*, Vol. 45, No. 9, pp. 1461-1463, 1521-3773

Okada, Y.; Akaba, R. & Chiba, K. (2009). Electrocatalytic Formal [2+2] Cycloaddition Reactions between Anodically Activated Aliphatic Enol Ethers and Unactivated Olefins Possessing an Alkoxyphenyl Group. *Organic Letters*, Vol. 11, No. 4, pp. 1033-1035, 1523-7052

Okada, Y.; Akaba, R. & Chiba, K. (2009). EC-backward-E electrochemistry supported by an alkoxyphenyl group. *Tetrahedron Letters*, Vol. 50, No. 38, pp. 5413-5416, 0040-4039

Okada, Y. & Chiba, K. (2010). Continuous electrochemical synthetic system using a multiphase electrolyte solution. *Electrochimica Acta*, Vol. 55, No. 13, pp. 4112-4119, 0013-4686

Okada, Y. & Chiba, K. (2010). Electron transfer-induced four-membered cyclic intermediate formation: Olefin cross-coupling *vs.* olefin cross-metathesis. *Electrochimica Acta*, Vol. 56, No. 3, pp. 1037-1042, 0013-4686

Okada, Y.; Nishimoto, A.; Akaba, R. & Chiba, K. (2011). Electron-Transfer-Induced Intermolecular [2 + 2] Cycloaddition Reactions Based on the Aromatic "Redox Tag" Strategy. *The Journal of Organic Chemistry*, Vol. 76, No. 9, pp. 3470-3476, 0022-3263

Part 2

Organic Electrochemistry

5

Electrochemical Reduction, Oxidation and Molecular Ions of 3,3´-bi(2-R-5,5-dimethy-1-4-oxopyrrolinylidene) 1,1´-dioxides

Leonid A. Shundrin
N. N. Vorozhtsov Institute of Organic Chemistry,
Siberian Branch of the Russian Academy of Sciences,
Russian Federation

1. Introduction

An interest in studying of stabilized radical ions of cyclic nitrones is related, first of all, to the problems of revealing possible routes of formation of stable nitroxide radicals from nitrones, which are widely used as spin traps. Some nitrones can undergo reversible one electron electrochemical oxidation to radical cations (RC), and their highly resolved ESR spectra can be obtained in solution at room temperature [1]. In 1992 L. Eberson proposed "Inverted Spin Trapping" mechanism of nitroxide radicals formation, which includes one electron oxidation of initial nitrone to its radical cation (RC) and subsequent reaction with nucleophile [2].

Potentials of the electrochemical reduction (PER) of *N*-oxides vary in rather wide limits: the PER of heteroaromatic *N*-oxides, *viz.*, phenazine, acridine, quinoxaline, and pyrazine derivatives, change from –0.85 V to –1.90 V (in DMF *vs.* saturated calomel electrode (s.c.e)) [3,4,5]. In aprotic media, the first step of reductive electrode processes of mono- or *N,N*-dioxides is a one electron process. However, the formation of cyclic *N*-oxide's radical anions (RA) stable at 298 K is observed very rare, and the first wave of their reduction is irreversible [4,5-6,7]. The reversible electrochemical reduction (ECR) of 1,1,3-triphenyl-*N*-oxide was mentioned [8]. However, the ESR spectrum of the corresponding RA was not detected down to –40 °C. In turn, the ECR of *N*-oxides of phenazine, acridine, quinoxaline, and pyrazine are characterized by the formation of corresponding RA, and the coupling constants with the ^{14}N nuclei and protons of the aromatic system were determined [3].

In contrast to some types of non nitrone N-oxides described above, ECR of nitrones is irreversible process, and the first step of reduction can be one [4] or two electron [9] by nature. PER of acyclic α-aryl-*N*-alkyl(aryl)nitrones and cyclic nitrones, *viz.*, pyrroline and isoindole derivatives, range from –1.80 V to –2.30 V (in DMF *vs.* s.c.e) [6-9]. Published data on the formation of long-lived RA of mono or dinitrones in solution at room temperature are absent.

3,3′-Bi(2-R-5,5-dimethy-1-4-oxopyrrolinylidene)-1,1′-dioxides (**1–4**, Scheme 1), which represent cyclic dinitrones with conjugated C=C bond, are formed by the smooth oxidation of the corresponding pyrrolinones (Scheme 1). Dinitrones **1-4** (DN) have the π-conjugation chain including both nitrone groups. The extension of the graph of the π-system of the corresponding derivatives compared to the initial pyrrolinones (Scheme 1 (b)) should enhance the electron withdrawing ability of DN **1–4** and, in all probability, favor the stabilization of their RA in enough extent to measure ESR spectra under ECR. Therefore, we studied the peculiarities of ECR and electrochemical oxidation (ECO) of compounds **1–4** in aprotic solvents by cyclic voltammetry. Solvent effects on ECR in MeCN:H_2O mixtures have been studied also.

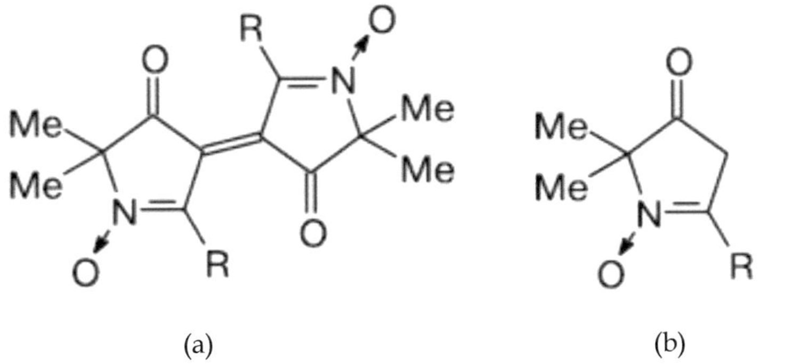

Scheme 1. Structures of 3,3′-bi(2-R-5,5-dimethy-l-4-oxopyrrolinylidene) 1,1′-dioxides ((R = CF_3 (**1**), Me (**2**), Ph (**3**), Bu^t (**4**)) (*a*) and corresponding initial 2-R-5,5-dimethy-4-oxopyrrolinone-1-oxides (*b*).

2. Experimental

Cyclic voltammograms (CV) of dinitrones **1–4** were measured on CVA-1BM electrochemical system (Bulgaria) equipped with a LAB-MASTER polyfunctional interface (Institute of Nuclear Physics, Novosibirsk, Russia), which enables one complete digital control of the system. Measurements were carried out in a mode of triangular pulse potential sweep in the range of sweep rates $0.1\ V{\cdot}s^{-1} < v < 50\ V{\cdot}s^{-1}$. A standard electrochemical cell with a working volume of 5 mL was connected to the system *via* the three electrode scheme and equipped with a salt bridge filling with a supporting electrolyte solution in DMF or MeCN to connect the working volume and reference electrode. The working electrode was a stationary spherical Pt electrode with a surface area of 8 mm^2, a Pt spiral was the auxiliary electrode, and a saturated aqueous calomel electrode (SCE) served as the reference electrode. The supporting electrolyte was Et_4NClO_4 (0.1 mol/dm^3) for aprotic solvents and $LiClO_4$ (0.1 mol/dm^3) for MeCN-H_2O mixtures. Oxygen was removed by passing argon through the working solution. The concentration of depolarizers was $1{\cdot}10^{-3}$ mol/dm^3.

ESR spectra of paramagnetic intermediates of the ECR and ECO processes were measured on a Bruker ESP-300 radio spectrometer equipped with a double resonator. Compounds **1–4** were oxidized and reduced in combination with ESR spectrometric measurements under anaerobic conditions at potentials of the corresponding first reduction or oxidation peaks at T = 298 K in a three electrode electrochemical cell for EPR measurements equipped with a Pt electrode. The space of the working electrode of the cell was placed into the front shoulder of the ESR spectrometer resonator. Numerical simulation of ESR spectra were performed according to the Winsim 2002 program with the SIMPLEX optimization algorithm.

To establish possible structures and of compounds **1**–**4**, as well as those of their molecular ions (radical anions (RA), radical cations (RC), dianions (DA) and dications (DC)) quantum chemical calculations of the corresponding species were carried out by semiempirical PM3 method (WinMOPAC 7.0 program set) using unrestricted (UHF) and restricted (RHF) Hartree-Fock approximations. Spin density distributions in RA and RC were calculated with DFT/PBE method ("Nature" program set [15]) for gas phase. For RA and RC of DN **2** in gaseous phase and in solution spin density distribution was calculated with (U)B3LYP method using 6-31+G* basis set. PCM model was used to describe the solvent (MeCN of H_2O). All calculations have been done with complete geometry optimization and symmetry restraints for the Me groups in position 5 of the pyrrolinone cycles and for substituents in position 2 according to their local dynamic symmetry (C_{3v}) for the CF_3, Me, and Bu^t groups and C_{2v} for Ph. No restraints were imposed on rotation of all substituents about the pyrrolinone cycles.

3. Results and discussion

3.1 Electrochemical reduction and oxidation of 3,3´-bi(2-R-5,5-dimethy-1-4-oxopyrrolinylidene) 1,1´-dioxides in aprotic solvents

Cyclic voltammograms of DN **1**–**4** in the region of negative potentials contain two reversible one electron reduction peaks (EE –process) corresponding to the formation of RA and DA (Fig.1, example for DN **1**). The replacement of the DMF [10] solvent by MeCN [11] does not qualitatively change the reduction processes. In MeCN, the potentials of reduction peaks are by approximately 0.1 V shifted to the region of more negative values compared to the potentials measured in DMF (Table1).

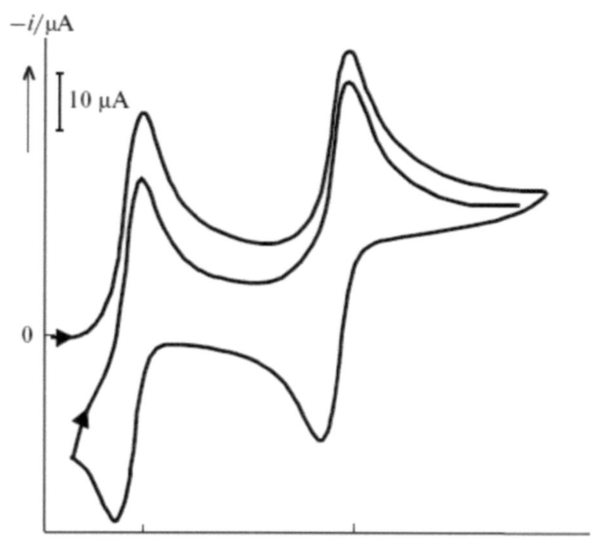

Fig. 1. Cyclic voltammogram of compound **1** (DMF, C=$1 \cdot 10^{-3}$ mol/dm^3, v = 0.08 V·s^{-1}, supporting electrolyte 0.1 *M* Et_4NClO_4).

The CV of DN **1–4** in the region of positive potentials are shown in Fig. 2. For the potential sweep rate $v < 5$ Vs^{-1}, the CV curve of **1** exhibits the single irreversible oxidation peak. It is most likely that the irreversibility is caused by a fast chemical reaction involving RC. In

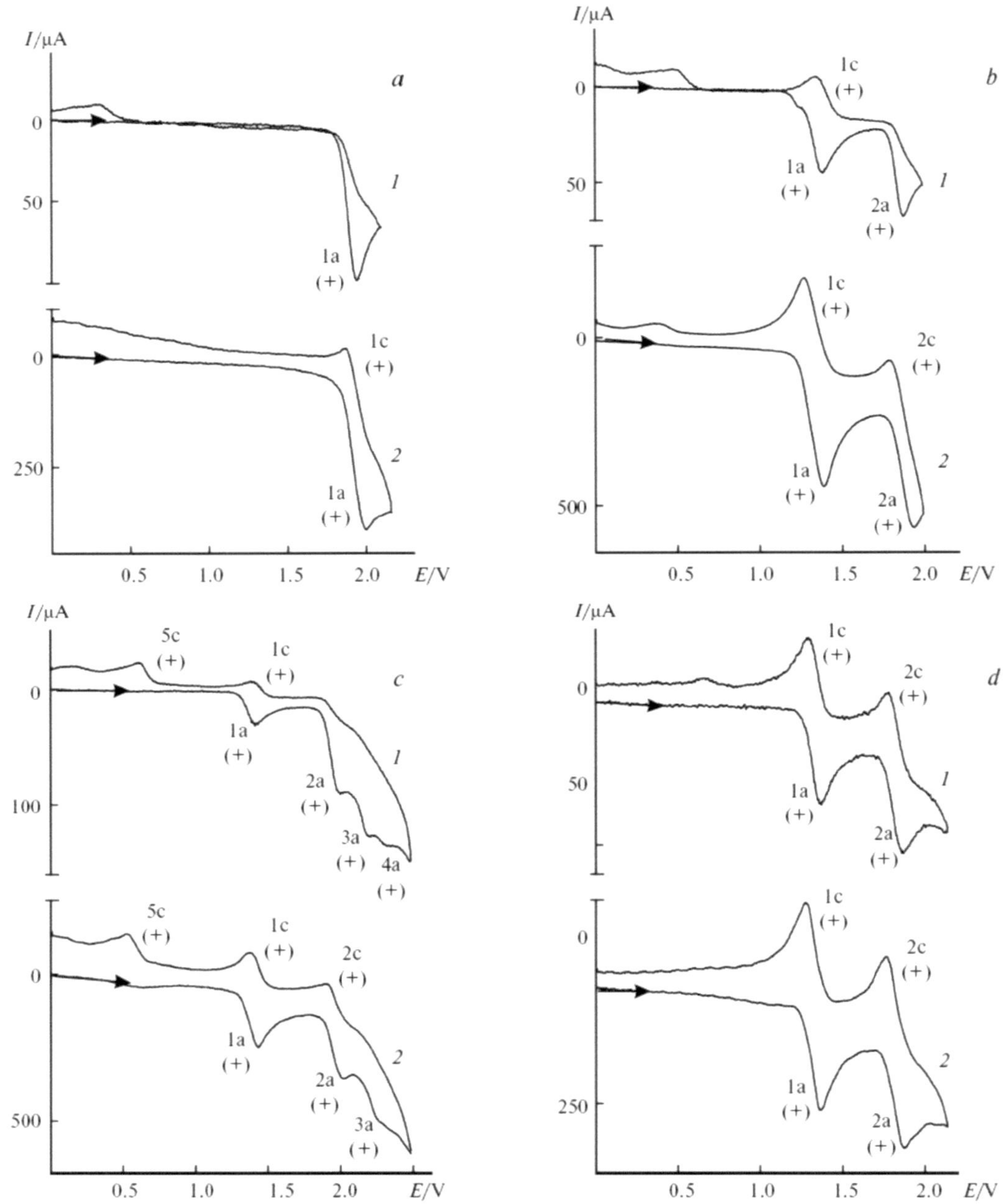

Fig. 2. Cyclic voltammograms of dinitrons **1–4** (*a–d*, respectively) in the region of positive sweep potentials and v = 0.1 (*1*), 5.0 V s^{-1} (*2*); $C_0 = 1{\cdot}10^{-3}$ mol/dm^3, MeCN, supporting electrolyte 0.1 M Et_4NClO_4.

		Oxidation[a]				Reduction			
№	R	$E_{(+)}^{1a}$	$E_{(+)}^{1c}$	$E_{(+)}^{2a}$	$E_{(+)}^{2c}$	$E_{(-)}^{1c}$	$E_{(-)}^{1a}$	$E_{(-)}^{2c}$	$E_{(-)}^{2a}$
1	CF_3	1.94 (1.98)[b]	- (1.88)	-	-	-0.18 **(-0.13)**[c]	-0.11 **(-0.06)**	-0.81 **(-0.83)**	-0.71 **(-0.76)**
2	CH_3	1.38 (1.39)	1.32 (1.30)	1.88 (1.90)	- (1.80)	-0.58 **(-0.50)**	-0.48 **(-0.43)**	-1.07 **(-1.14)**	-1.00 **(-1.07)**
3	Ph	1.41 (1.41)	1.36 (1.35)	1.98 (2.00)	- (1.89)	-0.55 **(-0.44)**	-0.47 **(-0.37)**	-1.05 **(-1.07)**	-0.95 **(-1.00)**
4	Bu^t	1.36 (1.36)	1.29 (1.26)	1.86 (1.87)	1.78 (1.77)	-0.52 **(-0.47)**	-0.45 **(-0.40)**	-1.00 **(-0.99)**	-0.93 **(-0.92)**

a At the Pt electrode *vs.* SCE in an 0.1 *M* solution of Et_4NClO_4 in MeCN, $C_0 = 1 \bullet 10^{-3}$ mol/dm^3, potential sweep rate 0.1 V s^{-1}.
b Peak potentials at a potential sweep rate of 5 V·s^{-1} are given in parentheses.
c Fist peak potentials in DMF.

Table 1. Peak potentials (V) of electrochemical reduction (MeCN, DMF) and oxidation (MeCN) of compounds **1**–**4**.

turn, the ECO of dinitrons **2** and **4** is characterized by two peaks, whereas four oxidation peaks are observed for **3**. The observed peaks are characterized here by their potentials in the region of positive ($E_{(+)}^{ij}$) and negative ($E_{(-)}^{ij}$) values, respectively (i is the number of peak, and j = a or c indicates the anodic or cathodic branch of the CV curve, respectively). The first oxidation peaks of all compounds are diffuse in nature, $I_{(+)}^{1a}\cdot v^{0.5}$ =const, where $I_{(+)}^{1a}$ is the maximum current of the first peak in the anodic branch of the CV.

For compounds **2**–**4**, the first peaks correspond to the reversible one electron process (ratio of currents of the anodic and cathodic branches $I_{(+)}^{1a}/I_{(+)}^{1c} \approx 1$ and $\Delta E = E_{(+)}^{1a} - E_{(+)}^{1c} = 0.06$ V).

At low sweep rates ($v < 4$ V·s^{-1}), the second oxidation peak of **2** is irreversible, and in the range 5 V·s^{-1} < v < 50 V·s^{-1} the CV of **2** exhibits two reversible one electron peaks (see Fig. 2, *b*). Noticeable instability of DC of **2** formed at the potential of the second oxidation peak is related, most likely, to proton elimination from the methyl group in position 2 (see also Ref. [12]). In fact, when the methyl groups are replaced by the *tert*-butyl groups, the ECO of dinitron **4** in MeCN at T = 298 K is an EE process with formation of stable RC and DC in the whole studied range of v (see Fig. 2, *d*). The peak corresponding to the formation of DC **3** is irreversible ($E_{(+)}^{2a}$, $v < 5$ V·s^{-1}, see Fig. 2, *c*), the ratio is $I_{(+)}^{2a}/I_{(+)}^{1a} = 2.57$, and two additional oxidation peaks, whose nature was not studied, are observed in the anodic branch. As in the case of methyl-substituted DN **2**, when v is increased to 5 V·s^{-1}, the $E_{(+)}^{2a}$ peak becomes reversible and one electron in nature ($I_{(+)}^{2a}/I_{(+)}^{1a} = 0.97$), and the $E_{(+)}^{4a}$ peak observed at low potential sweep rates disappears (see Fig, 2, c, curve 2). The irreversible (at all v) peak $E_{(+)}^{5c}$ in the cathodic branch of the CV (Fig.2, *c*) is attributed, most likely, to the ECO of the conversion products of DC **3**, because this peak is not observed for the potential sweep in the range 0 V < E < 1.7 V.

To establish the nature of the oxidation peak of trifluoromethyl substituted DN **1**, we measured the CV in the potential sweep region –1.0 V < E < 2.2 V (Fig. 3, *a*) and in the rate

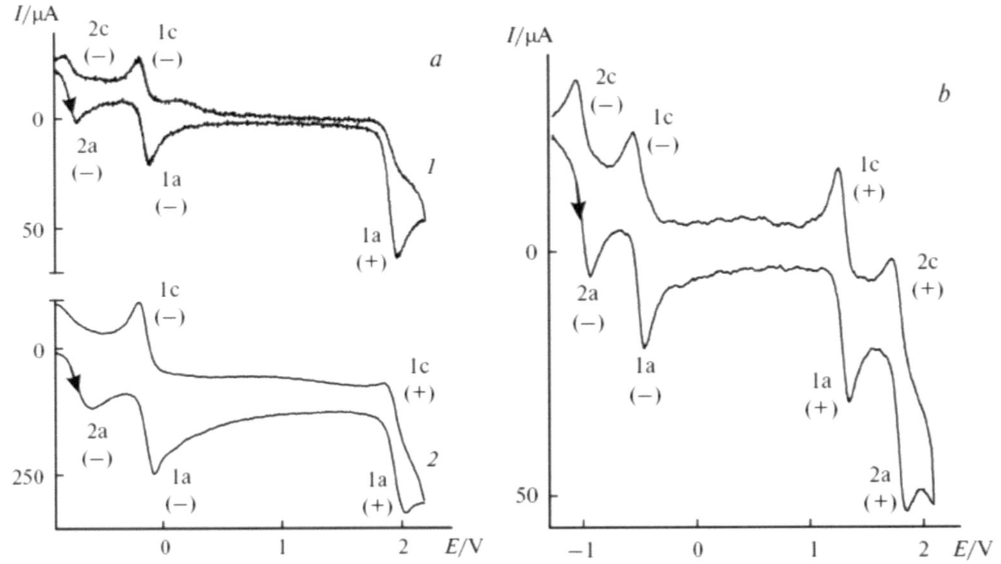

Fig. 3. Cyclic voltammograms of dinitrones **1** (*a*, v = 0.1 (*1*) and 5 V·s^{-1} (*2*)) and **4** (*b*, v = 0.1 V·s^{-1}) in oxidation and reduction areas of potential sweep in MeCN.

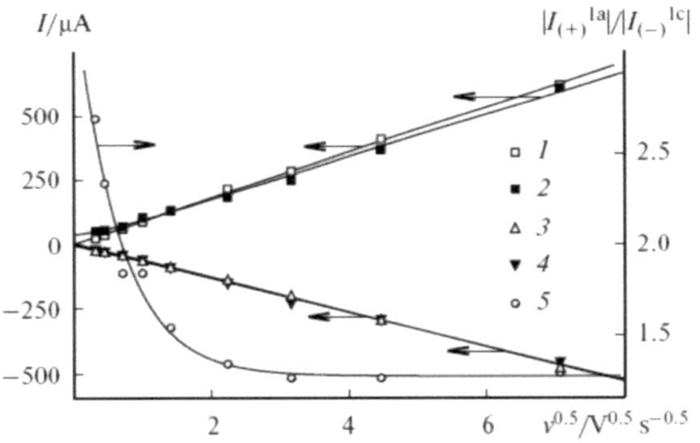

Fig. 4. Currents of the first oxidation ($I_{(+)}{}^{1a}$) and reduction ($I_{(-)}{}^{1c}$) peaks of DN **1** (*1*, *3*) and **4** (*2*, *4*) and the ratio $|I_{(+)}{}^{1a}| / |I_{(-)}{}^{1c}|$ (*5*) for DN **1** *vs.* $v^{0.5}$.

range 0.1 V·s^{-1} < v < 50 V·s^{-1}. For v = 0.1 V·s^{-1}, the ratio of currents of the observed oxidation peak and reversible one electron reduction peak is $|I_{(+)}{}^{1a}| / |I_{(-)}{}^{1c}|$ = 2.7. In the interval of 0.1 V·s^{-1} < v < 10 V·s^{-1}, the ratio decreases exponentially to 1.26 and remains constant up to v = 50 V·s^{-1} (Fig. 4). The plots of currents of the oxidation peak of DN **1** and its first reduction peak corresponding to the formation of RA **1** *vs.* potential sweep rate are shown in Fig. 4 along with similar plots for *tert*-butyl substituted DN **4**, whose all one electron peaks are reversible in both the negative and positive potential regions (see Fig. 3, *b*). The dependences of all currents of the reversible peaks on the potential sweep rate are well described by the classical equation:

$$I_{(\pm)}{}^{ij} = B + Av^{0.5}, \quad (1)$$

where $A = 0.4463 \cdot S_{el} \cdot (nF)^{3/2}(D/(RT))^{0.5} \cdot C_0$, $n = 1$ is the number of transferred electrons, S_{el} is the surface area of the working electrode, D is the diffusion coefficient of the substance, C_0 is the depolarizer concentration, and B is an empirical constant related to the perturbation of linear free diffusion. The parameters in Eq. (1) A, B, and r^2 (r^2 is the correlation coefficient) have the following values: DN **1**, $I_{(-)}{}^{1c}$, -67.56 μA·V$^{-0.5}$s$^{0.5}$, 7.73 μA, and 0.998; DN **4**, $I_{(-)}{}^{1c}$, -65.95 μA V$^{-0.5}$ s$^{0.5}$, -0.33 μA, 0.997, $I_{(+)}{}^{1a}$, 88.88 μA·V$^{-0.5}$s$^{0.5}$, 4.41 μA, and 0.998.

The dependence of the oxidation peak current of DN **1** on the potential sweep rate is described by the equation:

$$I_{(+)}{}^{1a} = B + Av^{0.5} + C\exp(-kv^{0.5}), \quad (2)$$

where C and k are empirical constants. The parameters in Eq. (2) for $I_{(+)}{}^{1a}$ are the following: $A = 88.33$ μA·V$^{-0.5}$s$^{0.5}$, $B = 6.28$ μA, $C = 35.33$ μA, $k = 1.48$ V$^{-0.5}$s$^{0.5}$, and $r^2 = 0.996$. The last term in Eq. (2) reflects the contribution of the electrode process associated with the oxidation of the RC **1** conversion products. This process seems to be rather slow in the CV time scale. Hence, for a relatively low increase in v (up to 5 V·s^{-1}), Eq. (2) can rapidly be reduced to Eq. (1), the oxidation peak of **1** becomes reversible and one electron, and the slope (A) of the plot of the peak current *vs.* $v^{0.5}$ is virtually equal to that of $I_{(+)}{}^{1a}(v^{0.5})$ for **4** (see Fig. 4). In the region of negative potentials, the slopes of the plots of the currents of the first reduction peaks of **1** and **4** ($I_{(-)}{}^{1c}$) are fairly close (see Fig. 4).

Thus, ECR of DN **1-4** is an EE process in aprotic solvents, whereas ECO is more complicated. The first step of oxidation of DN **1—4** is the one electron transfer to form RC **1—4**. For trifluoromethyl substituted DN **1**, the observed current of the first oxidation peak at low v contains an additional contribution from the electrode process related to the oxidation of the RC **1** conversion products. A similar situation is observed for the second step of oxidation of DN **3**. The second ECO step of DN **2** and **4** is one electron but only DC **4** is noticeably stable.

Note that both ECO and ECR of *tert*-butyl substituted DN **4** are EE processes accompanied by the formation of DA, RA, RC, and DC which are long-lived at $T = 298$ K (see Fig. 3, *b*). Perhaps, this is the first example of an organic heterocyclic compound for which four long-lived molecular ions were detected within one potential sweep cycle.

3.2 Electrochemical reduction of 3,3′-bi(2-R-5,5-dimethy-1-4-oxopyrrolinylidene) 1,1′-dioxides in MeCN:H_2O mixtures of various composition.

Among dinitrones **1—4**, methyl substituted DN **2** is most water soluble. In a saturated aqueous solution containing the supporting electrolyte $LiClO_4$ (0.1 mol/dm^3) at 25 °C, its concentration (~6·10^{-4} mol/dm^3) is sufficient for the direct CV study of ECR. Due to a lower solubility, the electrochemical reduction potentials of DNs **1, 3, 4** in water were determined by the extrapolation method, starting from the dependences of the reduction potentials on the water content in MeCN—H_2O binary mixtures. The same dependence was studied for comparison for DN **2**.

The transition from the Et_4NClO_4 supporting electrolyte (see Ref. [11]) to $LiClO_4$ shifts the potentials of the first cathodic peaks ($E_{(-)}{}^{1c}$) measured in MeCN by 0.1 V, on the average, toward less negative potentials. The diffusion nature and reversible one electron character of the first reduction peak are retained upon this transition. The region of reduction potentials accessible for measurement in MeCN—H_2O binary mixtures is 0.5 V > E > -1.1 V and restricted by the potential of water reduction on Pt under these conditions. When the water content in the binary mixture increases, the potentials of the first reduction peaks of compounds **1—4** shift toward less negative values (Fig. 5, example for DN **2**), reaching in water very low values (Table 2). In all cases, the first peaks are one electron and reversible in the whole range of compositions of MeCN—H_2O mixtures. The plots of the potentials of the first reduction peaks *vs.* MeCN content in the mixture are linear and described by the regressions in the form:

$$E_{(-)}^{1c} = E_{0(-)}^{1c} + A \cdot V_{MeCN}, \quad (3)$$

where V_{MeCN} is the volume fraction of MeCN. The regression parameters A and corresponding correlation coefficients r are given in Table 2. Unlike DN **2—4**, on going from MeCN to H_2O the potential of the first reduction peak of trifluoromethyl derivative **1** lies in the region of positive values *vs.* s.c.e.

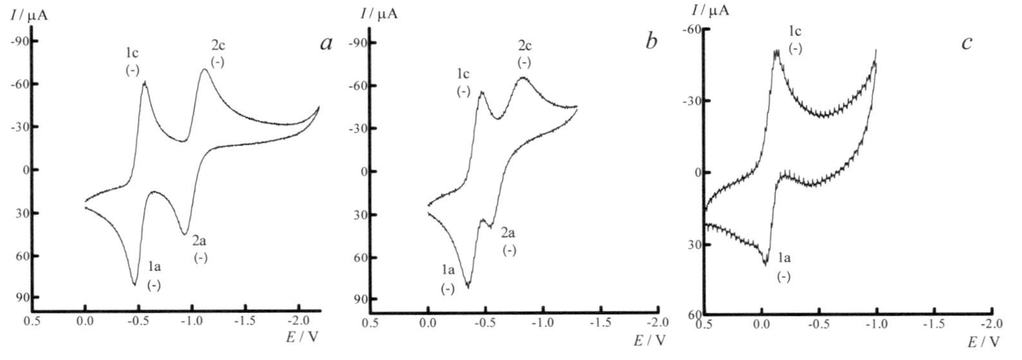

Fig. 5. Cyclic voltammograms of DN **2** in MeCN (*a*), MeCN—H_2O binary mixture (V_{MeCN} = 0.98) (*b*), and H_2O (*c*) (ν = 0.1 V s^{-1}, supporting electrolyte 0.1 *M* solution of $LiClO_4$).

Compound	$E_{0(-)}^{1c}$, V	A, V	r
1	0.19	-0.52	0.989
2	-0.13[b]	-0.35	0.998
3	-0.09	-0.33	0.998
4	-0.08	-0.32	0.998

a Potentials *vs.* SCE at the Pt electrode in a 0.1 *M* solution of $LiClO_4$; ν = 0.1 V· s^{-1}.
b The potential of the first reversible one electron reduction peak measured directly in H_2O at the Pt electrode is –0.14 V *vs.* SCE in a 0.1 *M* solution of $LiClO_4$ at ν = 0.1 V·s^{-1}.

Table 2. Parameters of the linear dependences of the first reduction peaks potentials of DN **1—4** ($E_{(-)}^{1c}$) on MeCN volume fraction in MeCN—H_2O binary mixtures[a] (see Eq. (3)).

The second reversible reduction peak, which is distinctly seen on CV of DN **1—4** in MeCN and related to the formation of the corresponding DA (Fig. 5, *a*) becomes quasi reversible (Fig. 5, *b*) upon the addition of even small amounts of H_2O (~2%), $E_{(-)}^{2c}$ shifts to the region of less negative values, and the limiting current of the second peak decreases appreciably compared to a similar value for the first peak with an increase in the water content. In the range $0 < V_{MeCN} < 0.5$, the second peak disappears (see Fig. 5, *c*). The result is unexpected: for the first time for compounds of the nitrone series we succeeded to observe the reversible one electron process in the first step of their electrochemical reduction in aqueous solutions and binary mixtures of the aprotic solvent—water type.

3.3 Radical anions and radical cations of 3,3′-bi(2-R-5,5-dimethy-1-4-oxopyrrolinylidene) 1,1′-dioxides

ESR spectra of the corresponding RA are detected under ECR at potentials of the first cathodic peaks ($E_{(-)}^{1c}$) [11] (Fig. 6). Corresponding isotropic hyperfine coupling constants (HFCC) are presented in Table 3. The character of the hyperfine structure of the ESR spectra of all RA generated in MeCN is the same as that for RA in DMF (Table 3, see ref. [10] also). Only a slight increase in the HFCC with ^{14}N nuclei is observed compared to the corresponding HFCC for RA in DMF. For RA **1** and **2**, all ^{19}F and ^{1}H nuclei of R substituents are spectrally equivalent and give septet hyperfine splitting with the binominal ratio of intensities of the components. At T = 298 K, no dynamic effects are observed, which are related to hindered rotation of the substituents in position 2 or torsional oscillations of the pyrrolinone moieties about the C=C bond when the solvent is replaced.

ESR spectra of the corresponding RC are observed under ECO of DN **2—4** at potentials of the first anodic peaks $E_{(+)}^{1a}$ (see Fig. 6). The nitrogen atoms of all RC are spectrally equivalent. A rather resolved hyperfine structure from protons of the substituents in position 2 of the pyrrolinone cycles is observed only for RC of methyl derivative **2** at T = 253 K, and the ESR spectra of RC **1** were not detected because of its instability (see Fig. 2, *a*). The HFCC of RC **2—4** are presented in Table 3.

Compound	R	Hyperfine coupling constants, mT	
		RA	RC
1	CF_3	0.462(^{14}N), 0.217(^{19}F) **0.444**(^{14}N)^{c}, **0.228**(^{19}F)	-
2	CH_3	0.414(^{14}N), 0.173(^{1}H) **0.408**(^{14}N), **0.174**(^{1}H)	0.141(^{14}N), 0.083(^{1}H)^{b}
3	Ph	0.425(^{14}N) **0.433**(^{14}N)	0.150 (^{14}N)
4	t-Bu	0.435(^{14}N) **0.424**(^{14}N)	0.188(^{14}N)

a The radical anions were generated in MeCN at the corresponding potentials of the first reduction peak $E_{(-)}^{1c}$, and the radical cations were generated at the potential $E(+)^{1a}$; the concentration of the depolarizer and Et_4NClO_4 was $1.5{\cdot}10^{-3}$ and 0.1 mol/dm^3, respectively.
b The hyperfine structure 5N×7H becomes resolved at T < 253 K.
c HFCC in DMF.

Table 3. Comparative characteristics of the ESR spectra^{a} of RA and RC **1—4** generated in MeCN and DMF (for RA) at 298 K.

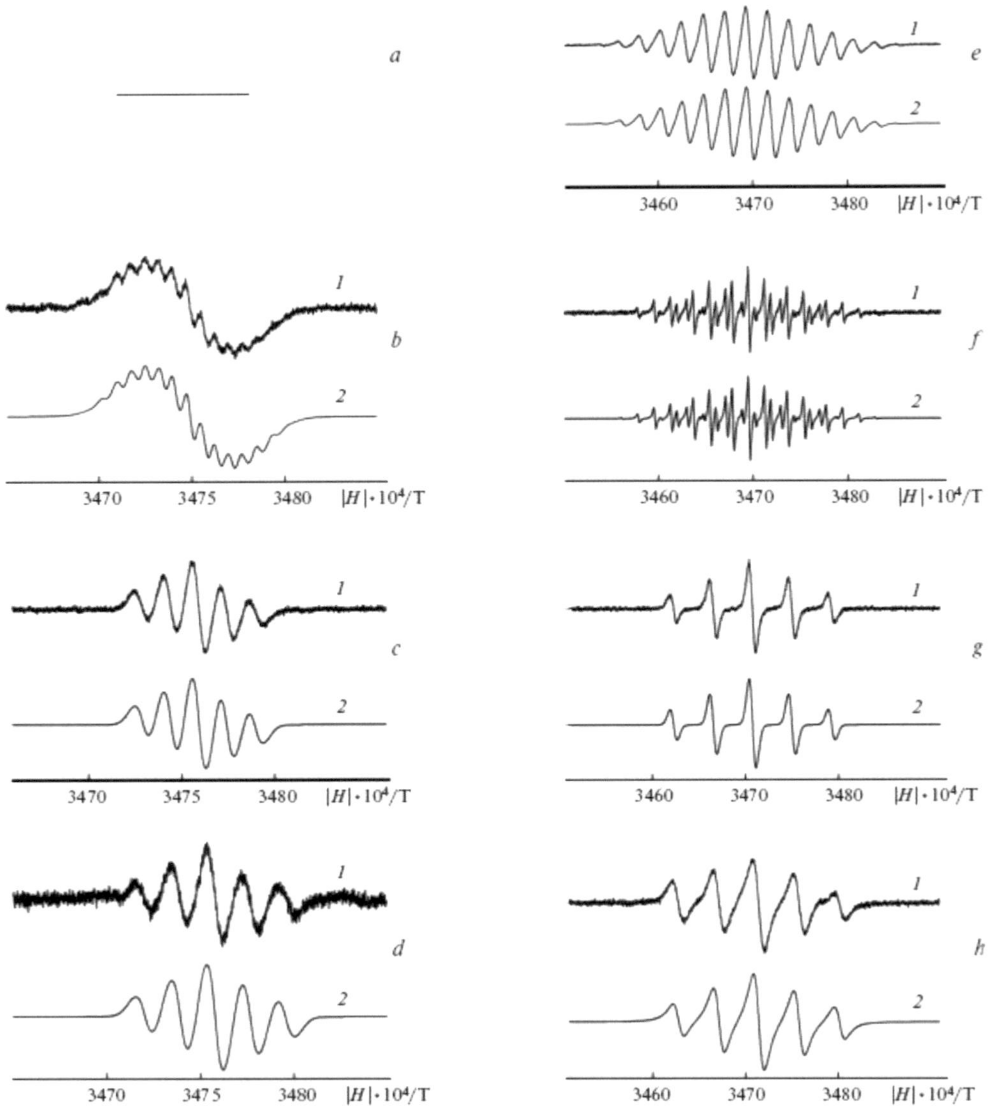

Fig. 6. ESR spectra of (*1*) RC (*a*–*d*) and RA (*e*–*h*) of dinitrones **1** (*a*, *e*), **2** (*b*, *f*), **3** (*c*, *g*), and **4** (*d*, *h*) generated by electrochemical oxidation and electrochemical reduction in MeCN (298 K) at the corresponding first peaks potentials and their numerical simulations (*2*).

On going from RA to the corresponding RC **2–4**, experimental values of nitrogen hyperfine splitting constant decreases, on the average, by 2.69 times (see Table 3), and the HFCC with the ^{1}H nuclei of the methyl groups (pair of RA **2**, RC **2**) decrease by 2.08 times. No HFCC with protons of the *tert*-butyl and phenyl groups are observed for both RA and RC **3** and **4**.

The ESR spectra of RA **1–4** were measured in MeCN–H_2O mixtures of various composition, which was characterized by the molar fraction of water (χ). Examples of the ESR spectra of RA **1**, **2**, **4** and their numerical simulations are given in Fig. 7. The maximum

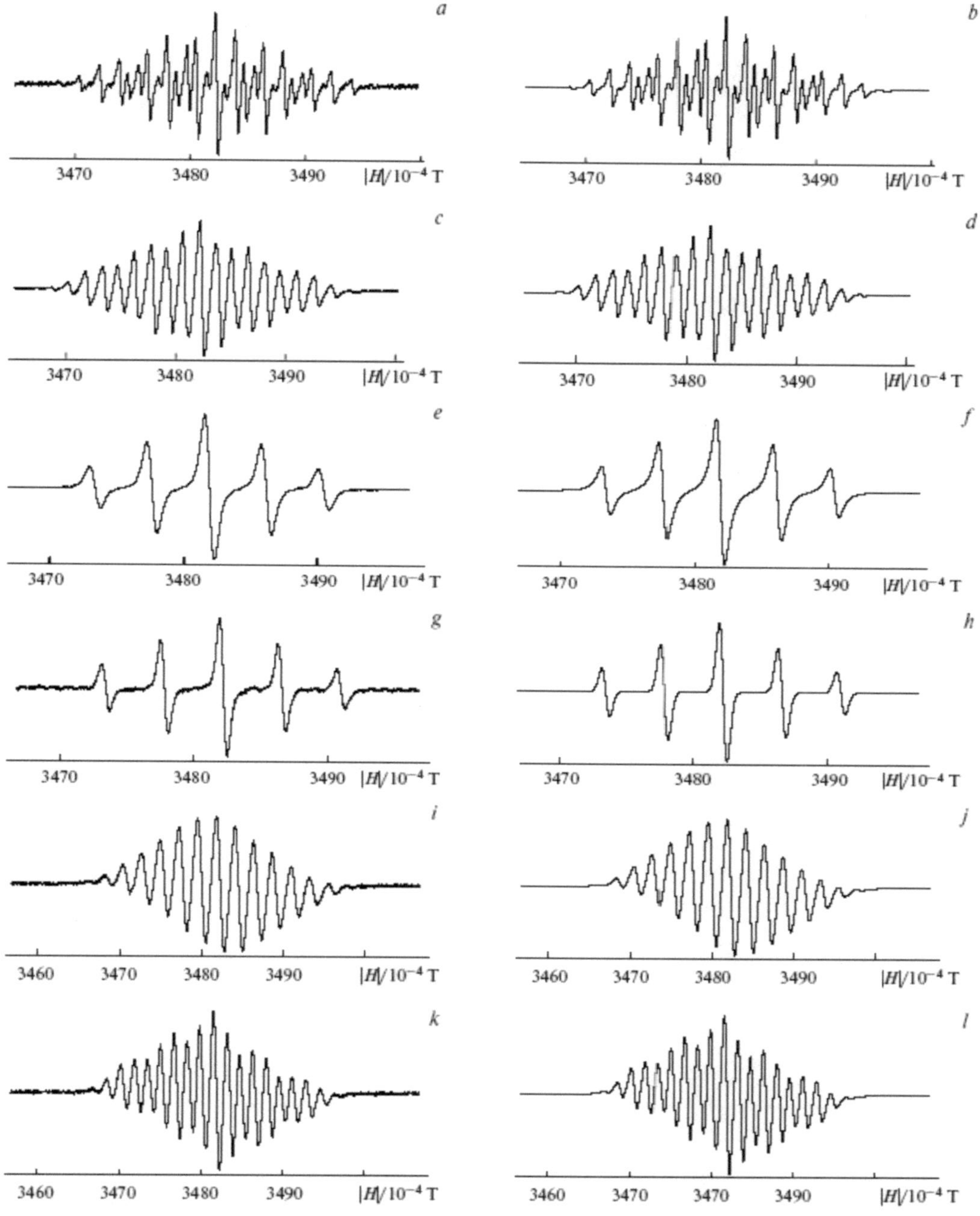

Fig. 7. The ESR spectra of RA **2** (*a* – *d*), **3** (*e* – *h*), and **1** (*i* – *l*) in MeCN and in MeCN – H_2O solvent mixture with maximum available content of H_2O: *a, c, e, g, i, k* are the experimental spectra, *b, d, f, h, j, l* are the numerical simulation of spectra; χ (molar fraction of H_2O) = 0 (*a, b, e, f, i, j*), 0.9 (*c, d*), 0.92 (*g, h*), and 0.5 (*k, l*).

value of χ in the mixture, at which the ESR spectra can be recorded, is not the same for nitrones **1–4** and is related to their different solubility [13]. For DN **2** possessing the highest solubility, the ESR spectra of the corresponding RA can be recorded within entire range of the MeCN–H_2O composition. The *g*-factors of RA **1–4** are close (2.0032±0.0004) and virtually do not depend on the mixture composition. The HFCC of RA **1–4** in MeCN and at the highest possible values of χ are given in Table 4. The well resolved hyperfine structure (HFS) on the ^{1}H and ^{19}F nuclei of substituents R are observed for RA **2** and **1** within entire available range of the mixture compositions. For RA **3** and **4**, only HFS on the ^{14}N nuclei are observed due to the remoteness of the substituents R protons, which is in agreement with the published data [11].

The dependencies of the HFCC for RA **1–4** on χ are shown in Fig. 8. For RA **1** and **2**, the nitrogen HFCC (a_N) increase with the increase of χ, the proton (a_H) and fluorine (a_F) HFCC constants change in opposite direction to the dependence of $a_N(\chi)$ (see Fig. 8, *a–c*). The solvation curves for all three types of the HFCC of RA **1** and **2** are well described by the empirical exponential functions of the form as follows:

$$a_{ki}(\chi) = a_{ki}(1) + S_{ki} \exp(-K_{ki}\chi), \qquad (4)$$

where k = ^{14}N, ^{19}F, ^{1}H, i is the RA's number. Parameters S_{ki} and K_{ki} for RA **1** and 2 are given in Table 5. The dependencies $a_N(\chi)$ for RA **3** and **4** sharply differ (see Fig. 8, *d*): for RA **3** (R = Ph) $a_N(\chi)$ has the S-like form qualitatively similar to the solvation dependencies of the nitrogen HFCC for RA of some *ortho*-substituted nitrobenzenes [14], whereas the solvent function $a_N(\chi)$ for RA **4** (R = Bu^t) has the minimum at χ = 0.3.

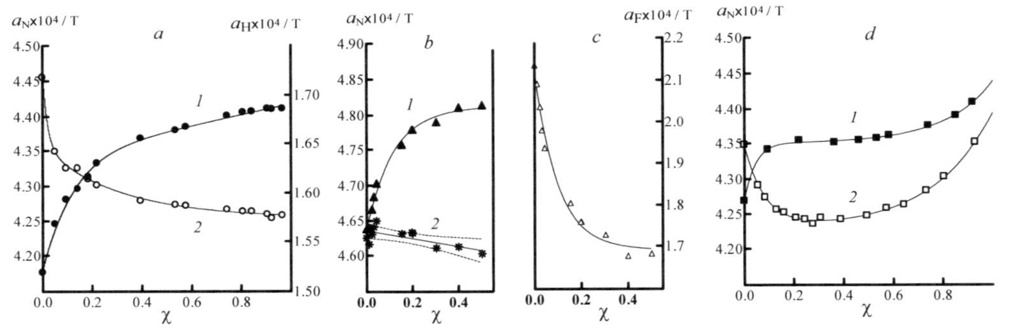

Fig. 8. Hyperfine coupling constants *versus* molar fraction of water (χ): (*a*) RA **2** (*1* – a_N, *2* – a_H); (*b*), (*c*) RA **1**, (*b*): a_N (*1*), the difference of experimental values and resolvation contribution for RA **1** ($a_N(\chi) - \Delta a_N^{solv}(\chi)$ (2), (*c*): a_F; (*d*): a_N for RA **3** (*1*), **4** (*2*).

R / *k*	CF_3 (**1**)		CH_3 (**2**)		Ph (**3**)		Bu^t (**4**)	
	χ=0	χ=0.50	χ=0	χ=0.97	χ=0	χ=0.93	χ=0	χ=0.92
^{14}N	4.638	4.814	4.176	4.406	4.269	4.409	4.348	4.351
^{1}H	—	—	1.718	1.581	—		—	
^{19}F	2.137	1.682	—	—	—		—	

Table 4. The HFCC (a_k×10^4 T) of RA of 3,3'-bi(2-R-5,5-dimethyl-4-oxopyrrolinylidene) 1,1'-dioxides **1–4** in MeCN–H_2O solvent mixtures.

R \ k	^{14}N			^{19}F			^{1}H		
	$a_{ki}(1)$	S_{ki}	K_{ki}	$a_{ki}(1)$	S_{ki}	K_{ki}	$a_{ki}(1)$	S_{ki}	K_{ki}
CF_3 (**1**)	4.813	-0.176	8.310	1.687	0.438	10.874	—	—	—
CH_3 (**2**)	4.408	-0.219	4.716	—	—	—	1.582	0.129	8.803

Table 5. Parameters of empirical solvent dependencies (4) for RA **1** and **2** ($[a_{ki}(1)]$, $[S_{ki}] = \times 10^4$ T=G).

3.4 Spacial, electronic structures of 3,3´-bi(2-R-5,5-dimethy-1-4-oxopyrrolinylidene) 1,1´-dioxides molecular ions and interpretation of hyperfine coupling constants solvent dependences for radical anions.

Semiempirical UHF and RHF calculations with PM3 Hamiltonian of the geometric parameters of the neutral species and molecular ions of **1**—**4** in the ground state show that all structures have the C_2 symmetry. In the molecular ions, both pyrrolinone cycles are planar in fact, and the angle between the cycles (θ_0) in optimized conformations increases with an increase in the effective volume of substituent R (Table 6.), whereas the θ_0 values are higher than those for the neutral species and increase with an increase in the ion charge.

According to the UHF/PM3 calculations [11] for both types of paramagnetic ions **2**—**4**, the ^{14}N nuclei in the pyrrolinone cycles should be spectrally equivalent, which is confirmed by experiment. The corresponding spin site occupancies of the 2s-AOs of the nitrogen atoms are the following: ($S_{2s\text{-}AO}$(RA **2**), $S_{2s\text{-}AO}$(RC **2**)) = (0.02166,–0.00181);

Com-pound	R	$\theta_{0(n)}$		$\theta_{0(RC)}$		$\theta_{0(DC)}$		$\theta_{0(RA)}$		$\theta_{0(DA)}$	
		UHF	RHF	UHF	RHF	UHF	RHF	UHF	RHF	UHF	RHF
1	CF_3	44.2	30.8	47.6	47.6	—	—	45.9	47.7	89.8	89.8
2	CH_3	38.2	30.1	36.6	36.7	39.3	42.7	37.0	37.4	41.4	43.9
3	Ph	63.6	37.2	47.8	48.7	58.2	73.9	67.7	56.0	86.6	80.6
4	Bu^t	—[a]	55.1	88.0	72.1	86.9	86.9	62.7	72.7	—[a]	—[a]

Table 6. Dihedral angles (θ_0/ deg) between the pyrrolinone cycles in neutral DN **1**—**4** and their molecular ions according to the data of PM3 calculations in the UHF and RHF approximations.

($S_{2s\text{-}AO}$(RA **3**), $S_{2s\text{-}AO}$(RC **3**)) = (0.02354,–0.00196); ($S_{2s\text{-}AO}$(RA **4**), and $S_{2s\text{-}AO}$(RC **4**)) = (0.02314,–0.00125). Therefore, the calculation predicts qualitatively the close constants in the series of RC **2**—**4** and RA **2**—**4** and a considerable decrease in the a_N constant on going from RA to RC due to the redistribution of spin density from nitrogen to oxygen atoms of nitrone groups.

This fact is confirmed by more sophisticated UB3LYP method with 6-31+G* basis set for the gaseous phase in the pair RA-RC of DN **2** as an example. Corresponding spin density distributions are shown in Fig. 9.

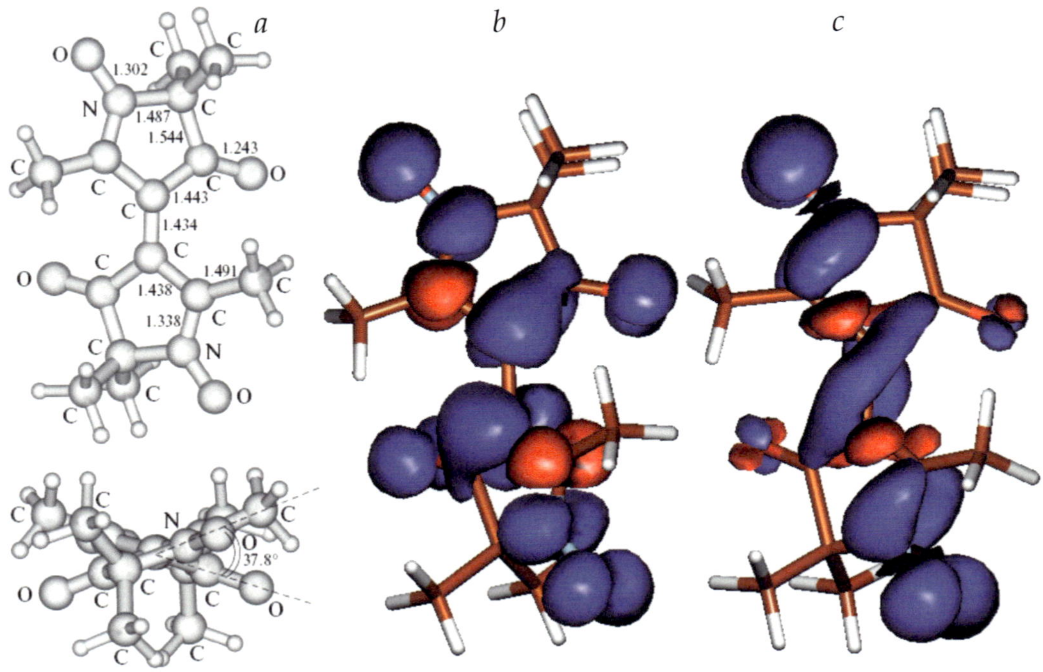

Fig. 9. The equilibrium conformation of RA **2** (*a*), distributions of the spin density in RA **2** (*b*) and RC **2** (*c*) according to the UB3LYP/6-31+G* data for the gaseous phase (blue color - positive sign of spin density, red - negative sign).

The angle θ_0 and HFCC values for RA **2** in the gaseous phase and with allowance for the medium using the PCM model are given in Table 7. According to the calculations, the HFCC with the protons of the methyl group in position 2 are negative, whereas with the ^{14}N nuclei, positive. Transition from the gaseous phase to the liquid causes insignificant increase of the angle θ_0 and nitrogen HFCC (Table 7) and decrease of the proton constant. On the increase of the dielectric permittivity of the medium from 36.64 (MeCN) to 78.39 (H_2O), the turning angle between the rings remains virtually unchanged.

The nitrogen HFCC calculated with allowance for the medium are close to the experimental values, whereas the proton HFCC are by 1.5 times larger than experimental. According to UB3LYP/6-31+G*/PCM calculations the change of the medium virtually has no effect on the values of both HFCC.

Let us suggest that starting from the results of calculation of RA **2** by the UB3LYP method, the shape of the experimental solvation dependence $a_N(\chi)$ for RA **2** is caused only by the change of the solvate cage upon increase of χ. In this case, the function

$$a_{N2}(\chi) - a_{N2}(0) = \Delta a_{N2} + S_{N2}\exp(-K_{N2}\cdot\chi) = \Delta a_N^{solv}(\chi) \quad (5)$$

Phase	ε^a	r_{solv} / Å^b	θ_0 / deg	a_N^{UB3LYP} ×10^4, T	$< a_H^{UB3LYP} >$×10^4, T^c
Gaseous	1	-	37.81	3.848	–2.705
MeCN (PCM)	36.64	3.720	42.48	3.962	–2.549
H_2O (PCM)	78.39	1.385	39.84	3.964	–2.544

a Dielectric constants in the PCM calculations.
b Effective radius of the solvent.
c Average dynamic value: $< a_H^{UB3LYP} > = (1/3)(\sum_{j=1}^{3} \rho_H(r_j))A_H$, where $\rho_H(r_j)$ are spin densities at the protons of the methyl groups at position 2, A_H = 159.22 mT is the atomic constant for the proton.

Table 7. The HFCC and angles θ_0 in the equilibrium conformations of RA **2** from the UB3LYP/6-31+G* calculation data in the gaseous phase and with simulation of the liquid phase in the framework of the PCM model.

where $\Delta a_{N2} = a_{N2}(1) - a_{N2}(0)$, can be approximately considered as general «resolvation contribution» into the dependence of $a_N(\chi)$ for all RA **1–4**, which is related only to the resolvation without conformational change. In such a case, for the RA of trifluoromethyl derivative **1** the following equality should take place:

$$a_{N1}(\chi) - \Delta a_N^{solv}(\chi) \approx const, \qquad (6)$$

which, in fact, is observed with the accuracy to 0.005 mT in the range $0 < \chi < 0.5$ (see Fig. 8, *b*). This means that for RA **1**, the angle θ is virtually constant, at least in the range of χ studied, whereas the absence of the structural contribution $a_N(\theta,\chi)$ into the solvent dependences of the nitrogen constants for RA **1** and **2** in the range of χ studied explains their similar shape (see Fig. 8, *a*, *b*).

We will evaluate the contributions $a_N(\theta,\chi)$ for RA **3** and **4** related to the change of θ starting from the calculations of the angular functions of the nitrogen constants by the DFT/PBE method [15] in the basis 3z* and experimental dependences of the constants with allowance for the resolvation contribution $\Delta a_N^{solv}(\chi)$. The results of calculation of angles θ_0 and corresponding HFCC in the equilibrium conformations of RA **1–4** are given in Table 8.

For RA **2**, the DFT/PBE calculations on average twice underestimate the nitrogen constant values as compared to the UB3LYP calculations (*cf.* with the data in Table 7), but in the series of RA **1–4** they correctly indicate the tendency of the change of a_N measured in MeCN despite small values of their absolute changes. The relations of experimental and calculated constants $R_i = a_{Ni} / a_{Ni}^{PBE}$ are given in Table 8.

Since all the RA possess different values of θ_0, it is convenient to introduce the value $\phi = \theta - \theta_0$ as an argument of the angular functions. According to the calculations in the range $-30° < \phi < 30°$, the angular functions of the nitrogen constants are linear:

$$a_{Ni}^{PBE}(\phi) = A_i + B_i\phi \qquad (7)$$

* 3z basic set {3, 1, 1/1} for H and {6, 1, 1, 1, 1, 1/1} for C, N, O, F [15].

RA	θ_0/deg	a_{Ni}^{PBE} ×10⁴/ T	$< a_{Fi}^{PBE} >^a$ ×10⁴ / T	$< a_{Hi}^{PBE} >^a$, ×10⁴/ T	R_i	A_i×10⁴ / T	B_i×10⁴ / T / deg	r^b
1	43	1.7 95	-1.715	—	2.512	—	—	—
2	38	1.584	—	-1.233	2.637	—	—	—
3	40.8	1.764	—	—	2.420	1.764	-0.006638	0.998
4	52	1.731	—	—	2.512	1.731	-0.005723	0.997

a The HFCC with ^{19}F and ^1H nuclei of CF_3 and Me groups are the dynamic average values.
b Correlation coefficients of the linear regressions (7).

Table 8. The HFCC, relationship of experimental and calculated nitrogen constant values (R_i), angles θ_0 in the equilibrium conformations of RA **1**−**4**, as well as parameters of the angular dependencies (7) from the DFT/PBE calculation data.

Parameters A_i, B_i of regressions (7) and the corresponding correlation coefficients are given in Table 8. Multipliers B_i are negative, *i.e.* an increase of θ leads to the decrease of the nitrogen HFCC and to the redistribution of the spin density involving the O atoms of the nitrone groups.

The structurally dependent solvation contributions for RA **3** and RA **4** are determined as follows:

$$a_{Ni}(\theta,\chi) = a_{Ni}(\chi) - \Delta a_{N}^{solv}(\chi) = R_i[A_i + B_i\phi_i(\chi)] \quad (8)$$

Direct calculations lead to the qualitatively similar functions $\phi_i(\chi)$ differing in the angular amplitude and behavior in the starting sections. They are shown in Fig 10. The maximum deviations of the angles for RA **3** and **4** are ~8 and ~22°, respectively. The maxima of $\phi_i(\chi)$ for both RA are very close and correspond to $\chi \approx 0.5$. An increase in the content of water in the mixture MeCN−H_2O leads to the different behavior of the angular functions $\phi_i(\chi)$ in the starting regions ($0 < \chi < 0.2$, see Fig. 11) and, therefore, to the domination of different contributions for RA **3** and **4**. In the solvent dependence $a_{N2}(\chi)$ for RA **3** in this range of molar fractions, the resolvation contribution is predominant, that leads to a sharp increase of a_N (see Fig.8, *d*), whereas at $\chi > 0.5$, the structural contribution dominates, that finally leads to an S-figurative general solvent function $a_{N2}(\chi)$. For RA **4** containing substituent of larger effective size (Bu^t) and possessing maximum noncoplanarity (θ_0 = 52°), the situation is reverse, and the function $a_{N3}(\chi)$ has the minimum (see Fig. 8, *d*). RA **1** and **2** contain substituents of small effective size (CF_3 and Me groups) and, according to the given estimates, do not undergo noticeable distortions, whereas the shape of functions $a_{N1,2}(\chi)$ is due to the redistribution of the π-spin density from the atoms C(2), C(2′) to the atoms N and O (see Fig. 10) of the nitrone groups upon increase in water content.

4. Concluding remarks

Thus, generalizing the results of our studies (see [10, 11, 13] also) one can assert, that we described the electrochemical behavior of the new class of heterocyclic compounds of the nitrone series with a very high electron withdrawing ability capable of forming long-lived molecular ions in aprotic media. The electrochemical behavior of the Bu^t substituted dinitrone is unique: the EE processes within one cycle of potential sweep in cyclic voltammogram

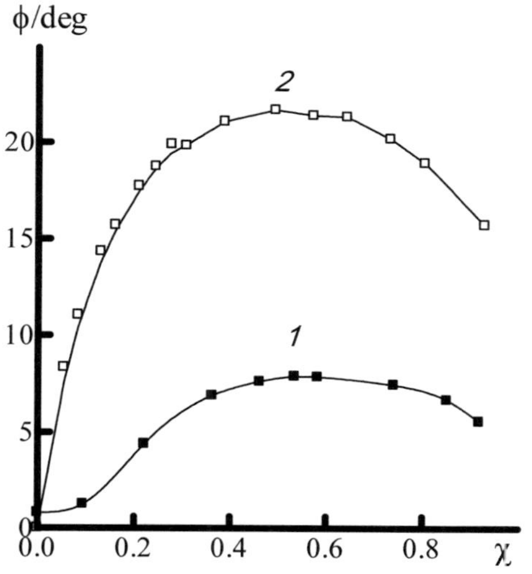

Fig. 10. The turning angle of pyrrolinone cycles for RA **3** (*1*) and **4** (*2*) *versus* molar fraction of water in the binary MeCN – H_2O mixtures.

were observed in both regions of negative and positive potentials with the formation of dianion, radical anion, radical cation, and dication, which are long-lived at $T = 298$ K.

Radical anions of dinitrones, which are long-lived in aprotic solvents and their mixtures with water have been described for the first time. The differences in solvent dependencies of hyperfine coupling constants with ^{14}N nuclei were explained based on the assumption on competition of resolvation and structural effects upon increase in H_2O content in the binary mixture MeCN-H_2O.

The corresponding values of the first peaks potentials of DN's **1-4** ECR in water are rather low and close to the redox potentials of the ferrocene derivatives, which are widely used in the recent time as redox-active labels in electrochemical detection of nucleic acid hybridization [16, 17] particularly, for the development of DNA microarrays with electrochemical detection of hybridization [18]. We believe that the dinitrone derivatives, which possess low potentials of electrochemical reduction and contain appropriate active groups (substituents) capable of binding with DNA, can be rather efficient redox-active labels for the electrochemical detection of DNA hybridization when the double DNA chain does not undergo irreversible destruction. Our recent study was devoted to the development of synthetic method of introduction such substituents in the described dinitrones for this aim [19].

4. Acknowledgment

We thank the Russian Foundation for Basic Research for financial support of this work (Project No. 10-03-00844-a).

5. References

[1] I.A. Grigor'ev, G.I. Shchukin, V.V. Khramtsov, L.M. Vainer, and V.F. Starichenko // *Russ. Chem. Bull., Int. Ed.*, 1985,. 34, 2169.

[2] Eberson L. // *J. Chem. Soc. Perkin Trans. 2.*, 1992, 1807-1813.

[3] K. Nishikida, T. Kubota, H. Miyazaki, and S. Sakata, // *J. Magn. Reson.*, 1972, 7, 260.

[4] H. Miyazaki, T. Kubota, and Y. Matsuhisa, *Bull. Chem. Soc. Jpn.*, 1981, 54, 3850.

[5] H. Miyazaki, T. Kubota, and M. Yamakawa, *Bull. Chem.Soc. Jpn.*, 1972, 45, 780.

[6] I.M. Sosonkin, V. N. Belevskii, G. N. Strogov, A. N. Domarev, and S. P. Yarkov, *Zh. Org. Khim.*, 1982, 18, 1504 [*J. Org. Chem. USSR*, 1982, 18, 1313 (Engl. Transl.)].

[7] T. H. Walter, E. E. Bankroft, G. L. McIntire, E. R. Davis,L. M. Gierasch, H. N. Blount, H. J. Stronks, and E. V. Janzen, *Can. J. Chem.*, 1982, 60, 1621.

[8] K. Ashok, P. M. Scaria, P. V. Kamat, and M. George, *Can. J. Chem.*, 1987, 65, 2039.

[9] G. L. McIntire, H. N. Blount, H. J. Stronks, R. V. Shetty, and E. V. Janzen, *J. Phys. Chem.*, 1980, 84, 916.

[10] L. A. Shundrin, V. A. Reznikov, I. G. Irtegova, and V. F. Starichenko // *Russ. Chem. Bull., Int. Ed.*, 2003,. 52, 939.

[11] L. A. Shundrin, I. G. Irtegova, A. D. Rogachev, and V. A. Reznikov // *Russ. Chem. Bull., Int. Ed.*, 2005, 54, 1178.

[12] A.S. Morkovnik and O. Yu. Okhlobystin // *Chem. Heterocycl. Compd.*, 1980, 16, No. 8, 777.

[13] L. A. Shundrin, N. V. Vasil'eva, I. G. Irtegova, V. A. Reznikov //*Russ. Chem. Bull., Int. Ed.*, 2007, 56, 1273.

[14] L. A. Shundrin, V. F. Starichenko, L. N. Shchegoleva, V. D. Shteingarts // *J. Sruct. Chem.*, 2003, 44(4), 592.

[15] D. N. Laikov, *Chem. Phys. Lett.*, 1997, 281, 151.

[16] T. S. Zatsepin, S. Yu. Andreev, T. Gianik, and T. S. Oretskaya, // *Russ. Chem. Rev.*, 2003, 72, 537.

[17] A.Anne, A. Bouchardon, and J. Moiroux, *J. Am. Chem. Soc.*, 2003, 125, 1112.

[18] S. Abramowitz, *J. Biomed. Dev.*, 1999, 1, 107.

[19] I.A. Khalfina, N. V. Vasil'eva, I. G. Irtegova, L. A. Shundrin, and V. A. Reznikov // *Russ. J. Org. Chem., 2010, 46, 399.*

6

Electron Transfer Kinetics at Interfaces Using SECM (Scanning Electrochemical Microscopy)

Xiaoquan Lu, Yaqi Hu and Hongxia He
Key Laboratory of Bioelectrochemistry & Environmental Analysis of Gansu Province, College of Chemistry & Chemical Engineering, Northwest Normal University, P. R. China

1. Introduction

Today, there is barely an aspect of our lives that is not touched fundamentally by Chemistry. We know that Chemistry is in us, because our body is composed of atoms and molecules, and functions through the extremely intricate patterns of their interactions. However, Chemistry is also around us, in natural phenomena such as photosynthesis, and in the artificial products and materials that sustain the development of our civilization: medicines, fertilizers, plastics, semiconductors, etc. Moreover, the most important global problems-those relating to food, human health, energy and the environment-cannot be solved without the aid of Chemistry.

Chemistry occupies a central position among scientific disciplines, and provides the main links between Biology and Physics. In addition, Materials Science, Chemical Engineering, Earth Sciences, Ecology, and related areas are largely based on Chemistry. And, in the near future our medical problems-and perhaps also our feelings, thoughts, and emotions-will be described and discussed on a molecular (i,e., a chemical) basis. Chemistry, in fact, is far more than a discrete scientific discipline, since its methods, concepts, and practitioners are penetrating virtually all fields of science and technology. Chemistry can, therefore, also be regarded as a trans-disciplinary science that provides an essential means and a fundamental language to understand fully these other scientific disciplines. A key feature of chemistry is chemical reactivity, and of all the chemical reactions electron transfer is undoubtedly the most important. First, electron transfer is the key step in a number of biological processes that have enormous relevance to life, such as photosynthesis and respiration. Second, it is a fundamental feature of many processes of vast technological impact, such as information storage (photography) and energy conversion (batteries). Third, electron transfer is a type of reaction that is amenable to detailed experimental investigation and accurate theoretical descriptions. But perhaps most importantly, the ubiquity of electron transfer in Chemistry and related fields has helped to demolish-and is still demolishing-the arbitrarily created barriers that until now have subdivided Chemistry into its separate organic, inorganic, physical, and analytical branches. In this respect, electron transfer encourages research at the cross-roads of different disciplines where scientific and technological progress is more likely to occur. During the past decade, our knowledge of electron transfer has grown at an astonishing rate, and it was for this reason that a need was recognised for a unified view of the field.

2. Development of SECM

Scanning electrochemical microscopy (SECM) has grown to be a powerful analytical technique for probing a wide range of interfacial processes with high spatial and temporal resolution, including heterogeneous electron-transfer reactions, molecular transport across membranes, adsorption/desorption processes, corrosion processes, the activity of living biological cells and charge-transfer at liquid/liquid interfaces. In fact, SECM is a scanning probe technique that is based on faradaic current changes when a tip is moved across the surface of a sample. This technique is useful in obtaining topographic and surface reaction kinetic information. The charge transfer is of highly importance in bioenergetics and is an active field of research. Much attention has been devoted to studying the heterogeneous electron-transfer reactions, charge transfer at liquid/liquid interfaces and molecular transport processes. SECM has recently been shown to be a valuable technique because it overcomes many typical problems, including the effects of the resistive potential drop in solution (iR drop), charging current, and allows separation of ET and IT processes at the interface between two immiscible electrolyte solutions (ITIES). Hence, analytical measurements can be performed in the interfacial region. The theory and application of SECM have been reviewed from different facets recently. In this chapter, we will discuss the basic principles of the SECM technique and chiefly focus on the latest applications of SECM to studies of charge transfer at the solid/liquid interfaces and liquid/liquid interfaces. Our group has utilized the SECM to investigate the kinetics of heterogeneous electron transfer at the modified liquid/liquid interface, liquid/solid interface by porphyrin and its derivatives.

3. Principle of SECM

3.1 Mode of operation

The illustration of an SECM instrument shown in Fig. 2 includes bipotentiostat, inchworm positioner (3D piezo positioner), electrochemical cells and data acquisition system. More detailed description of this instrument was given in literatures, such in ref (A.J. Bard, 2001). Several modes of operation of the SECM can be realized including feedback (FB) mode, tip generation/substrate collection (TG/SC) mode, substrate generation/tip collection (SG/TC) mode, penetration mode, ion transfer feedback mode, equilibrium perturbation mode, potentiometric detection mode, constant current mode, constant high mode, constant distance mode, reverse imaging mode, redox competition mode and direct mode, surface interrogation mode, microreagent mode, standing approach mode. Fig. 1 shows the mode on the SECM, the Simple model for the ET of Photosystem II in the chloroplast and three modes for SECM operation.

3.1.1 Amperometric feedback mode

The feedback mode has been widely employed to study the kinetic and imaging. In the feedback mode, the ultramicroelectrode (UME) which serves as the working electrode in a three or four-electrode system is usually called the tip. The sample that is usually called the substrate may serve as a second working electrode. The electrodes are immersed in a solution containing redox mediator (e.g., a reduced species, R). When a sufficiently positive potential is applied to the tip, O is electro-generated at a diffusion-controlled rate from solution species R to the UME Fig.1, as follow reaction (1):

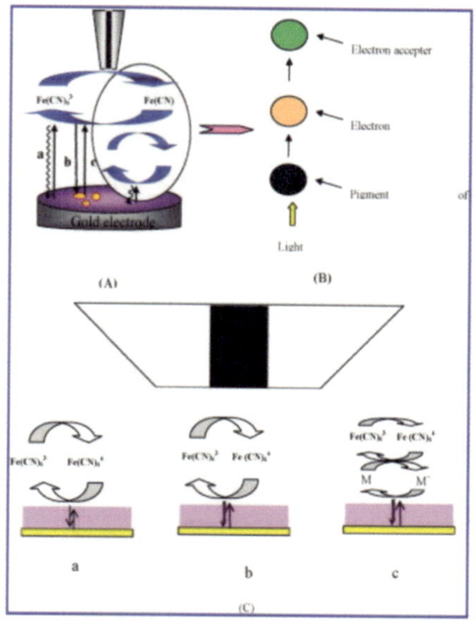

Fig. 1. (A) The Mode on SECM; (B) the Simple Model for the ET of Photosystem II in the Chloroplast; (C) Three Modes for SECM Operation: (a) Electrons of the Redox Mediater Directly Transfer through the Monolayer, Pinholes, and Defects; (b) between the Monolayer-Attached Redox Species and the Underlying Au; (c) Bimolecular Reaction between the Attached SAMs and the Redox Mediator (X.Q. Lu et al., 2010).

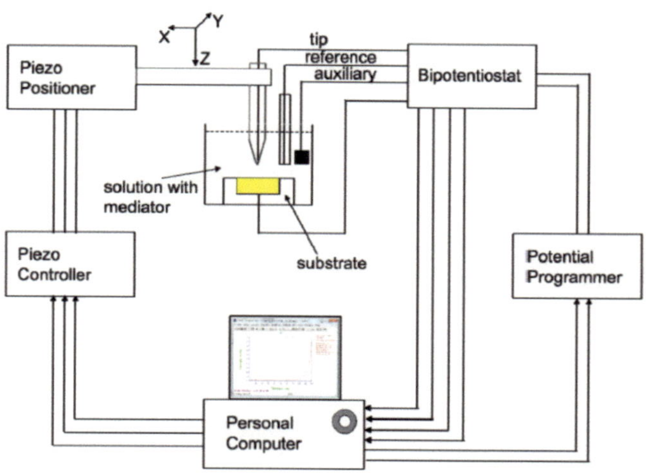

Scheme 1. Diagram of the scanning electrochemical microscopy (SECM) apparatus

$$R - ne^{-} \rightarrow O \tag{1}$$

If the tip is far (i.e., greater that several tip diameters) from the substrate, the steady-state current, $i_{T,\infty}$, for a disk-shaped tip, is given by

$$i = 4nFDca \tag{2}$$

where n is the number of electrons involved in the electrode reaction, F is the Faraday constant, D is the diffusion coefficient of R, c is the concentration of the electroactive reactant, R, and a is the radius of the ultramicroelectrode tip. When the tip is brought near a substrate, the tip current, i_T, is perturbed by hindrance of diffusion of R to the tip from the bulk solution and by reactions that occur at the substrate surface. If the tip is close to a conductive substrate, the O species formed in reaction (1) can diffuse to the substrate and be reduced back to R. When this process occurs, the flux of R to the tip is increased, and $i_T > i_{T,\infty}$, This phenomenon is termed "positive feedback" (Fig. 3a). On the other hand, when the substrate surface is an insulator or a surface where the tip generates product, O, does not react, $i_T > i_{T,\infty}$. This decrease in current is called "negative feedback" (Fig. 3b). If positive feedback measurements are carried out under diffusion-controlled condition, chronoamperomtric feedback characteristics are found to be sensitive to theratio of the diffusion coefficients of the O/R couple. But when a steady-state is established, the normalized current becomes independent of the ratio of diffusion coefficients and depends only on the tip/substrate distance. Moreover, the smaller the tip/substrate distance (d) is, the larger i_T is. The value of $i_{T,\infty}$ gives the normalizing factor for current: $i_T / i_{T,\infty}$. The approach curve is presented in from $i_T / i_{T,\infty}$ versus L (where L = d/a, and d is the distance between the tip and the substrate), which provides information on the kinetics of the process at the substrate. In addition, from a high-quality approach curve, one can obtain the size and shape of polished nanoelectrodes.

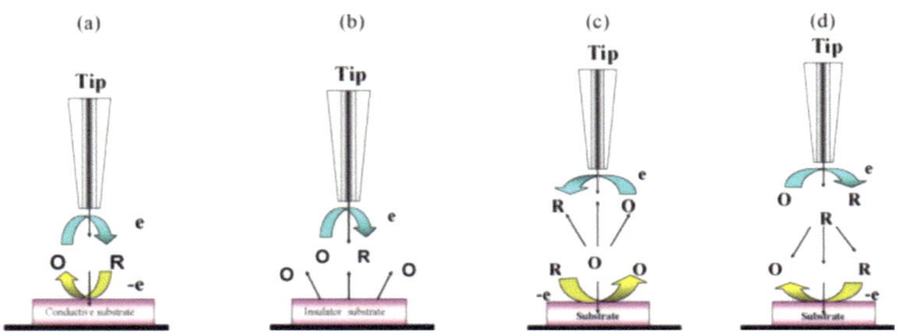

Fig. 2. Schematic diagrams of feedback mode and generation–collection mode. (a) The positive feedback mode at a conductive substrate; (b) the negative feedback mode based on hindered diffusion by insulating; (c) substrate generation/tip collection (SG/TC) mode; (d) tip generation/substrate collection (TG/GC) mode.

3.1.2 Generation/collection (TG/SC) mode

In the generation/collection (G/C) mode, both tip and substrate can be used as work electrodes, one work electrode generates some species which are collected at the other electrode. There are two significantly different G/C modes: the substrate generation/tip collection (SG/TC) mode and the tip generation/substrate collection (TG/SC) mode. The G/C mode is more sensitive because the background signal is very weak. The information about localized quantitative kinetic studies of electrode processes based on the G/C mode has been studied (R.D. Martin, 1998; A.J. Bard, 2004). The ratio of diffusion coefficients of the mediator couple may be accurately measured (R.D. Martin, 1998). Here, take the SG/TC mode as an example to explain. SG/TC is useful for measurements of concentration profiles and fluxes of either electrochemically inactive species (e.g., alkali metal ions) or species undergoing irreversible oxidation/reduction (e.g., glucose). In SG/TC experiments, the tip transmits within a thick diffusion layer generated by the substrate. If an amperometric tip is used, the substrate diffusion is disturbed when the reaction takes place on its surface. Therefore, a potentiometric tip is usually employed in the SG/TC mode because it does not consume species and it can simultaneously diminish the disturbance to the concentration profile of electroactive species generated or consumed at the substrate. G/C imaging can be used to make high-resolution chemical concentration maps of corroding metal surfaces, biological materials, and polymeric materials.

In addition, there are still ion-transfer feedback mode, penetration mode, equilibrium perturbation mode and potentiometric detection, which are developed to investigate variously electrochemical processes.

Ion transfer feedback mode was recently developed technique by Shao and Mirkin (Y.H. Shao etal., 1997; Y. Shao., 1998). In this mode, a micropipet filled with solvent (e.g., water) immiscible with the outer solution (e.g., organic) serves as an SECM tip. The tip current is due to transfer of an ion between the pipet and the outer solution. When such a tip approaches a macroscopic liquid/liquid or liquid/membrane interface, the IT process depletes the concentration of the transferred ion near the phase boundary. If the second phase (e.g., another aqueous solution) contains the same ion, the depletion results in the IT across the interface. The tip current depends on the rate of the interfacial IT reaction, which can be extracted from approach curve. This mode is potentially useful for studies of IT reactions at and high-resolution electrochemical imaging of the liquid/liquid and liquid/membrane interfaces, and also used to study chemical properties of non-electric activity substances.

The equilibrium perturbation mode is different from other working modes, the tip is used to perturb an equilibrium at the substrate (interface) by depletion of a component of the solution. In this process, the tip current is sensitive to variation of solution equilibrium. It was used to adsorption/desorption, dissolution of ionic solids, solubility and partition equilibrium. In addition, this method can also be used to study micro-structure that cannot apply to penetrate.

The mode of potentiometric detection was applied ion selective electrode as a SECM probe to measure the varied ion concentration, and dynamic information was accessed through varied potential caused by varied ion concentration. This mode is employed to study ions that generated in enzymatic reaction and membrane permeability.

The constant current mode, constant height mode and constant distance mode of SECM imaging were useful mode for extracting important information of sample surface. In these modes, the images are obtained by scanning the tip in x, y plane above the substrate and recording the variations in z-coordinate or the tip current in solution containing either oxidized or reduced form of a redox mediator. In the constant-current mode, the current was used to control the tip-substrate separation. This constant current was unusually seted as a 70~80% of the steady-state tip current for insulator substrate or 140~150% of the steady-state tip current for conductive substrate. The tip was then scanned over the sample in the x and y directions while the tip position in the z axis was changed to maintain the tip current within 1% of the set point. A home-built digital proportional-integral-derivative (PID) loop controller using tip current as input signal and tip position as output signal was constructed to control the vertical position of the tip during constant-current imaging. This mode was especially useful in mapping the biochemical activity of a living cell, large areas of nonflat and tilted substrates and sample consists of electroactive and nonelectroactive material. However, this mode is different from constant distance imaging of SECM, and the latter is most commonly imaging mode whereby the tip is positioned at a fixed height over the surface and scanned in the x-y plane, and the tip current versus the lateral tip positions were recorded to obtain SECM images. This work mode was potentially useful to study the enzymatic activity, modified substrate, corrosion processes, and surface catalysis and so on. In both imaging modes, it is noteworthy to identify the best initial current or set the most appropriate initial tip-substrate distance. If the set is inappropriate, this can lead to ambiguous interpretation of SECM images or tip crash and sample damage in experiment. In the constant height mode, where the tip is scanned in one plane of z-coordinate, the working distance will thus vary over rough samples. This working mode is used for studying numerous on single living cells and imaging of enzyme activity (T. Matsue etal., 2003).

Apart from these imaging modes, the reverse imaging mode (RIM) is also an important imaging method, which has been developed by White group in recent years. This is most useful for investigating transport across biological membranes (e.g., skin) and artificial membrane in situations where the SECM tip can access only the exterior membrane surface. Moreover, it is also useful in characterizing interfacial molecule or ion transfer kinetics because of the shape of RIM images is qualitatively sensitive to the rate of interfacial molecule transfer.

The redox competition mode of scanning electrochemical microscopy (RC-SECM) was first introduced by Schuhmann (K. Eckhard, 2006; W. Schuhmann, et al., 2007; W. Schuhmann, et al., 2009). It is a bipotentiostatic experiment, in which the SECM tip competes with the sample for the very same analyte. If the tip and substrate are very close to active catalyst sites, the tip current decreases. This mode was applied to investigate the electrochemical process for energy production and corrosion reactions occurring metal surface.

The direct mode of SECM is based on probing a faradaic current which flows between an ultramicroelectrode and an ionic conducting material, and has been applied successfully to microfabrication and topographical imaging of sample surface with submicrometer spatial resolution.

The surface interrogation mode of scanning electrochemical microscopy (SI-SECM) is a new in situ electrochemical technique based on scanning electrochemical microscope (SECM) operating in a transient feedback mode for the detection and direct quantification of adsorbed species on the surface of electrodes. It was used to study the adsorption phenomena in electrocatalysis and evaluate the rate constant of different redox mediators.

In the microreagent mode (W.B. Nowall et al., 1998; D.O. Wipf R et al., 2003; S. G. Denuault et al., 2005) of SECM, a reaction at the SECM tip changes the local solution composition to provide a driving force for chemical reactions at the surface. Electrolytic generation of an oxidizing agent at the SECM tip can precisely etch metal and semiconductor surfaces. Nowall et al. (W.B. Nowall et al., 1998) utilized this mode to generate micro-patterns of biotin on the electrode surface and imaged the modification with a fluorescence microscope after grafting an avidin labelled fluorophore.

The standing approach mode of SECM, in which the tip is moved vertically to first approach and then retracted from the sample surface at each measurement point, is a useful method for simultaneous electrochemical and topographic imaging of living cell and cell surface. It can avoid the damage of sample in the process of detection and image, because the tip-sample distance is controlled by feedback mechanism.

Alternating current impedance feedback and imaging of membrane pores have been investigated under alternating current scanning electrochemical microscopy (AC-SECM). AC-SECM is a versatile technique for imaging of interfacial impedance properties with high lateral resolution Scanning electrochemical microscopy double potential step chronoamperometry (SECM-DPSC) has been developed and examined experimentally. This powerful technique supplements earlier equilibrium perturbation transient SECM methods by providing the study of the kinetics of irreversible transfer and reversible transfer processes across the ITIES.

All in all, the technology of SECM has a variety of operating modes, but it is difficult to investigate the complex systems by one of those modes in practical application. Usually, it was used by integrated application of various modes or even in conjunction with other analytical techniques to study complex problem.

3.2 Imaging surface reactivity

An important advantage of SECM is the capability to image the surface topography or reactivity by using amperometric and potentiometric probes. Two modes are used to map the surfaces of targets by SECM: the constant height mode and the constant distance mode. Constant height images are obtained by moving a probe laterally in the x and y direction above the substrate surface and monitoring the tip current, i_T, as a function of tip location. This approach is good for relatively large tip electrodes. However, when smaller tip is used in the hope of attaining higher resolution, scanning in the constant height mode becomes more difficult. The tip is possible to crash because of a change of sample height or greatly increase in surface tilt. This imaging mode is usually suitable for a flat surface or for SG/TC mode with a probe positioned far from the surface. The constant distance mode is a useful tool to provide independent information about the sample topography and monitor local electrochemical activity/conductivity in proximity of solid/liquid interfaces. In contrast, the constant distance mode can protect a tip from crashing at surface protrusions. Thus, for high resolution, SECM must be performed in the constant distance mode, as is often used with the scanning tunneling microscope (STM), where the distance is adjusted by a feedback loop to the z-piezo to maintain i_T constant. The simplest approach to constant distance imaging is the constant current mode, where the tip current is used as a feedback signal. A SECM image depending on the sample topography or reactivity can be obtained by constant current mode. Constant current imaging is straightforward when the substrate surface consists of only insulating or only conducting material. The image containing both insulating and conducting substrate is also possible using the tip position modulation (TPM) mode or the picking mode. TPM and picking modes can be used to control the tip-to-sample distance. TPM uses a small-amplitude vertical modulation of the tip position. An important aspect of this technique is that the phase of the TPM current is different for a conducting or insulating surface. In the picking mode, a surface-induced convective current is caused when the tip rapidly approaches from a large tip–substrate separation, which minimizes the chance of tip–substrate crashes during lateral motion. In addition, the constant-impedance mode is used for distance control. Tip positioning is based on applying a high-frequency alternating voltage between the SECM tip and a counter electrode, measuring the AC current response of this system using a lock-in amplifier or frequency response analyser, and thus determining its impedance. In contrast to the constant current imaging, ac impedance imaging does not require the presence of an electroactive species in the solution. The tip impedance-based constant-distance mode has the important advantage that topography and faradaic current can be measured simultaneously. Constant distance imaging has also been showed by using a shear-force-based feedback mechanism with an optical detection system, which was developed for the positioning of an ultra small optical fiber tip in scanning near-field optical microscopy (SNOM). Using this technique, a flexible SECM tip is vibrated laterally at its resonant frequency using a piezoelectric buzzer for agitation. But the instrumentation of optical shear force-based approach is complicated. Therefore, non-optical shear-force-based detection method was developed as a highly sensitive, simple approach. It is easier to use and the operator is not limited to special electrochemical chambers and transparent solutions. Furthermore, another important approach to control tip to sample distance is the combination of atomic force microscopy (AFM) by using specially designed AFM cantilevers acting simultaneously as a force sensor for topographical AFM imaging and an UME for electrochemical SECM imaging. Because the SECM response depends on the rate of heterogeneous reaction at the substrate, it can be used to image the local surface reactivity. The resolution attainable with the SECM largely depends upon the tip radius, and the distance between tip and sample. The kinetics of several rapid heterogeneous electron-transfer reactions were measured with nanometer-sized SECM tips. Pt nanoelectrodes as SECM probes were successfully applied to improve the quality of the imaging resolution. Therefore, the resolution of the topographic imaging can be improved by reducing the size of the tip or using constant distant imaging. In addition, microelectrode arrays were used to map the surface topography or reactivity because microelectrode arrays take all the advantages of microelectrodes, including rapid response time, steady-state diffusion, and small iR drop. The applications of SECM to the imaging of surface and electrochemical system have been widely studied. It can indicate the formation process of the self-assembled monolayers (SAMs) on gold electrode at different periods and provide much electrochemical information. There is a great deal of works in this domain involving the imaging of surface reactivity and the assessment of surface reaction kinetics.

4. The kinetics of heterogeneous electron-transfer reactions

The quantitative SECM theory has been developed for various heterogeneous and homogeneous processes and different tip and substrate geometries. Especially, with the introduction of various efficient algorithms, more exact formulas have been produced by different groups. The focus of this section is on general analytical formula and numerical simulation for scanning electrochemical microscopy.

The kinetics of heterogeneous electron transfer has been the focus of considerable research activity. SECM is a powerful approach for measuring the kinetics of heterogeneous electron transfer. The kinetics of heterogeneous electron transfer can be determined with high lateral resolution while scanning a tip parallel to the surface. Distance-dependent measure ments provide quantitative information on sample properties. Here, we only briefly presented how to extract the heterogeneous ET rate constant for approach curves. The main quantitative operation is obtained from the feedback mode. All values of tip current are normalized by the steady-state current ($i_{T,\infty}$) given in Eq. (2) for a disk-shaped tip electrode.The electron transfer rate can be obtained according to the corresponding formulas. These expressions are written in terms of a normalized current $I_T = i_T/i_{T,\infty}$, and a normalized distance, L = d/a. The following equations can be used to extract the rate of heterogeneous reaction occurring at a substrate:

$$I_T^k = I_S^K(1 - \frac{I_T^{ins}}{I_T^c}) + I_T^{ins} \quad (3)$$

$$I_S^K = \frac{0.78377}{L(1+1/\Lambda)} + \frac{[0.68 + 0.3315\exp(-1.0672/L)]}{[1 + F(L,\Lambda)]} \quad (4)$$

Where I_T^c , I_T^k and I_T^{ins} represent the normalized currents for diffusion-controlled regeneration of redox mediator, finite substrate kinetics, and insulating substrate, respectively, and I_S^K is the kinetically controlled substrate current, $\Lambda = k_f\, d/D$, where k_f is the apparent heterogeneous rate constant (cm s^{-1}) and D is the diffusion coefficient, and $F(L,\Lambda) = (11/\Lambda + 7.3)/(110 - 40L)$. The analytical approximations for I_T^c and I_T^{ins} are give by following:

$$I_T^C = \frac{0.78377}{L} + 0.3315\exp\left(\frac{-1.0672}{L}\right) + 0.68 \quad (5)$$

$$I_T^{ins} = \frac{1}{\{0.15 + 1.5385/L + 0.58\exp(-1.14/L) + 0.0908\exp[(L - 6.3)/1.017L]\}} \quad (6)$$

By fitting an experimental current/distance curve to theory (Eqs. (3)–(6)), the rate of an irreversible heterogeneous reaction can be obtained. The tip shape affects the approach curves because of proportional differences in lateral and normal diffusion to the tip. The steady-state and the transient currents at a hemispherical UME operating in the feedback mode of the SECM have been simulated using the alternating direction implicit finite difference method (ADIFDM).NThe steady-state current, $i_{T,\infty}$, at a simple hemispherical electrode is given by:

$$i_{T,\infty} = 2\pi nFDc^* a \quad (7)$$

Therefore, the approximations for the positive feedback and the negative feedback of hemispherical electrode are different from those of a disk electrode. The simulated positive and negative feedback curves can be accurately described (with an error of less than 1%), respectively :

$$I = 0.873 + \ln(1 + L^{-1}) - 0.20986\exp\left[\frac{-(L - 0.1)}{0.55032}\right] \quad (8)$$

$$I = 0.39603 + 0.42412L + 0.09406L^2 \quad (9)$$

For $0.1 \le F \le 2$

In addition, SECM for localized quantitative kinetic studies of electrode reactions based on the tip generation–substrate collection operation mode was presented. The substrate current is monitored with time until it reaches an apparent steady-state value. For an irreversible first-order electrode reaction at the substrate, the analysis includes transient and steady state simulations performed using an explicit finite difference method (FDM). Transient responses, steady-state polarization curves, and TG-SC approach curves can be used to obtain substrate kinetics.

The accuracy of equation (3) is claimed by the authors as within 1-2% for $0.1<L<1.5$ and $0.01<k<1000$ for a disk microelectrode with Rg=10. This theoretical formula was applied to extract the heterogeneous rate constants of complex reaction, such as charge transfer in liquid-liquid interface , electron transfer (ET) through self assembled monolayers, mediated ET in living cells, charge transport across the polymeric film enzyme mediator kinetics studies at modified substrate.

Fernandez et al. (A.J. Bard et al., 2005) has studied heterogeneous catalytic reactions using SECM feedback operating and digital simulation. They found that the tip current was strongly dependent on the stoichiometry of the substrate reaction relative to the tip reaction and extracted the rate constant of the catalytic reaction by fitting equation (10)

$$I = \frac{A_1 - A_2}{1 + \left(\frac{\kappa}{A_3}\right)^{A^4}} + A_2 \quad (10a)$$

$$A_1 = -0.00245 + 0.907\left[1 - \exp\left(-\frac{L}{0.99}\right)\right] + 3.328 \times 10^7\left[1 - \exp\left(-\frac{L}{3.637 \times 10^9}\right)\right] \quad (10b)$$

$$A_2 = -0.0028 + 0.972\left[1 - \exp\left(-\frac{L}{0.48}\right)\right] + 0.4447\left[1 - \exp\left(-\frac{L}{5.05}\right)\right] \quad (10c)$$

$$A_3 = \frac{0.261}{\frac{0.214}{L} + 1} + 0.22\exp\left(-\frac{L + 0.034}{1.195}\right) + \left[1 - \exp\left(-\frac{L + 0.034}{0.236}\right)\right]^{3.48} \quad (10d)$$

$$A_4 = 0.7826 + 0.22\left[1 - \exp\left(-\frac{L}{1.384}\right)\right] \quad (10e)$$

where κ is the normalized rate constant of reaction on the substrate as defined $\kappa = k_a/D$. Equation (10) was useful in treating experimental data but it doesn't have physical meaning and claimed by the authors with error less than 0.5% in the range $10^{-3} < \kappa < 10$ for an Rg =5 disk UMEs.

Until now, some research groups are the rate constant from experimental data by fitting equation (3) but they ignored the R g of the electrode used in experiment is smaller 10. Therefore, the inaccuracies and unreliability results was obtained. A new available equation (11) proposed by Lefrou (C. Lefrou et al., 2008) in 2008 used in the kinetic information extraction of SECM steady-state feedback experiment data. The accuracy of equation (3) is claimed by the authors as within ±2.5% for any kinetics in the condition of L>0.1 for a disk microelectrode with R g smaller than 20.

$$I_T = I_T^c\left(L+\frac{1}{\Lambda}\right)+\frac{I_T^{ins}-1}{\left(1+2.47R_g^{0.31}L\Lambda\right)\left(1+L^{0.006Rg}\Lambda^{-0.236Rg+0.91}\right)} \tag{11}$$

Where $I_T^c(L+1/\Lambda)$ is a simple translation of I_T^c calculated by equation (12b)same L along the tip-substrate distance(L) axis and more detailed description showed in literature (C. Lefrou, 2007).

I_T^{ins} given by equation (12a) d $\Lambda = k_a L/D$ where k is the apparent heterogeneous rate constant and D is the diffusion coefficient of the redox mediator. When the Rg is equal to 10 as same as previously published literature, the new equation suitable for processing data of L>0.01 and all Λ was produced and described as following: (11)

$$I_T = I_T^c\left(L+\frac{1}{\Lambda}\right)+\frac{I_T^{ins}-1}{\left(1+5.6L\Lambda\right)\left(1+L^{0.177}\Lambda^{0.64}\right)} \tag{12}$$

$$I_T^{ins} = \frac{\frac{2.08}{R_g^{0.358}}\left(L-\frac{0.145}{Rg}\right)+1.585}{\frac{2.08}{R_g^{0.358}}(L+0.0023Rg)+1.57+\frac{\ln Rg}{L}+\frac{2}{\pi Rg}\ln\left(1+\frac{\pi Rg}{2L}\right)} \tag{12a}$$

$$I_T^c = \alpha(Rg)+\frac{\pi}{4\beta(Rg)\arctan(L)}+\left(1-\alpha(Rg)-\frac{1}{2\beta(Rg)}\right)\frac{2}{\pi}\arctan(L) \tag{12b}$$

In equation (12), the definition of parameters consistent with the previous equation (11). From long-term development perspective, these equations are not the ultimate perfect because of variable complex tip geometry for SECM research system. More accurate analytical equation will be generated by numerical simulation after we founded the mathematical model that was very close to research conditions. And it will be the developmental direction of SECM technology.

5. Much work about numerical simulation for SECM quantitative research

Numerical simulation is a well developed and efficient tool to treat complicated systems like microelectrodes in diffusive or convective fields. The microelectrode has been largely simulated using various numerical methods such as finite element method, finite difference method, alternating direction implicit finite-difference method, boundary element method, conformal mapping transformations and so on. The commonly general of numerical simulation includes the following steps: (i) a numerical model of the electrochemical experimental system is set up within a computer Specifically, this step is to build diffusion equation (Fick's second law) and find its corresponding initial and boundary conditions. (ii) According to the actual situation, an efficient and high-accuracy calculation method has been selected from the methods mentioned in previous. (iii) Write your own program and calculated or using commercially program (e.g. COMSOL Multiphysics (FEMLAB), PDEase, MATLAB) calculated. (iv) When the simulation is completed, large amounts of data are obtained. And then, one can extract from them numerical equation of current functions, concentration profiles, potential transients and so on. Comparing it with experimental results, one can obtain large numbers of useful information about complicated electrochemical problem at irregular microelectrodes and predict the results of difficult to research using general experimental methods. In addition, multidimensional integral equations (MIE) and their numerical solution can be used as an effective theoretical treatment method for microelectrode electrochemistry. Here, we mainly overview the application of numerical simulation in scanning electrochemical microscopy.

Since microelectrode is applied in electrochemical research, the simulation method has been changing from one-dimensional to multidimensional space simulation, such as in consideration of microelectrode arrays, scanning electrochemical microscopy and so on. Two-dimensional finite element method with exponentially expanding grid was first used in simulation study of SECM by Kwak and Bard (A.J. Bard et al., 1989). The theoretical current-distance curves for ring microelectrode with various R g and a/b ratios approaching either a conductive or an insulating substrate were simulated by Lee et al. (.J. Bard et al., 2001) using a commercial program PDEase2D, and they determined the size and shape of ring microelectrodes combined with experimental curves. Adaptive finite element (AFE) algorithm was introduced for use in electrochemistry and used this method to simulate the electron-flux over an arbitrarily geometry shaped SECM tips. And then they explained the simulation current deviates from linear approximation for very small distance ($L<10^{-1}$) using 'thin-layer loss'. Using finite element analysis program PDEase2D, Zoksi et al (A.J. Bard et al., 2008) studied and analyzed theoretical SECM approach curves for UMEs with different conical geometries approaching either a conducting or insulating substrate. This research showed that the magnitude of either positive or negative feedback current decreases with increasing the aspect ratios H of finite conical electrode (i.e. H=h/a, the ratio of the cone height (h) to its base radius (a)). This study provided a sensitive method for extract geometric parameters including h, a and R g of etched conical electrode for H<3. A heterogeneous electron-transfer (ET) process at an unbiased conductor was studied theoretically and experimentally using SECM coupled with COMSOL Multiphysics software based on finite element method. Zhu et al. (Z. Ding et al., 2008) quantitatively analyzed variations in surface reactivity by fitting probe approach curves to curves simulated with the similar simulated method. Edwards and co-works (P.R. Unwin et al., 2009) used finite element method to investigate approach curves to a planar surface for an electrolyte-filled Scanning ion conductance microscopy (SICM) micropipet with relatively large semiangle and Rg and explored the influence factors of SICM surface topography on the current response.

A lot of research work for analytical equations for quantitative steady-state SECM experience has been carried out using conformal map transformations in recent years. Conformal mapping transformations are powerful tools for the modeling of some electrochemical systems. This method is to transfer the original space into another better space using complex function (D. Britz. 2005), and it can be applied to extracting analytical equations from digital simulations of the microdisk SECM tip for steady-state feedback mode approach curves. The analysis provides a theoretical proof that the extraction of the kinetics parameters from experimental approach curves for quasi-reversible or irreversible kinetic reaction at the bare substrate or modified substrate. Cornut et al. (R. Cornut et al., 2010) proposed accurate analytical approximations to study microring electrode SECM approach curves in pure positive and negative feedback modes using commercially software COMSOL Multiphysics combined with MATLAB.

6. SECM measurement of the bimolecular ET reactions through molecular monolayers

A new experimental and theoretical treatment to measure the rates of ET reactions through self-assembled monolayers was developed. The bimolecular ET rate constant between a solution-base redox probe and redox centers attached to the SAMs can be determined. the general equation of ET reaction as following:

$$R + M_{OX} \underset{k_{red}}{\overset{k_{ox}}{\rightleftarrows}} O + M_{red} \tag{13}$$

When it comes to chlorophyll electrochemistry, ET could be studied by the bimolecular reaction model and the relative rate constants. The process is described as above:

This represents the bimolecular reaction at the solid/solution interface; could be used as an electron donor, reducing matter Red_2 could be an electron acceptor, and the rate constant (k_{ox}, k_{red}) is the bimolecular oxidation-reduction rate constant (mol^{-1} cm 3 s^{-1}).

When the monolayer absorbed on the substrate contains an electroactive group, M, the regeneration of the mediator occurs via a bimolecular reaction with these surface-bound redox centers with a rate constant k_{BI}. The rate of the tunneling ET is governed by the relative contributions from the forward and backward reactions, kf and kb, which are both dependent on the value of substrate potential Es. If bimolecular ET reaction is irreversible and the electron tunneling through the monolayer is quasi-reversible, the effective rate constant obtained from fitting an experimental approach curve to theory, k_{eff}, should be equal to:

$$k_{eff} = \frac{k_{BI} k_f \Gamma}{k_{BI} c + k_f + k_b} \tag{14}$$

Two limiting cases can occur:(a) are the formal potentials of the SAM bound redox species and redox mediator in solution, respectively. c is not very small, and Es is not much more positive than Eads). Eq. (14) becomes

$$k_{eff} \cong \frac{k_f \Gamma}{c} \tag{15}$$

In this case, the overall rate of ET is a function of kf and c. The effective rate constant is controlled largely by the rate of tunneling through the monolayer, kf, and is inversely proportional to c. (b) and c is small or Eq. (14) becomes

$$k_{eff} \cong k_{BI}\Gamma \tag{16}$$

Here, the concentration and potential will not effect on k_{eff}. The overall rate of ET is limited by the bimolecular ET rate constant, k_{BI}.

7. SECM measurement of charge-transfer processes at the liquid/liquid interface

Electrochemistry at the interface between two immiscible electrolyte solutions gains more and more interest due to its wide range of applications. There is great interest in heterogeneous charge transfer (CT) reactions at the ITIES. The ITIES can serve as an SECM substrate and be probed directly by the UME. In most cases, these studies have involved four electrode systems, where the current is measured when a potential is applied across the interface. With the SECM, all electrodes are contained in a single phase, and a constant potential drop across the ITIES is controlled by the composition of the two liquid phases. Using the SECM, one can quantitatively separate interfacial ET from IT and study both processes independently. These charge transfer reactions can be classified into three main categories: (1) electron transfer, (2) simple ion transfer, (3) facilitated ion transfer.

7.1 Electron-transfer (ET) process

In a typical SECM experiment at the ITIES, studies of ET have employed the feedback mode; a tip is placed in the upper liquid phase containing one form of the redox species (e.g., the reduced form, Red_1). If the tip is held at a sufficiently positive potential, the oxidation of Red1 at the tip occurs so as to produce the oxidised form of the species, Ox_1. When the tip approaches to the interface, ET between the tip-generated species in phase 1 (Ox_1) and the redox-active species in phase 2 (Red_2) occurs at the interface and regenerates the original species in phase 1 (Red_1) via the bimolecular redox reaction (17)

$$Ox_1 + \mathrm{Re}\,d_2 \xrightarrow{K_{12}} Ox_2 + \mathrm{Re}\,d_1 \tag{17}$$

Because i_T increases with a decrease in d, the effect of positive feedback of the regenerate species enhances the steady state current. The kinetics of such a reaction at the interface can be evaluated from the SECM approach curve. In SECM measurements, a non-polarisable ITIES is poised by the concentrations of the potential-determining ions providing a constant driving force for the ET process. To maintain electroneutrality in both phases, the transfer of common ions (eitheranions or cations or both) following the ET reaction occurs.

In SECM studies of ET reactions at the ITIES, there are four main steps of the over process that can influence the tip current:

1. mediator diffusion limiting current in phase 1 between the tip and the ITIES, i_T^c
2. limiting current of heterogeneous electron-transfer reaction, i_{ET},
3. diffusion limiting current of the redox species in phase 2, i_d,
4. charge compensation current by IT at the interface, i_{IT}.

The electrical current across the ITIES (iS) caused by this multistage serial process can be expressed as Eq. (18) .

$$\frac{1}{i_s} = \frac{1}{i_T^C} + \frac{1}{i_{ET}} + \frac{1}{i_d} + \frac{1}{i_{IT}} \tag{18}$$

Any of these stages can be rate-limiting, but the concentration of Red_2 is usually made sufficiently high to avoid the complication of diffusion effects in the bottom phase.

Of more interest in studies of the kinetics of ET reactions is the driving-force dependence of the rate constant. The relationship between the rate constant of ET, kf, and the energy of activation, | G can be written as follows:

$$k_f = const \exp(-\frac{\Delta G^{\neq}}{RT}) \tag{19}$$

At low driving force, the potential dependence of the rate constant appears to follow the Bulter–Volmer (B–V) theory. A Butler–Volmer type approximation can be employed:

$$\Delta G^{\neq} = -\alpha F(\Delta E^o + \Delta\phi) \tag{20}$$

where α and F are the transfer coefficient and the Faraday constant, respectively. ΔE^0 is the difference between the standard potentials of the two-redox couples in both phases. However, at high driving force, a plot of log k_{12} versus driving force is parabolic, which is consistent with Marcus theory inverted region behavior. The Marcus theory predicts that the rate constant of an ET reaction increases when the driving force is low and decreases when the driving force is high. The inverted region of Marcus theory can be observed by a plot of log k_{12} versus the driving force. According to Marcus theory, the energy of activation for an ET reaction is given by:

$$\Delta G^{\neq} = (\frac{\lambda}{4})(\frac{(1 + \Delta G^o)}{\lambda})^2 \tag{21}$$

where is the reorganization energy, and is given by

$$\Delta G^o = -F(\Delta E^o + \Delta\phi) \tag{22}$$

Therefore, the activation energy of the ET reaction depend parabolically on ΔG^o. At high driving force, the kinetics gradually comes into the Marcus inverted region and agrees with Marcus theory.

7.2 Ion transfer (IT) processes

Ion transfer reactions are numerous and of importance for many biological and chemical processes, such as drug delivery, transmembrane signaling, and phase-transfer catalysis. There are two different types of IT reactions that may be studied in SECM measurements, i.e., simple (unassisted) ion transfer and facilitated ion transfer. The highly ordered porous anodic aluminumoxide (AAO) was modified at the interface between two immiscible electrolyte solutions (ITIES) to recover the process of ion transfer through the ion channels revised Scheme 3 (X.Q. Lu et al., 2010).

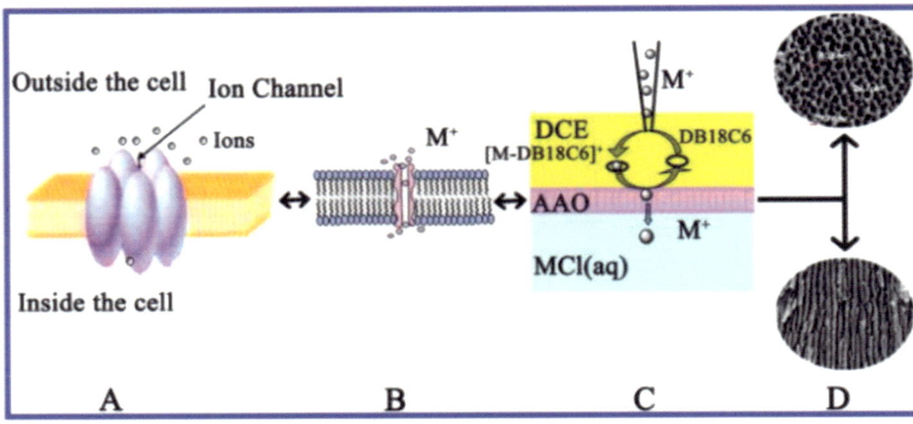

Scheme 2. (A) Ions through the Ion Channels; (B) Simple Model for Single Ion through the Ion Channel; (C) Model for a Three-Electrode to Investigation of Ion Transport Traversing the "Ion Channels" by (SECM); (D) Scanning Electron Micrographs of AAO

7.2.1 Simple ion transfer

Simple IT reactions are of great importance for biological system. It can be induced in two different ways: one is to apply external potential across the interface; another is to deplete the concentration of the common ion in one of two liquid phases near the ITIES. The reaction of simple IT can be showed as follows:

$$M^{+}(o) \rightarrow M^{+}(W) \text{ (at the pipet tip)} \tag{23}$$

$$M^{+}(W) \rightarrow M^{+}(o) \text{ (at the ITIES)} \tag{24}$$

In this case, both the tip and the bottom phase contain the same ion at equilibrium. M^+ is transferred from organic phase into the aqueous filling solution inside a pipet. When tip depletes concentration of this common ion in the top solvent near the ITIES, the ions transfer across the ITIES, which produces positive feedback if the bottom phase contains a sufficiently high concentration of M^+.

7.2.2 Facilitated ion transfer

SECM is a powerful tool to probe a facilitated ion-transfer (FIT) at ITIES. The kinetics of FIT reactions has been investigated by SECM combining with micropipets or nanopipets. A micropipet filled with a solvent immiscible with the outer solution can serve as a tip electrode. The facilitated iontransfer by interfacial complexation (TIC) at the micropipet tip is:

$$M^{+}(W) + L(o) \rightarrow M^{+}L + (o) \text{ (at the pipet tip)} \tag{25}$$

When the concentration of M^+ inside the pipet is much higher than that of L in the organic phase and the tip is biased at a sufficiently positive potential, the reaction at the tip orifice is controlled by the diffusion of L to the interface. Subsequently, the interfacial dissociation reaction at the bottom water/oil (w/o) interface occurs. When the tip approaches the bottom (aqueous) phase, M+ is released from the complex and transferred into the aqueous solution and L is regenerated to its neutral form:

$$ML^{+}(o) \rightarrow M^{+}(W) + L(o) \quad \text{(at the ITIES)} \tag{26}$$

Although the mass-transfer rate for IT process is similar to that for heterogeneous ET process, the charge transfer processes at the tip and bottom ITIES are due to ions rather than electrons. Moreover, the tip is a pipet whose type and size strongly affect the attainable information. The smaller the pipet is, the higher mass-transfer rate that we can obtain is. The nanopipet electrodes can be fabricated and used as SECM tip (P.R. Unwin et al., 1997; A.J. Bard et al., 1999; G. Denuault et al., 1999; G. Denuault et al., 1999; G. Wittstock et al., 2002). If the ITIES formed at the pipette tip is polarizable, the driving force for the IT process is provided by the potential applied between the micropipette and the reference electrode. If the interface between the organic (top) and water (bottom) layer is non-polarizable, the Gavani potential difference ($\Delta_O^W \varphi$) at the interface is determined by the ratio of concentrations of the common ion (e.g., ClO^{4-}) in the two phases.

8. Applications

8.1 Solid/liquid interface

8.1.1 ET

The study of long-distance bridge-mediated electron transfer is an important problem in electroanalytical chemistry. Self-assembled monolayers are monomolecular layers which are spontaneously formed upon immersing a solid substrate into a solution containing amphifunctional molecules. SAMs of alkanethiols have been widely used as the bridging moieties because of well-ordered, close-packed nature of the monolayer and the almost unlimited possibility of introducing functional groups in alkanethiols. Although transient electrochemical techniques, such as cyclic voltammetry, electrochemical impedance spectroscopy and chronoamperometry have been most commonly used for measuring the rates of ET through alkanethiol monolayers, these methods are affected by the resistive potential drop and double layer charging current. SECM is a useful technique for the investigation of electron transfer, eliminating problems with double layer charging and other transient contributions.

Liu et al. (A.J. Bard et al., 2004) developed a new experimental and theoretical methodology to investigate long-distance ET across molecular monolayers by SECM. The developed model can be used to independently measure the rates of ET mediated by monolayer-attached redox moieties and direct ET through the film as well as the rate of a bimolecular ET reaction between the attached and dissolved redox species by SECM. The upper limits for the electron tunneling and bimolecular rate constants measurable by the developed technique are ~$10^8 s^{-1}$ and ~5×10^{11} mol^{-1} $cm^3 s^{-1}$, respectively. They determined a tunneling decay constantof 1.0 per methylene group. Recently, Holt studied the electron-transfer kinetics of cytochrome c electrostatically immobilized onto a COOH-terminated alkanethiol self-assembled monolayer on a gold electrode by the technique of SECM (K.B. Holt, 2006). Approach curves were recorded with ferrocyanide as a mediator at different substrate potentials and at different coverages of cytochrome c. k_{BI} =2×10^8 mol^{-1} $cm^3 s^{-1}$ for the bimolecular ET and k° = 15 s^{-1} for the tunneling ET are measured by the the moretical treatment proposed by Liu et al. Moreover, the kinetics of ET was also found to depend on the immobilization conditions of cytochrome c. The tunneling ET rate constant, k°, was found to decrease on covalent binding of the protein to them SAM and increase when mixed CH_3/COOH-terminated MLs were used.

Thiols and mercaptoalkanoic acid are often chosen in SAMs, because they can covalently link to gold surfaces by Au−S linkages to form highly ordered SAMs. The electron ransfer could be affected by the structure of the probe molecules, substrate potential and the property of the solution. Cannes et al. measured the kinetics of ferrocenemethanol electron transfer at a mono- and a bilayer-modified gold electrode.As the apparent heterogeneous rate onstant obtained at the tetradecanethiol/Au(C_{14}SH/Au) electrode, $k^{o,mono}$ (0.02cm s^{-1}) is larger than the value obtained at the phospholipids/tetradecanethiol/Au(POPC/C_{14}SH/Au) electrode, $k^{o,bi}$ (0.0007cm s^{-1}),the redox process is easier at the C_{14}SH/Au electrode than at the POPC/C_{14}SH/Au electrode(A.J. Bard et al., 2003). Recently, Lu etal. (X.Q. Lu et al, 2006; X.Q. Lu et al, 2006) investigated electrochemical characterization through a self-assembled monolayer of thiol-end-functionalized tetraphenyl-porphines(SH-TPP) and metal tetraphenylporphines(SH-MTPP). Different potentials on substrate controlled from 500 to 1100 mV were studied (A.J. Bard et al., 2003). The effect of substrate potential indicates that the oxidation of the porphyrins at the more positive potential is helpful to ET due to a similar bimolecular reaction between the porphyrin ring with positive charge and the probe molecules with negative charge. The reaction equation was provided as follows (X.Q. Lu et al., 2010):

$$[H_2MPTPP]^+ + Fe(CN)_6^{4-} \Leftrightarrow H_2MPTPP + Fe(CN)_6^{3-} \tag{23}$$

The equation above can also be applied on the bimolecular reaction model for partial ET of photosystem II in the chloroplasts.

In addition, the electron transport ability of the SAMs decreased because of increased the length of the thiol-end-link spacer (alkyl group). With the insertion of metallic ions, the electron transport ability of the SAMs of SH-MTPP increased in comparison with that of the SAMs of SH-TPP (X.Q. Lu et al., 2006). The interaction between metal ions and SAMs has been studied by Burshtain and Mandler D. Mandler. 2005). They used both cyclic voltammetry and SECM techniques to investigate the binding of metal ions or protons by ω-mercaptoalkanoic acid SAMs on a gold electrode. It showed the degree of complexation affected the total charge of the monolayer and the feedback current of a charged mediator. The effect of pH on the charge transfer kinetics was also discussed. When the pH increases, the feedback current changes from positive to negative. Recently, the immobilization of ferrocenyl (Fc)-dendrimers was reported at molecular printboards via multiple host-guest interactions and the electrochemically induced desorption by SECM.

8.1.2 Immobilized enzymes

In recent years, biological samples and targets with biocatalytic activity attract a lot of interest; SECM has become a very useful technique for the study of biological systems. In particular, SECM has been devoted to studying the activity of immobilized enzymes on patterned interfaces in order to attempt to structure biosensors, chip-based assay and enzyme immunoassays (EIAs). SECM can be used in the FB mode for imaging enzyme activity at interfaces, but the G/C mode is employed frequently because enzyme kinetics are often too slow for feedback measurements. Compare FB with G/C mode, FB shows higher lateral resolution but lower sensitivity than G/C mode. However, a new detection scheme that combines the advantages of the FB and G/C modes was proposed. This method allows high sensitivity and lateral resolution in SECM imaging. Glucose oxidase (GOx) catalyzed reaction of glucose with dissolved oxygen is often investigated in biosensor research. The oxidation of glucose catalyzed by the immobilized GOx is shown below:

$$\beta - D - glu\cos e + O_2 \xrightarrow{GOx} D - gluconolactone + H_2O_2 \quad (24)$$

The progress of this biocatalytic can be followed by detecting the hydrogen peroxide production or by detecting the decrease of oxygen concentration. Hydrogen peroxide is formed in several oxidase-catalyzed oxidations of substrates by oxygen. High currents are measured above regions of H_2O_2 production, which provides a map of GOx activity. The current increases significantly only over the immobilized GOx structures, due to the high local concentration of produced H_2O_2 during enzymatic oxidation of glucose. The glucose oxidase monolayer immobilized with an electropolymerized polyphenol film was imaged by using a µm diameter meso-porous platinum microelectrode operating in the G/C mode. The concentration distribution inside immobilized enzyme containing layers was investigated. Liquid enzyme layer immobilized with flat Cellophane membrane or cross linked polyacrylamide gel membrane containing entrapped enzyme as biocatalytic media was used in the SECM imaging. Recently, the development of chemical sensors has been focusing on the function complex multienzyme reaction layer. Csóka et al. (G. Nagy et al, 2003) studied the concentration profiles of oxygen and hydrogen peroxide recorded inside the two different enzyme layers with the SECM. The reaction layer of the amperometric enzyme sensor was made of two different parts: one was for interference elimination and the other for selective molecular recognition and measurement.

In FB imaging, a quantitative detection limit can be obtained, which indicates that the mediator concentrations and tip radius affect the limit of detection

$$k_{cat}\Gamma_{enz} \geq 10^{-3}\frac{D_{\mathrm{Re}d}C_{\mathrm{Re}d}}{a} \quad (25)$$

The left side of Eq. (25) summarizes the enzyme-dependent terms: the rate constant, kcat, and the enzyme surface concentration, enz In general, kcat is characteristic of a particular combination of enzyme, mediator, and substrate. In the case of enzyme-loaded films, enz should be replaced by the product of enzyme volume concentration in the film and the film thickness. The right side of Eq. (25) presents the experimentally controllable conditions: diffusion coefficient of the mediator DRed, mediator concentration cRed, and the tip radius a. It indicates that low mediator concentration and large tip radius decrease the limit of detection.

Similarly to Eq. (25), there is a quantitative detection relation for the active of surface immobilized enzymes in G/C mode:

$$k_{cat}\Gamma_{enz} \geq \frac{c'D}{r_s} \quad (26)$$

where k_{cat} is the minimum catalytic reaction rate at substrate saturation, Γ_{enz} mol cm^{-2} is the surface coverage of enzyme, r_s is the radius of the enzyme modified spot, c` and D are the detection limit for the species observed at the tip and the diffusion coefficient of the detected component, respectively. This relation indicates that the sensitivity of the experiment to kinetics is increased for large sample radius rs, low c`, low D, and high surface coverage of enzyme.

SECM was used to study biocatalytic reactions inside the enzyme layer of a biosensor. Concentration changes of substrate, products and pH changes in the reaction layer were probed by amperometric and potentionmetric tip. Many enzyme reactions and metabolic processes of microorganisms can cause pH changes. The pH sensitive antimony (Sb) tip was selected to check pH changes. The current–distance curves can be useful to give information about an optimal reaction layer thickness. These curves for both H_2O_2 and O_2 concentration profiles showed the products have local maximum inside reaction layer approximately 200μm far from the substrate/enzyme layer boundary.

Pyrroloquinoline quinone (PQQ)-dependent glucose dehydrogenase (GDH) is of interest because of its high activity and independence of dissolved oxygen in catalyzing the transfer of electrons from glucose to an electron mediator. It can be studied by SECM using FB mode or G/C mode. Because galactosidase (Gal) which is not an oxidoreductase cannot be imaged in the conventional FB mode, it was studied in the G/C mode. Zhao and Wittstock (G. Wittstock et al, 2004) investigated a multienzyme system composed of immobilized Gal and PQQ-dependent GDH with a new detection method which combined the advantages of the FB and G/C modes and allowed high sensitivity and lateral resolution in SECM imaging. Gal and GDH were immobilized on separate batches of paramagnetic microbeads.In conventional GC imaging of Gal, p-aminophenyl- β -D-galactopyranoside (PAPG) is added to the solution, generating p-aminophenol (PAP) which can be oxidized to p-quinoneimine (PQI) under diffusion controlled conditions at the tip. The electrochemically produced PQI diffuses to the microspot, where it is reconverted into PAP by immobilized GDH in the presence of d-glucose. Therefore, SECM feedback loop was formed when PAP diffused back to the UME. The UME current, i_T, is caused by the G/C contribution from Gal and the feedback amplification by GDH. After the addition of glucose to the assay solution, a clear increase in the current strength was observed above the spot. Moreover, the peak current observed in the combined mode is about 1.8 times stronger than that observed in the conventional GC mode. Yamada et al. (Yamada, et al. 2005) have used a share force-based tip–substrate position SECM system to image a platinum-patterned array electrode and a diaphorase/albumin coimmobilized glass surface. A standing approach mode to avoid contact between the tip and the substrate was used, which repeated the approach and retraction at each data point of the surface, to obtain simultaneously both a current image and a topographic image.

8.1.3 Electrocatalytic activity

Catalyzed heterogenous reaction has been extensively investigated by SECM. Different electrode materials and pH of the solution have different catalytic activities. Hydrogen peroxide decomposition catalyzed by different solid surfaces which were catalase-modified nylon surface and Pt nanoparticles immobilized glass was performed by using an amalgam

Au–Hg disk tip. Rate constants for decomposition reaction on immobilized catalase and Pt nanoparticles were measured at different pH values. Catalase-modified nylon surface exhibited obvious activity in the pH range 5 < pH < 10, with a maximum at around pH 7. The value of ks is between 1 × 10^{-3} and 3 × 10^{-3} cm s^{-1}. The activity of immobilized Pt particles (5–10 nm average diameter) showed a maximum at pH 11.9 in the pH range 10 < pH < 13 with an apparent rate constant (ks = 5 × 10^{-3} cm s^{-1}). Zhou et al. (A.J. Bard et al., 2000) reported the SECM feedback behavior of the H^+/H_2 mediator system at 10 mM concentrations of strong acid at different substrates and demonstrated that the feedback of the redox couple depended on the catalytic activity of the substrate surface for hydrogen oxidation. The prohibitive effect of adsorbed anions (Br^- and I^-) and reduction products of NO_3^- for catalytic H_2 oxidation on the Pt surface was also studied. The kinetics of hydrogen oxidation on Pt, Ir, and Ru substrates in acidic media was investigated. The rate constants for hydrogen oxidation on Pt (k° = 0.22 cm s^{-1}), Ir (k° = 0.25 cm s^{-1}) and Rh (k° = 0.010 cm s^{-1}) were obtained by fitting approach curves. The SECM can be used to quantitatively map the rate constant for the hydrogen oxidation reaction. Recently, a screening method was presented that quantitatively detects protons at a surface using the SECM. The variation in reactivity of the platinum catalyst gradient toward the hydrogen oxidation reaction was measured directly as a function of spatial position. The local rate constant value was proportional to the local platinum surface coverage. A tip reaction was employed to measure the activity of PtxRuy and PtxRuyMoz electrodeposited catalyst toward the hydrogen oxidation reaction in the absence and presence of a monolayer of carbon monoxide. The activity and the onset potential of various catalysts toward the hydrogen oxidation reaction determined by performing screening studies as a function of composition electrode potential. The use of SECM can effectively weaken the hydrogen oxidation activity and poison (CO) tolerance of the catalysts. Bard (A.J. Bard et al, 2003) detected and imaged electrocatalytic activity for the oxygen reduction reaction (ORR) in acidic medium of different electrode materials by operating in a modified TG/SC mode. When the tip placed at 30 └m from the substrate was moved in the x-y plane, activity-sensitive images of heterogeneous surfaces, e.g., with Pt and Au electrodes, were obtained from the substrate current. Moreover, the images were obtained by scanning an array of Pt and Ru spots supported on glassy carbon (Fig. 5). It showed Ru metal had weaker ORR activity than Pt. The electroactive of trodes can be sufficiently discriminated by SECM-image. Later, the heterogeneous kinetics for the ORR has been studied by TG/SC mode. The electrocatalysis of O_2 reduction at copper (II)-poly-l-histidine complex was studied and the activity of arrays of Cu^{2+}-poly-his complex spots of various compositions were imaged. After the electrocatalytic activity for the ORR of different electrode materials were studied by TG/SC mode, Fernandez et al. (A.J. Bard et al, 2005) reported the electrocatalytic activity of bimetallic (and trimetallic) materials for the ORR in acidic medium by using SECM in a new rapid-imaging mode.

8.2 Liquid / liquid interface

8.2.1 ET

The kinetics of heterogeneous electron transfer depends on the driving force at the ITIES. At low overpotential, the kinetics of ET obeys the Bulter–Volmer (B–V) theory. As the driving force is increased, the Marcus inverted region appears. The existence of a Marcus inverted

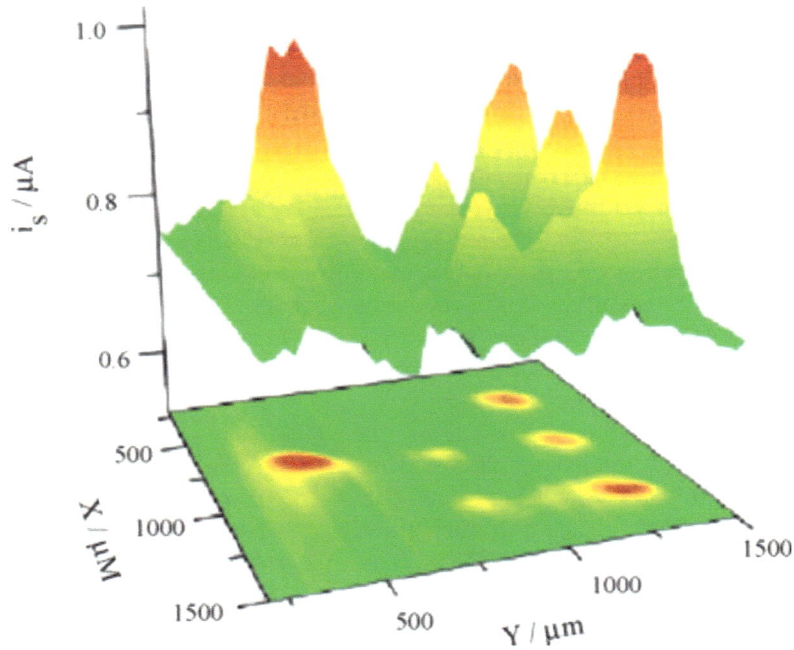

Fig. 3. ORR images obtained by the TG-SC mode of an array of Pt (left spot and right row) and Ru (middle row) spots supported on glassy carbon: scan rate 600μms^{-1}; d = 30nA; i_T = 210 nA; ES = 0.1 V. (A.J. Bard et al, 2003)

region at a lipid modified ITIES or unmodified ITIES was demonstrated by SECM. The previous SECM experiments were often performed with a high concentration of reactant in phase 2 compared with that of the mediator in phase 1. Barker et al. (A.L. Barker et al, 1999) developed a theoretical model that allowed the use of a relatively low concentration of the reactant in the second phase. When the concentration ratio Kr of aqueous to organic reductant ($K_r = C^{*,W}_{\mathrm{Re}d_2} / C^{*,O}_{\mathrm{Re}d_1}$) is decreased, the interfacial ET rate decreases. This approach is beneficial in studying the fast kinetic of the ET reaction. Recently, Li et al. (P.R. Unwin et al, 2007) illustrated how the use of small values of K_r is optional for fast kinetics studied and the effect of Galvani potential on the ET rate constants of the reaction between TCNQ$^-$ and Ru(bipy)$^{3+}{}_3$ at the ITIES. The ET reaction between electrogenerated ZnPor$^+$ (oxidized form of zinc-21H, 23H-tetraphenylporphine) and different redox species at different water/organic solvent interfaces were studied. The redox reactions at the tip and ITIES are as follows:

$$ZnPor(o) - e^- \rightarrow ZnPor^+(o) \quad \text{(at the tip)} \tag{27}$$

$$ZnPor^+(o) + \mathrm{Re}d(W) \rightarrow ZnPor(o) + Ox(W) \quad \text{(at the ITIES)} \tag{28}$$

where the redox couple represents $Ru(CN)_6^{3-/4-}$,$Mo(CN)_8^{3-/4-}$, $Fe(CN)_6^{3-/4-}$, $W(CN)_8^{3-/4-}$, $Fe(EDTA)^{-/2-}$,$Ru(NH_3)_6^{3+/2+}$, $Co(Sep)^{3+/2+}$, and $V^{3+/2+}$.

Although the solvent used for the $Fe(CN)_6^{4-}$ study was different from that of the other three reactants, the overall trend is consistent with the prediction of Marcus theory. Moreover, the effect of aqueous ionic strength was studied by the addition of salt to the aqueous by using approach curves of SECM .

Recently, Sun et al. (Y.H. Shao et al, 2003) demonstrated the existence of an inverted region by the use of SECM with a three-electrode setup and variable concentration ratio of redox species in both phases. For the ET reaction between ZnPor in DCE and $Fe(CN)_6^{3-}$ in aqueous phase, at lower applied potential (0.54- E_O^W <440 mV, Ewo is the externally applied potential), the increasable trend was obvious. When the applied potential (E_O^W) is given at >440 mV, the values of log k_{12} were increasedslowly and reached a maximum at about 500 mV. When the driving force exceeded 600 mV, the rate constant of an ET reaction diminished (Fig. 6a). The reorganization energy (λ = 47.2 kJ mol− 1) was obtained from th best-fitting of the experimental curve. The ET reaction between TCNQ in 1, 2-dichloroethane (DCE) and $Fe(CN)_6^{3-}$ in the aqueous phase was also studied (Fig. 6b). The reorganization energy is 28.9 kJ mol^{-1}. Compared with the studies of the forward reaction between TCNQ in DCE and $Fe(CN)_6^{4-}$ in the aqueous phase, the dependence of the back ET rate constants on $\Delta_W^O\phi$ for the oxidation of $TCNQ^-$ in DCE by aqueous $Fe(CN)_6^{3-}$ was studied using SECM with a fixed concentration of THAP (tetra-n-hexylammonium perchlorate) in DCE.

An interface between two immiscible electrolyte solutions is considered as the simplest model for biomembranes. There is a great deal of interest in understanding charge-transfer at modified interfaces. Amphiphilie lipid molecules dissolved in an organic phase spontaneously form a monolayer film at the water/organic interface. Although the lipids make the CT at the L/L interface more complicated, the lipid-modified interface may be a great help to study the electrical properties of membrane-bound components that control ion-transfer and electron-transfer across biological membranes. The ET reactions at lipid modified L/L interfaces have been investigated by SECM. The ET rate depends on the lipid concentration and the chain length. On the one hand, the ET rate through the ITIES decreases when the lipid concentration increases, but the ET rate reaches a limiting value and doesn't change for lipid concentrations higher than 50 Mm because of a complete lipid monolayer formed at the ITIES. On the other hand, the ET rate increases as the lipid chain length increases. The blocking effect of the lipid monolayerm makes the ET rate significantly decreased. In addition, the extent of the blocking effect apparently depends on the driving force for ET process. The dependence of log k_f on driving force for ET between $Fe(CN)_6^{4-}$ and $ZnPor^+$ was linear with aslope of=0.59 at lower overpotential ($\Delta E^o + \Delta_O^W \varphi \leq 630nm$). At a larger overpotential, the ET rate for Co (II) sepolchrate and V^{2+} decreased with the increasing driving force, which was consistent with Marcus theory inverted region behavior.

To investigate the bionic ET between dopamine (DA) and ferrocene (Fc) at the water/1, 2-dichloroethane (w/DCE) interface, the recently proposed three-electrode setup is adopted based on a glassy carbon electrode and it can be used to study the CT at a L/L interface with different phase ratios.

For an ET process, the solvent effects have recently aroused much attention, such as the effects of dielectric constants and viscosities, which mainly influence on its activation Gibbs energy and pre-exponential factor. The dependence of rate constant of the ET reaction at the interface upon the solution viscosity is evaluated using SECM. Liu and Mirkin (M.V. Mirkin et al, 1999) have investigated the effect of solvent dynamics on the rate of ET from ZnPor to $Ru(CN)_6^{3-}$ using various organic solvents. The ET rate was found to be essentially independent of the potential drop across the interfacial boundary when the organic redox reactant is a neutral species. Miao et al. (A.J. Bard et al, 2002) studied the heterogeneous electron-transfer kinetics for the oxidation of ferrocenemethanol ($FcCH_2OH$) over the whole composition range of dimethyl sulfoxide (DMSO)-water solutions of different viscosities (η) containing 50.0 mM $(CH_3)_4NClO_4$ (TMAP) at a Pt microelectrode by the SECM technique. Bai et al. (Y.H. Shao et al, 2003) reported the effect of solution viscosity on heterogeneous ET reaction kinetics between TCNQ and ferrocyanide. The k_{12} dependence on viscosity was explained. The ET rate constant was shown to be inversely proportional to the aqueous solution viscosity and the solvent longitudinal relaxation time, and directly proportional to the diffusion coefficient of the electroactive species. TLCV is demonstrated to be a useful means for investigating the kinetics of heterogeneous consecutive ET. (revised Scheme 3 and Fig.4) X.Q. Lu. et al. (X.Q. Lu et al., 2010) study two-step electron transfer of ZnTPP/$[Fe(CN)_6]^{4-}$, indicating that the Bulter-Volmer (B-V) theory is suitable for the consecutive electron transfer. This system provides an interesting example of how the relatively new theory of TLCV is used to probe more complex biological redox chemistry.

8.2.2 Ion transfer (IT) at the ITIES

Heterogeneous IT reactions are essential for many biological and technological systems. Especially, simple IT reactions are numerous and of great importance for biological systems. Wei et al. were the first to apply the SECM for investigating the IT process at the ITIES induced by ET. They demonstrated IT limitation on the overall kinetics in two cases: the organic phase without electrolyte and the effect of the concentration of the potential determining ion (TEA^+) on the shape of feedback curves. SECM is possible to probe simple (unassisted) IT reactions and map ion-transfer reactivity of the interfacial boundary with ion-transfer feedback mode of the SECM and basing some ions (mainly Na+ and K+), play a fundamental role in the excitability of the nerves and muscles because they are transported through the ionic channels of their cells, the ion transport traversing the "Ion Channels" was investigated by Scanning Electrochemical Microscopy (SECM). The ion-transfer feedback mode of the SECM that is used for imaging of solid/liquid and liquid/liquid interfaces has

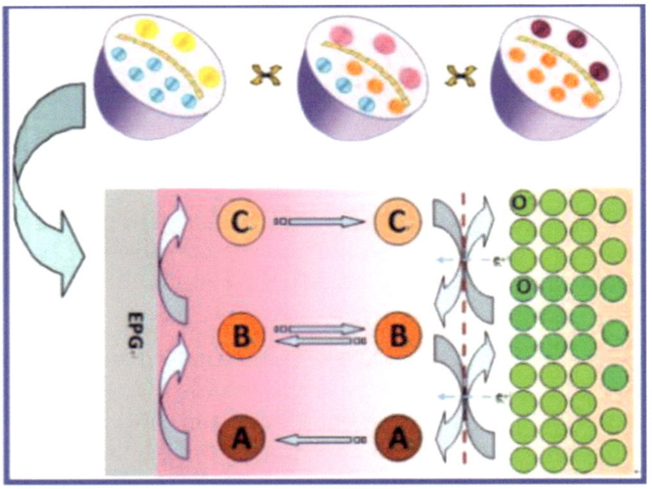

Scheme 3. Simplified Two-Step Electron Transfer Process of Respiratory Chain in the Mitochondria and the Corresponding Reaction at the Electrode

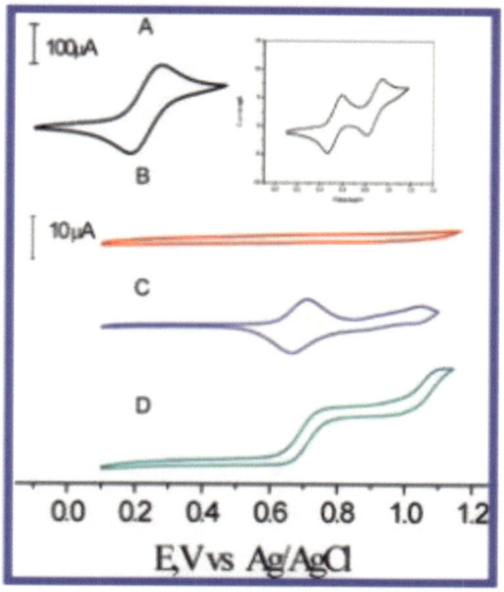

Fig. 4. Voltammetric observation of electron transfer between ZnTPP and $K_4Fe(CN)_6$. (A) Cyclic voltammogram for 10 mM $K_4Fe(CN)_6$ at an uncoated EPG electrode: (C) Cyclic voltammogram with the electrode covered with 1.5 µL of NB containing 1 mM ZnTPP and supporting electrolyte (0.01 M $TBAClO_4$).

some advantages, including the availability of nanometer-sized pipet tips and the possibility to work without redox mediator in solution. Most IT reactions at the ITIES are often very fast and hard to measure. A mass-transfer rate sufficiently high for measurements of rapid IT kinetics under steady-state conditions can be obtained using nanometer-sized pipets based ITIES. Cai et al. measured the kinetic parameters for two rapid simple ion transfer (IT) reactions—the transfers of tetraethylammonium (TEA^+) and tetramethylammonium (TMA^+) ions between DCE and water. An excellent agreement of the k° values was obtained for the forward (from DCE to water) and reverse (from water to DCE) transfer of TEA^+. Ion transfer across L/L interface facilitated by various types of ionophores has been extensively investigated for different cations. The process of facilitated K^+ transfer by dibenzo-18-crown-6 (DB18C6) from water into DCE has been studied extensively in the past. The reactions studied in the system can be described as follows:

$$K^+(W) + DB18C6(DCE) = \left[K^+DB18C6\right](DCE) \quad \text{(at the pipet tip)} \tag{29}$$

$$\left[K^+DB18C6\right](DCE) = K^+(W) + DB18C6(DCE) \quad \text{(at the ITIES)} \tag{30}$$

The tip current was produced by the facilitated transfer of K^+ from the aqueous solution inside the pipette into DCE assisted by DB18C6. With the concentration of K^+ inside apipette is much higher than that of DB18C6 in DCE phase, the tip current is limited by diffusion of DB18C6 to the pipet orifice. When the tip approaches the aqueous layer, DB18C6 is regenerated to its neutral form by interfacial dissociation and diffuses back to the tip to establish the feedback effect. The kinetics of these processes has been determined using

SECM combined with nanopipet and a three-electrode arrangement. A new method in which an SECM can be combined with a polarized L/L interface was introduced. Recently, the group of Shao (Y.H. Shao et al, 2006) has investigated the dependence of rate constants (k_f) on the potential drop wo°of facilitated ion-transfer reaction at the W/DCE interface. The FIT rate constants kf were found to be dependent upon the driving force. When the driving force was low, the dependence of ln kf on the driving force was linear with a transfer coefficient of about 0.3 in the case of a facilitated Na^+-transfer reaction, which followed theclassical Butler–Volmer theory. At higher driving, the Marcus inverted region was observed in the facilitated Li^+-transfer system In addition, it was reported that the ion transfer across the micro-water/nitrobenzene interface supported at the tip of a micropipet could be facilitated by a functionalized fullerene derivative using cyclic voltammetry (CV) and osteryoung square wave voltammetry (OSWV).

The kinetic information about IT reactions at planar bilayer lipid membranes (BLMs) was extracted by SECM with a Pt microelectrode as the tip, where membrane permeability to conventional redox mediators was studied. The voltammetric K^+-selective micropipet electrodes were used in the SECM feedback and generation–collection experiments to probe K^+ transfer through gramicidin channels in BLMs. Afterwards, Ti(I) can be used as a surrogate for K(I), so that ion transport across gramicidin channels can be monitored using an amperometric UME instead of an ion-selective probe. An apparent heterogeneous rate constant (k_{het} = 2.8 (± 0.1) × 10^{-4} cm s^{-1}) for the transport of Tl(I) through the gramicidin to the tip was obtained . The charge transfer coupling (CTC) processes at L/L interface are common phenomena, which play an important role in many chemical and biological systems.Moreover, the processes of coupling ET–IT and IT–IT have been probed by SECM.

8.2.3 Molecular transfer

Studies of molecular transfer processes of non-charged species at the L/L interfaces can be performed by SECM. The effect of fatty alcohol monolayers on the rate constant for Br_2 transfer across the water/air (w/A) interface has been investigated using SECM-DPSC. The intracellular transfer of oxygen is of importance in bioenergetics and the diffusion of O_2 through membranes is an active area of research .The adequate supply of oxygen to cells is necessary for the growth of multicellular systems; lack of oxygen can cause cell death or necrosis. Scanning electrochemical microscopy induced transfer (SECMIT) was introduced by the Unwin's group. SECMIT is a development of the SECM equilibrium perturbation method. It can be used to characterize reversible phase transfer processes at a wide variety of interface. In other words, the SECM tip can be used to locally deplete the concentration of an electroactive species near the ITIES and induce the transfer of this species from the second liquid phase. It was shown to be a valuable method for probing the dynamics of partitioning of electroactive solutes between two immisciblen phases. For the special case of slow interfacial transfer, the theory is conceptually similar to simple (unassisted) IT. Recently, SECMIT was applied to investigate the transport of molecular oxygen across phosphatidylcholine monolayers adsorbed at the interface between a buffered aqueous phase and DCE. The influence of different lipids, a series of 1, 2-diacyl-sn-glycero-3-phosphocholines (C14:0, C16:0, C18:0), on the dioxygen transfer process acrossn the interface was studied, which demonstrated that a monolayer of C18:0 formed a barrier for the transport of a small molecule (O_2) from DCE to water but there is relatively small effect on dioxygen transfer across C14:0 and C16:0 monolayers at ambient temperature. The interfacial rate constant decreased with increasing concentration of C18:0. This inhibited behaviour was explained in terms of a simple energy barrier mode. Cannan et al. (S. Cannan et al., 2004) have reported the kinetics of oxygen transfer across an L-R-phosphatidylethanolamine, distearoyl (DSPE) monolayer spread at three different interfaces by using SECM combined with a Langmuir trough. At all three interfaces, oxygen transfer was diffusion controlled when the monolayer was in the liquid expanded state, but the rate of transfer decreased with increasing surface pressure in the liquid condensed state. For the decane/water interfaces, oxygen transfer was diffusion-controlled up to surface pressures of approximately 20 mN m^{-1} for the thin layer and 40 mN m^{-1} for the thicker layer. At higher pressures, the rate of oxygen transfer decreased rapidly, especially at the thick decane interface. Moreover, the rate constant for oxygen transfer at thick oil/water was higher than that at thin oil/water or air/water for any given surface pressure. The phenomenon showed oxygen had an easier permeation at a thick oil/water interface where the hydrocarbon tail region of the phospholipid was less ordered (Fig. 6).

In recent years, the applications of the SECM have greatly increased in numerous fields due to the fabrication of miniaturized tips and new combinations with other techniques, for instance, with atomic force microscopy, optical microscopy (OM), single-molecule fluorescence spectroscopy (SMFS), electrochemical quartz crystal microbalance (EQCM), Langmuir trough technique, electrochemical scanning tunneling microscopy (ECSTM), scanning force microscopy (SFM) and surface plasmon resonance imaging (SPR-i) These combinations with SECM make the spatial resolution, accuracy, sensitivity further improved. In addition, more information about the topographic, optical, kinetic, or photoelectrochemical properties in situ at various interfaces can be obtained. Obviously, these combinations are to be a trend.

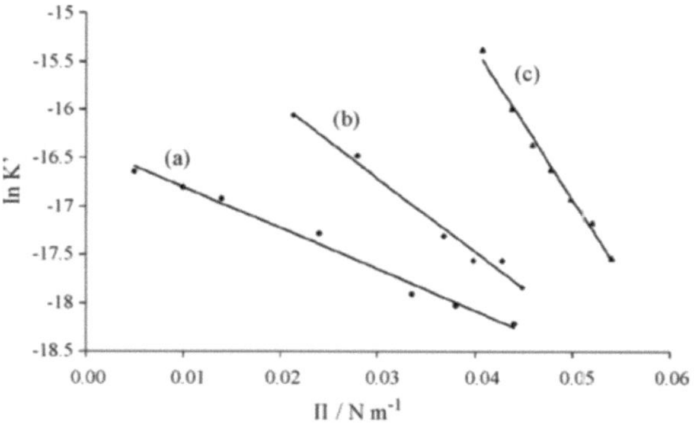

Fig. 5. Plot of ln k`vs. *Π* for the transfer of oxygen across a DSPE monolayer spread at the interface between (a) air/water, (b) thin decane film/water, and (c) thick decane/water (S. Cannan et al., 2004).

8.3 Photoinduced charge transfer at the solid/electrolyte interface

The use of mesoporous oxide films as a substrate to anchor the dye molecules allows sunlight to be harvested over a broad spectral range in the visible region. Similarly to chlorophyll in the green leaf, the dye acts as an electron transfer sensitizer. Upon excitation by light, it injects an electron into the conduction band of the oxide, resulting in the separation of positive and negative charges. Charge transfer from photoexcited dyes into semiconductors was discovered more than a century ago in a famous experiment. He observed that the photoelectric effect reported earlier by Becquerel on silver plates was enhanced in the presence of erythrosin dye. A few years before, Vogel in Berlin had associated dyes with the halide semiconductor grains to make them sensitive to visible light. This led to the first panchromatic film, able to render the image of a scene realistically in black and white. However, the clear recognition of the parallelism between the two procedures, a realization that the same dyes in principle can function in both systems, and a verification that their operating mechanism is by injection of electrons from photoexcited dye molecules into the conduction band of the n-type semiconductor substrates date to the 1960s. In subsequent years the idea developed that the dye could function most efficiently if chemisorbed on the surface of the semiconductor. The concept emerged of using dispersed particles to provide a sufficient interface, then photo-electrodes were employed. Finally, the use of nanocrystalline $Ti0_2$ films sensitized by a suitable molecular dye provided an important technological breakthrough. These mesoporous membranes have allowed in effect for the first time the development of a regenerative photoelectrochemical cell based on a simple molecular light absorber, which attains a conversion efficiency commensurate with that of silicon-based photovoltaic devices, but at a much lower cost.

The photoinduced electron-transfer (PET) processes are of vital importance in the versatility of biological and chemical systems, which exhibited extensive research prospects, such as photocatalysis, photo-to-electric conversion, photosynthesis, and photo-induced supramolecular electron-transfer (ET). Photoelectrochemical properties of porphyrin compounds, including porphyrin-fullerene dyad and ferrocene-porphyrin-fullerene triad, were investigated by self-assembled monolayers (SAMs) on nanostructured substance and electrode surfaces. With the development of SECM techenology, it has been demonstrated to be an effective means of determining electron transfer (ET) kinetics on polymer films at electrodes, Langmuir monolayers at air-water interfaces, liquid-liquid interfaces, phospholipid bilayers, the traditional metal electrode-electrolyte solution interfaces, thiol-porphyrin self-assembled monolayers, as well as the semiconductor/electrolyte interface (SEI). As well all known, almost each-step ET occurring in nature can not proceed without light, indicating the studying on PET has important practical implications. Because porphyrin molecules were an important class of conjugated organic molecules for light harvesting in photosynthesis, and exhibited ultrafast electron injection, slow charge-recombination kinetics, high absorption coefficients and good chemical stability, the carboxylic groups on some porphyrins can spontaneously bind to TiO_2 nanoparticles to obtain the excellent photosensitized material (H.X. Ju et al., 2011). By fitting experimental feedback curves to theory ones, the heterogeneous rate constant (k_{eff}) is estimated, meanwhile, the dependence of k_{eff} with light source wavelength and intensity is determined (X.Q. Lu et al., 2011).

9. Scope

The scanning electrochemical microscopy has been proven to be a powerful instrument for the quantitative investigation and surface analysis of a wide range of processes that occur at interfaces. SECM has the advantage that measurements are carried out under steady-state conditions, eliminating the resistive potential drop in solution, double layer charging current and some difficulties that are frequently associated with the other traditionally electrochemical techniques. In addition, the SECM technique has provided a great advantage of easy separation of electron transfer and ion transfer processes. These advantages of SECM make it become a ver- satile technique for determining electron-transfer kinetics, micropatterning and imaging of cellular activities of single cells. The development of UME and the combination of SECM and other techniques show much higher spatial resolution and precision. With the improvement of lateral resolution and sensitive detection, SECM will extend a wide variety of applications and show a great potential as analytical and microfabrication tool, especially in life sciences and in the material.

10. References

[1] A.J. Bard, M.V. Mirkin. *Scanning Electrochemical Microscopy*. Marcel Dekker, Inc., New York, 2001.
[2] R.D. Martin, P.R. Unwin. (1998). *Anal. Chem*. 70: 276.
[3] A.J. Bard, J.L. Fernandez. (2004). *Anal. Chem*. 76: 2281.
[4] Y.H. Shao, M.V. Mirkin. (1997). *J. Electroanal. Chem*. 439: 137-143.
[5] Y. Shao, M.V. Mirkin. (1998). *J. Phys. Chem. B*. 102: 9915-9921.
[6] T. Matsue., D. Oyamatsu, Y. Hirano, et al. (2003). *Bioelectrochemistry*. 60: 115-121.
[7] K. Eckhard, X. Chen, F. Turcu, W. Schuhmann. (2006). *Phys. Chem. Chem. Phys*. 8: 5359-5365.
[8] W. Schuhmann., K. Karnicka, et al. (2007). *Electrochem. Commun*. 9: 1998-2002.
[9] W. Schuhmann., L.Guadagnini, et al, (2009). *Electrochim. Acta*. 54 (669): 3753-3758.
[10] W.B. Nowall., D.O. Wipf., W.G. Kuhr. (1998). *Anal. Chem*. 70: 2601-2606.
[11] D.O. Wipf R., C. Tenent, et al. (2003). *J. Electrochem. Soc*. 150: E131-E139.
[12] S. G. Denuault., A.G. Evans., et al. (2005). *Electrochem. Commun*. 7: 135-140.
[13] A.J. Bard., J.L. Fernandez., C. Hurth. (2005). *J. Phys. Chem. B*, 109: 9532-9539.
[14] R. Cornut., C. Lefrou. (2008). *J. Electroanal. Chem*. 621: 178-184.
[15] C. Lefrou. (2007). *J. Electroanal. Chem*. 601: 94-100.
[16] A.J. Bard., J. Kwak. (1989). *Anal. Chem*. 61: 1221-1227.
[17] A.J. Bard., Y. Lee, S. Amemiya. (2001). *Anal. Chem*. 73: 2261-2267.
[18] A.J. Bard., C.G. Zoski., B. Liu. (2004). *Anal. Chem*., 76: 3646-3654.
[19] Z. Ding., R. Zhu., et al. (2008). *Anal. Chem*. 80: 1437-1447.
[20] P.R. Unwin., M.A. Edwards, et al. (2009). *Anal. Chem*. 81: 4482-4492.
[21] D. Britz, *Digital Simulation in Electrochemistry*, Springer Berlin Heidelberg, 2005.
[22] R. Cornut., M. Mayoral., D. Fabre., J. Mauzeroll. (2010). *J. Electrochem. Soc*. 157: F77-F82.
[23] X.Q. Lu., T. X. Wang., et al. (2011). J. Phys. Chem. B., 115, 4800–4805.
[24] P.R. Unwin., C.J. Slevin., J.V. Macpherson. (1997). *J. Phys. Chem. B*. 101: 10851-10859.
[25] A.J. Bard., A.L. Barker., et al. (1999). *J. Phys. Chem. B*, 103: 7260-7269.
[26] G. Denuault., Q. Fulian, A.C. Fisher. (1999). *J. Phys. Chem. B*, 103(799): 4387-4392.
[27] G. Denuault., Q. Fulian., et al. (1999). *J. Phys. Chem. B*. 103: 4393-4398.
[28] G. Wittstock., O. Sklyar., (2002). *J. Phys. Chem. B*. 106(801): 7499-7508.
[29] S. A.J. Bard., E. Creager., et al. (2004). *J. Am. Chem. Soc*. 126: 1485.
[30] K.B. Holt. (2006).*Langmuir*. 22: 4298.
[31] A.J. Bard., C. Cannes., et al. (2003). *J. Electroanal. Chem*. 547: 83.
[32] X.Q. Lu., L.M. Zhang., et al. (2006). *Chem. Phys. Chem*. 7: 854.
[33] X.Q. Lu., M.R. Li., et al. (2006). *Langmuir* 22: 3035.
[34] D. Burshtain, D. Mandler. (2005). *J. Electroanal. Chem*. 581, 310.
[35] B. Csóka, B. Kovács, G. Nagy. (2003). *Electroanalysis*. 15, 15.

[36] C. Zhao, G. Wittstock. (2004). *Angew. Chem.* Int. Ed. 43, 4170.
[37] T. Koike., H. Yamada, et al. (2005). *Anal. Chem.* 77,1785.
[38] J.F. Zhou, Y.B. Zu, A.J. Bard, J. Electroanal. Chem. 491 (2000) 22.
[39] J.L. Fernandez, A.J. Bard. (2003). *Anal. Chem.* 75, 2967.
[40] X.Q. Lu., W. T. Wang., et al., (2010). J. Phys. Chem. B., 114, 10436-10441
[41] J.L. Fern ´andez, D.A. Walsh, A.J. Bard. (2005). *J. Am. Chem. Soc.* 127, 357.
[42] A.J. Bard., A.L. Barker, et al. (1999). *J. Phys. Chem. B.* 103, 7260.
[43] F. Li, A.L. Whitworth, P.R. Unwin. (2007). *J. Electroanal. Chem.* 602.
[44] Y.H. Shao., P. Sun, et al. (2003). *J. Am. Chem. Soc.* 125, 9600.
[45] B. Liu, M.V. Mirkin. (1999). *J. Am. Chem. Soc.* 121, 8352.
[46] W.J. Miao, Z.F. Ding, A.J. Bard. (2002). *J. Phys. Chem. B.* 106, 1392.
[47] X. Q. Lu., Sun P., et al., (2010) Anal. Chem., 82, 8598-8603.
[48] Y.H. Shao., Y.M. Bai, et al. (2003). *Electrochim. Acta.* 48, 3447.
[49] Y.H. Shao., P. Sun., et al. (2002). *Angew. Chem.* Int. Ed. 41, 3445.
[50] S. Cannan, J. Zhang, F. Grunfeld, P.R. (2004). *Unwin, Langmuir.* 20, 701.
[51] H.X. Ju., W.T. Wen., et al. (2010). *Anal. Chem.*, 82, 8711-8716.
[52] X.Q. Lu., W.T. Wang., et al. (2011). *Chem. Commun.*, 47, 6975-6977.

Part 3

Electrochemical Energy Storage Devices

7

Studies of Supercapacitor Carbon Electrodes with High Pseudocapacitance

Yu.M. Volfkovich, A.A. Mikhailin,
D.A. Bograchev, V.E. Sosenkin and V.S. Bagotsky
*A. N. Frumkin Institute of Physical Chemistry and Electrochemistry,
Russian Academy of Sciences, Moscow,
Russia*

1. Introduction

During the last decades different new capacitor types were developed based on electrochemical processes. According to Conway [1] an electrochemical capacitor is a device in which different quasi-reversible electrochemical charging/discharging processes take place and for which the shape of the charging and discharging curves is almost linear, similarily to those in common electrostatic capacitors [1-13]. Electrochemical capacitors can be classified as film-type (dielectric), electrolytic and supercapacitors.

Electrolytic capacitors based on aluminium foils and liquid electrolytes are well-known for many decades. In them a thin film (thickness in the order of micrometers) of aluminum oxide prepared by electrochemically oxidizing the Al foils serves as dielectric film. Their specific energy is of the order of some hundredths Wh/L.

Electrochemical supercapacitors (ESCs) can be subdivided into electrical double-layer capacitors (EDLCs), pseudo-capacitors, and hybride-type capacitors. Historically the first ESCs which were developed were EDLCs. Up to now they remain the most important ESC version. The first prototypes of EDLCs were developed in the 1970s in Russia by N. Lidorenko and A. Ivanov [14] and also in Japan under the names "molecular energy accumulators" and "*Ionistors*".

The double-layer capacitor (EDLC) comprise two porous polarizable electrodes. The accumulation of energy in them proceeds through dividing positive and negative electrical charges between the two electrodes while maintaining a potential difference U between them. The electrical charge on each electrode depends on the electrical double-layer (EDL) capacity. Due to the very low thickness of this layer (tenths of a nanometer) the capacity value referred to unit of electrode's surface area is much higher than for electrolytic capacitors. Hence the term "*supercapacitor*" was introduced.

In order to achieve high capacity values in EDLCs highly dispersed carbonaceous electrodes with a high specific surface area of 1000-3000 m^2/g are used, such as activated carbon (AC) or activated carbon cloths (ACC), nanofibers, nanotubes graphene sheets. The specific energy density of such capacitors reaches values of 1-20 Wh/L.

For a EDLSC with ideal polarizable electrodes the energy A delivered through a single discharge can be represented as:

$$A = (1/2)\, C\, [(U_{max})^2 - (U_{min})^2], \qquad (1)$$

where C is the average electrode's capacity, and U_{max} and U_{min} are the initial, resp. the final values of discharge voltage. For full discharge until $U_{min} = 0$ the maximal discharge energy will be: $A = A_{max} = (1/2)C[(U_{max})^2]$.

The development of EDLCs was induced by a necessity for rechargeable power sources with higher enegy values, and power capabilities, and much better cyclability properties than those for existing storage batteries. Among the most remarkable features of EDLCs are excellent cyclability (hundred of thousands charge/discharge cycles as compared with hundreds of cycles for storage batteries) and the possibility to deliver for short periods high power and current densities, and also the possibility to be used at high and low temperatures (up to +60°C and down to -50°C). The cycling efficiency (ratio of energy consumed during charging and delivered during discharge) is about 92-95 %. Taking into account these properties, very promising are the following fields of applications for EDLCs: in ICE vehicles (in parallel with storage batteries) for starting purposes, delivering the necessary initial peak power (especially at low temperatures) and thus increasing the battery life time; and also in all-electric and in hybride vehicles for energy recuperation during slow-down and braking.

In *pseudocapacitors* electrical charges are accumulated mainly as the result of fairly reversible redox reactions (faradaic pseudo-capacity). Many such reactions are known in which oxides and sulfides of transition metals RuO_2, IrO_2, TiS_2 or their combinations take part.

One of the important achievements of modern electrochemistry is the development of *electron-conducting polymers*. Electrochemical reactions in systems with conjugated double bonds such as polyanilin, polythiophene, polypyrolle, polyacethylene and others, are reversible and can be used in supercapacitors. Such processes are called electrochemical doping or dedoping the polymers with anions and cations. The electron conductivity during doping is due to the formation of delocalized electrons or electron-holes and their migration in the system of polyconjugated double-bonds under the influence of an applied electrical field. The use of some electron-conducting polymers as supercapacitor electrodes is based on the high reversibility of doping and dedoping reactions and the high conductance values of such polymers. The specific energy values of pseudocapacitors are fairly high 10-50 Wh/L and their cyclability reaches hundred thousands cycles. A disadvantage of supercapacitors based on transition metal oxides and sulfides is their high price. A disadvantage of those based on electron-conducting polymers is their insufficient stability.

Recently some *hybride-type supercapacitors* were developed in which different types of electrodes are used. In [15-18] capacitors were investigated in which the positive electrodes were based on metal oxides, and the negative electrodes on activated carbon (e.g. the system NiOOH/KOH/AC or the system PbO_2/H_2SO_4/AC. In both of these two systems as positive electrodes conventional electrodes from alkaline, resp. lead-acid storage batteries are used. An advantage of the hybride-type supercapacitors as compared with their analogues, with symmetrical AC, is their higher U_{max} value and correspondingly their higher specific energy (up to 10-20 Wh/kg). The lower U_{max} value for symmetric AC systems in comparison with the hybride versions is due to the fact that at not very high anodic potentials (0.9-1.0 V) an oxidative carbon corrosion is observed while hydrogen evolution potential on the the hybride's negative AC electrode remains at very negative values up to -0.8 V). An advantage of hybride supercapacitors over the corresponding storage batteries is the much higher cycling capability, and the possibilities of faster charging and of easier hermetically sealing. Hybride PbO_2/H_2SO_4/AC capacitors are used in wheelchairs and in electrical motor-buses.

In [19] a hybride supercapacitor with a RuO_2 positive electrode and an AC negative electrode was described. A high specific energy of 26.7 Wh/kg was reported.

An important feature of supercapacitors in comparison with storage batteries is the possibility of rapid charging and discharging in very broad time intervals ranging from less than one second to several hours. Correspondingly supercapacitors can be subdivided into power units with high values of specific power and into energy units with high specific energy values.

Power supercapcitors allow carrying out the charging and discharging processes in very short time periods (from fractions of a second to minutes) and obtaining herewith high power characteristics from 1 to 5 kW/kg in concentrated aqueous solutions with high specific conductivity. Measurements for highly dispersed carbon electrodes in the energy capacitor operation modes usually yield the specific charge values in the range of 40 to 200 C/g [20, 21]. In the case of carbon materials, the limiting capacity obtained in [22] was 320 F/g due to a considerable contribution of pseudocapacitance of reversible surface group redox reactions (thus, they are not pure EDLCs anymore). In [23] high power characteristics (above 20 kW/kg) were obtained for electrodes based on single-wall carbon nanotubes (SWCNT). Such high power values can be explained by the regularity of the SWCNT's pore structure. As can be seen from Fig. 1 representing a SEM picture of this material, the pores are neither corrugated nor curving, thus providing for a high conductivity and, therefore, for a high power.

The electrodes used in energy-type ECSCs are often electrodes on which rather reversible faradaic processes occur. Such electrodes include electrodes based on electron-conducting polymers (polyaniline, polythiophene, polypyrrole etc.) and also electrodes based on some oxides of variable-valency metals (oxides of ruthenium, iridium, tungsten, molybdenum, zirconium etc.) [1, 2]. These electrodes feature different limitations for practical application, such as expensiveness, insufficient cyclability due to degradation processes etc. Using nonaqueous electrolytes in ECSC with electrodes based on highly dispersed carbon materials allows obtaining high (up to 3-3.5 V) charging voltage values, which significantly enhances the energy but limits the power of capacitors due to low conductivity of these electrolytes [1, 2, 24]. Aqueous alkali solutions allow obtaining rather high power values, but the low operating voltage range (about 0.8 V) decreases the energy characteristics of ECSC. Aqueous electrolytes with the highest conductivity are sulfuric acid solutions with concentrations from 30 to 40 wt. %. Besides, the working voltage range in the region of reversible processes proves to be above 1 V due to the relatively low corrosion activity towards carbon as compared to other aqueous electrolytes.

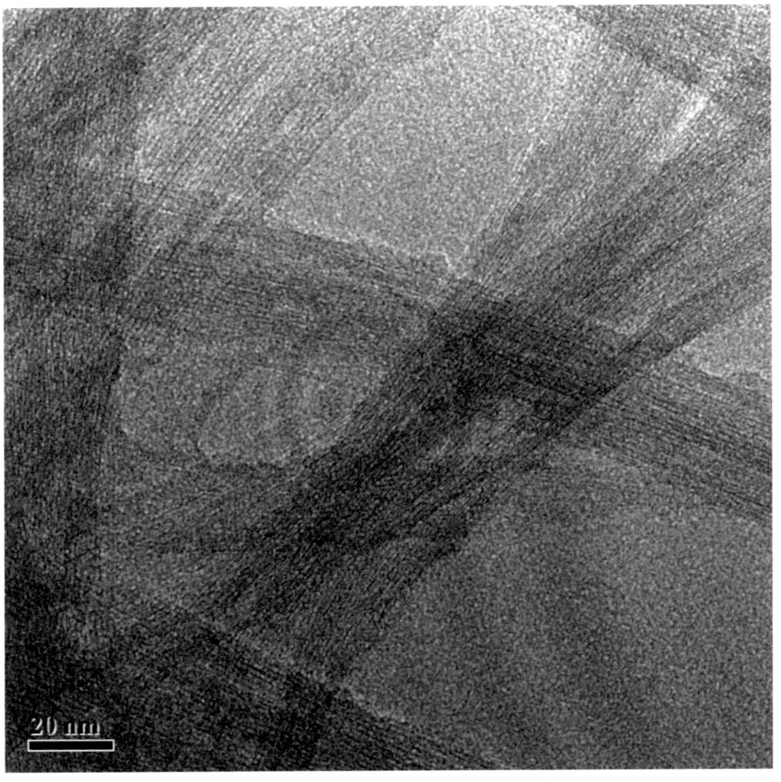

Fig. 1. SEM image of single-wall carbon nanotubes (SWCNT)

In [25], a very high maximum amount of electricity of 1150 C/g was obtained with electrodes based on ADG type activated carbon with the specific surface area of 1500 m^2/g. This value was reached after deep cathodic charging to the potentials of -0.3 to -0.8 V RHE. On the basis of these data and on the basis of other various experiments it was assumed that such a high amount of electricity is obtained as a result of hydrogen intercalation into AC carbon limited by solid-phase diffusion. These experiments are as follows: a very slow (20 h and more) deep charging process, memory effect under potential scanning in the positive direction, absence of correlation between the specific surface area and specific capacitance values, linear dependence of current maximums on the square root of the potential sweep rate, and some other experimental data. It was concluded in this paper that the most probable limiting amount of electricity is that corresponding to the formation of a compound C_6H similar to the compound of C_6Li for negative carbon electrodes in rechargeable lithium-ion cells. However that formation of the C_6H compound according to Faraday's law requires consumption of 1320 C/g that was not obtained in [25].

Therefore, the aim of the present investigation was to reach the maximum possible capacitance values on AC and also a more detailed study of mechanisms of the processes occurring under deep cathodic charging. Besides that, the aim of the investigation was to develop a mathematical model of these processes and to compare it with experimental data.

There are only few papers on mathematic modeling of processes in ECSC. In [26], an EDLC model was developed taking into account EDL charging, potential distribution in a porous electrode due to ohmic energy losses and its porous structure. The calculation results agreed well with experimental galvanostatic charging-discharge curves. In [27, 28] an operation theory was developed for electrodes based on electron-conducting polymers used in supercapacitors. The theory takes into account EDL charging, potential distribution in a porous electrode, electrochemical kinetics, intercalation of counterions in the polymer phase, and a nonsteady-state solid-phase diffusion of counterions in this phase. Simulation of the experimental discharge curves allowed obtaining the values of the parameters of processes occurring in the electrode: the solid-phase diffusion coefficient, EDL specific capacitance, and exchange current density of the electrochemical reaction. In [29], a model of a SWCNT electrode was developed taking into account EDL charging, hydrogen electrosorption-desorption, and kinetics of hydrogen electrooxidation-electroreduction according to the Volmer theory. In [30] an AC-based electrode impedance model accounted for EDL charging and intercalation processes, using statistical thermodynamics. However, in [29, 30] there was no comparison between the theory and experiment

2. Experimental methods

The following electrochemical methods were used in this work: cyclic voltammetry, galvanostatic, and impedance techniques. The electrochemical impedance spectra were obtained using an electrochemical measurement system consisting of a Solartron 1255 frequency analyzer, Solartron 1286 potentiostat, and a computer, and also using a FRA impedance meter. Measurement of CVs and galvanostatic curves was carried out using Solartron 1286 and PI-50 potentiostats. Prolonged cell cycling was achieved using a Zaryad 8k cycling device. Apart from the impedance technique, the dc charged electrode resistance was measured as follows. After deep anodic charging, the cell for charging was disassembled and the working electrode was removed. Further, in order to remove the acid residue, the electrode was dried in contact with microporous acidproof filter paper. Then it was placed into a special four-electrode measurement cell and pressed between two liners of foil of thermally expanded graphite (TEG); copper disks with copper current leads were placed on the rear sides of this set. Further, the cell was sealed and the dependence of voltage on current was measured, from which the electrode resistance was calculated. An important feature of the method is that the cell with copper current leads must be assembled in a predetermined period of time (in our case, 3 min), as a deeply charged electrode undergoes gradual oxidation in air. An advantage of this method over impedance analysis is that it allows eliminating the polarization contribution into resistance as in the case of the impedance technique. Due to these limitations, both of these methods were used.

The main electrochemical measurements were carried out in a specially designed teflon filter-press cell with carbon current leads that allowed performing studies in a wide range of potentials. Its scheme is shown in Fig. 2,a and its photograph is presented in Fig. 2,b. Fig. 3 shows schematically the electrochemical group of this cell representing a matrix system, in which the electrolyte is in the pores of the electrode and separator. The separator is clamped between two ring-shaped acidproof rubber gaskets that are chosen depending on the electrode thickness to provide reliable separation of the working and auxiliary electrodes. A fine-texture Grace-type separator (polyethylene with silica gel) was used in the cell. The capacitance of the AC auxiliary electrode was much higher than that of the working electrode, which allowed eliminating interaction of electrodes through the evolved gas. Application of graphite current leads instead of the usual metallic (most often, platinum) ones allowed reaching high negative potentials (up to -1 V RHE), which in its turn allowed obtaining ultrahigh capacitance values (see below).

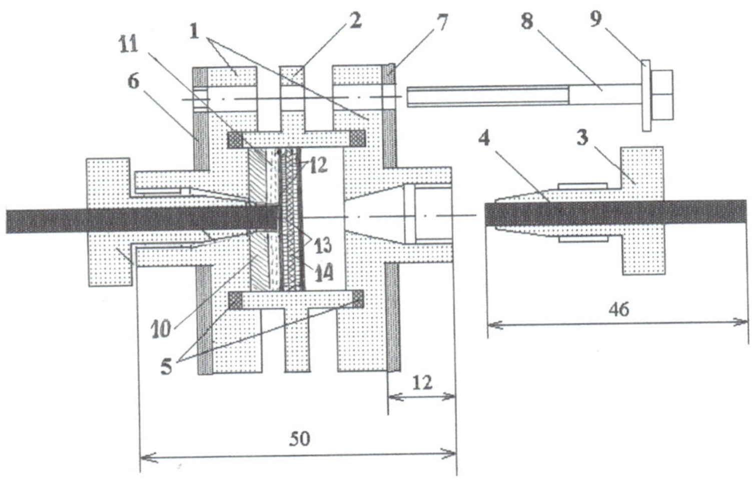

Fig. 2.a Electrochemical cell. The principal details: 4 - a graphite current lead, 12 - current distribution layer of thermal expanded graphite, 13 - electrodes, 14 - separator.

Fig. 2.b Photograph of the electrochemical cell

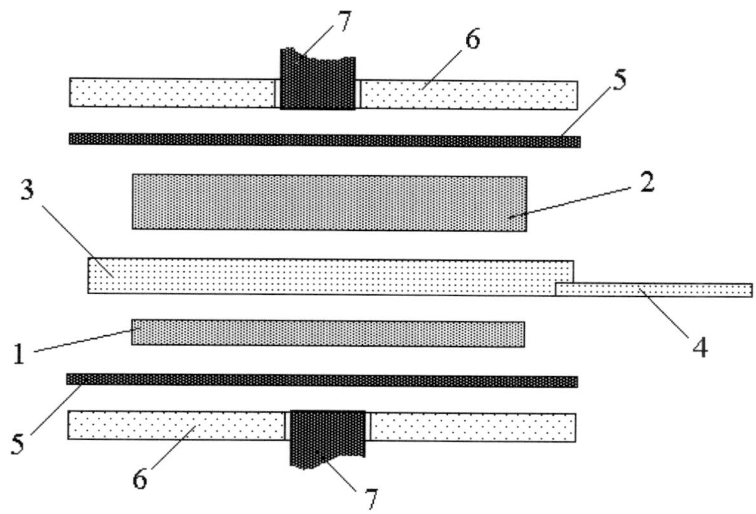

Fig. 3. Electrochemical group of the Teflon cell.

The electrolyte used in this study was mainly concentrated sulfuric acid with concentrations from 30 to 60%. The electrode used was a CH900-20 activated carbon cloth (ACC) (Japan). The electrodes had a surface area of 3 cm^2 , a thickness (in the compact state under the pressure of 0.5 MPa) of about 0.4 mm, and a mass of about 0.06 g.

Studies of the porous structure and hydrophilic-hydrophobic AC properties were carried out using the method of standard contact porosimetry (MSCP) technique [32] with evaporation of octane and water. The method is based on the laws of capillary equilibrium. If two (or more) porous bodies partially filled with a wetting liquid are in capillary equilibrium, the values of the liquid's capillary pressure in these bodies are equal. The value of the capillary pressure according to the Laplace equation is $p^c = 2\sigma \cos\theta / r$, where r - is the pore radius, σ - the surface tension, θ - the wetting angle. In this method the amount of a wetting liquid in the test sample (V_t) is measured; Simultaneously, the amount of the same wetting liquid (V_s) is measured in a standard specimen of known porous structure. The liquids in both porous samples are kept in contact. After some time a thermodynamic equilibrium is reached. The measurements are performed for different overall amounts of the liquid $V_0=V_s+V_t$. During the experiment this overall amount is changed by gradual evaporation of the liquid. The MSCP with appropriate standard samples can be used to measure pore sizes in the range from 1 to $3 \cdot 10^5$ nm. The MSCP has several substantial advantages over mercury porosimetry and other porosimetric methods. It has the possibility:

- to investigate materials with low mechanical strength (for example clothes), frail materials and even powders;
- to measure samples at fixed levels of compression and/or temperature, i.e., under conditions in which they are commonly used in different devices;
- to use for measurements the same liquid (for example water or aqueous solutions) as that, used in real devices (i.e. leading to the same swelling degree of the sample);
- no use toxic materials such as mercury.

One of the most pronounced advantages of the MSCP is the possibility to investigate the wetting (hydrophilic/hydrophobic or liophilic/liophobic) properties of porous materials. Primarily, the MSCP measures the distribution of pore volume vs. the capillary pressure p^c, i.e. vs. the parameter $r^*=r/\cos\theta$ (henceforth, this parameter is called effective pore radius). For partially hydrophobic materials (for which $\theta>0$) the porosimetric curves measured with water are shifted towards higher values of r^* with respect to the curves measured with octane that wets most materials almost ideally ($\theta=0^0$). The value of this shift for a certain value of pore volume V_n and of the corresponding pore radius r_n allows to determine the wetting angle of water for pores with the radius r_n:

$$\cos\theta = r_n / r^*. \tag{2}$$

For porous materials the wetting angles $\theta(r)$ for pores of different size can be different.

3. Experimental results and discussion

Fig. 4 presents integral pore radius distribution curves measured using MSCP with octane and water for the CH900-20 AC cloth. As follows from this figure, this cloth has a very wide pore spectrum: from micropores with radii r ≤ 1 nm to macropores with r > 100 μm, i.e. in the range of more than 5 orders of magnitude. It is of interest that this cloth contains micropores and also macropores with r > 1 μm, but there are practically no mesopores with 1 nm < r < 100 nm. Micropores provide high full specific surface with S_f = 1520 m²/g and hydrophilic specific surface S_{phi} = 870 m²/g. According to [32], porometric curves are measured using octane for all pores and using water only for hydrophilic pores. As follows from these curves, the full porosity (by octane) was 86%, the hydrophilic porosity was 78.5%, and the hydrophobic porosity iwas 7.5 %.

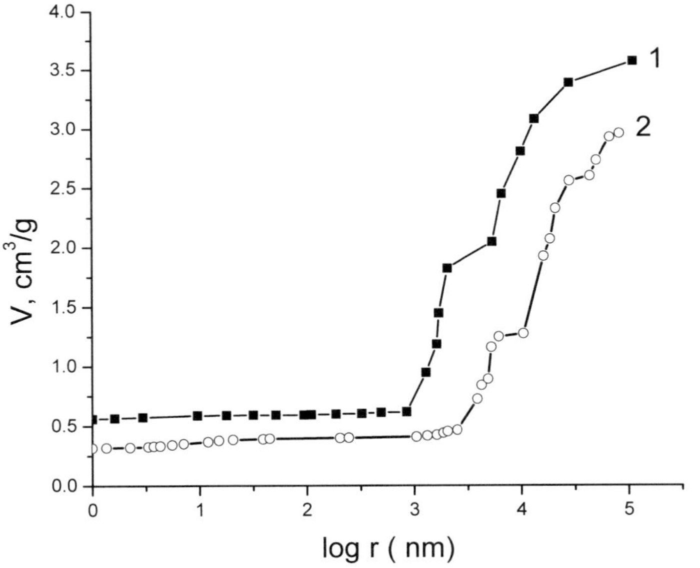

Fig. 4. Integral pore radius distribution curves measured in octane (1) and water (2) for a CH900-20 AC cloth.

Fig. 5 presents the dependence of the wetting angle θ on the pore radius calculated from Fig. 5 according to [32]. As may be seen, the θ values for the whole pore radius range are close to 90°. Therefore, though most of the pores are hydrophilic with θ < 90°, they are still badly wetted. This and also the presence of fully hydrophobic pores may point to high amounts of graphite (graphene) impurities that are practically hydrophobic. The complex curve character in Fig. 5 is largely due to a nonuniform distribution of surface groups in the pores of different radii.

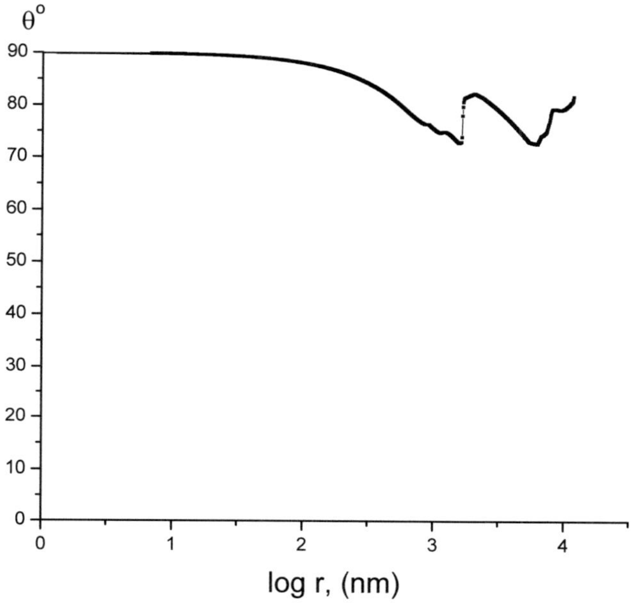

Fig. 5. Dependence of wetting angle θ on the pore radius for the CH900-20 AC cloth.

In this work, studies of AC were carried out in a large range of potentials from -0.8 to 1.0 V RHE. It is convenient to present the CV (especially those measured at different potential sweep rates w), graphically in the form of capacitance-voltage curves constructed in the coordinates of differential capacitance (C) vs. potential (E), where $C = dQ/d\tau = I\, d\tau\, /dE = I/w$, I is the current, Q is the amount of electricity, $w = dE/d\tau$, τ is the time.

Fig. 6 compares the cyclic capacitance-voltage curves measured in 48.5% H_2SO_4 at different potential sweep rates in two ranges of potentials: in the reversibility range (from 0.1 to 0.9 V) and in the deep charging range (from -0.8 to 1 V). As follows from curve 4 measured in the reversibility range, only electric double layer (EDL) charging occurs here, while pseudocapacitance of redox reactions of surface groups is very low in this case. This distinguishes the CH900-20 ACC from ADG AC in which a considerable contribution is introduced by the pseudocapacitance of fast redox reactions of surface groups [25]. As follows from curve 4, the value of the EDL capacitance is approximately 160 F/g. Taking into account that S_{phi} = 870 m²/g, we obtain C_{DEL} = 18.4 μF/cm² for unit of the true hydrophilic carbon surface ara. This value is close to the classical EDL capacitance value of platinum [33]. Much lower C_{EDL} values presented in [34], in our opinion, are explained by the fact that the values of the full specific surface area measured using the BET technique were used, while it is known that carbon materials have both hydrophilic and hydrophobic pores. The MSCP used in this work allows to obtain the values of the hydrophilic surface area S_{phi} and thus the C_{DEL} values per unit of the real working surface area.

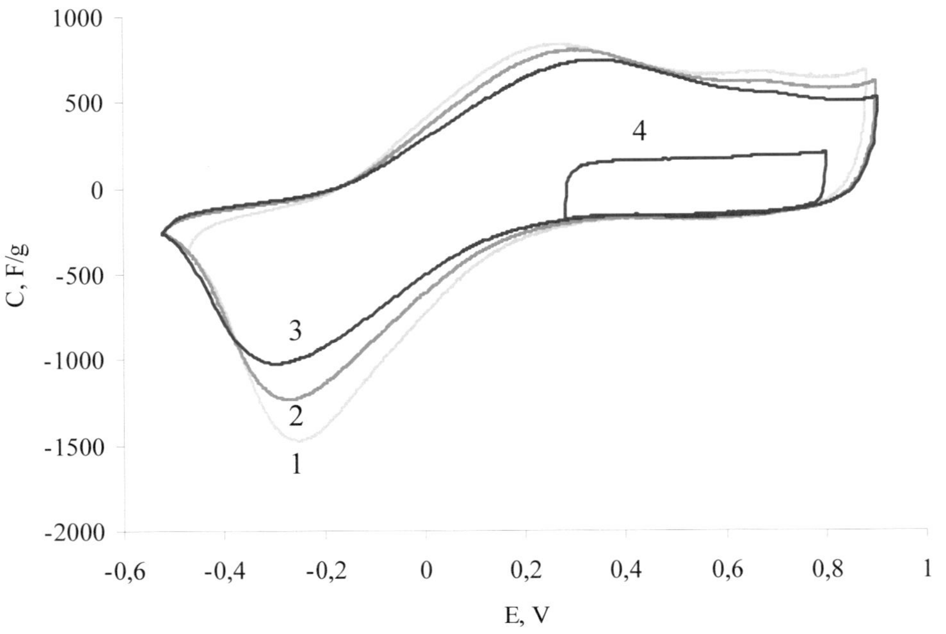

Fig. 6. Capacitance-voltage cyclic curves for CH900-20.

In the range of negative potentials (< -0.1 V) (curves 1, 2, 3), faradaic processes with a very high pseudocapacitance are observed. As may be seen, the EDL capacitance decreases in the range of negative potentials. This is probably due to the partial surface blocking by adsorbed particles. There are two pronounced maximums in the anodic branches of curves 2, 3 measured at low w values, which points to the probable occurrence of two slow processes. There is a single deep maximum corresponding to a very high amount of electricity under deep cathodic charging of AC in catodic branches of curves 1, 2, 3.

The method of galvanostatic curves was used to measure the amounts of electricity after long-term charging at negative potentials. Fig. 7 presents the dependence of the amount of electricity Q under discharge on the charging time at potential E = -250 mV in 40.3% H_2SO_4. As may be seen, the Q value grows very fast at very low charging times of seconds and minutes and continues increasing further for many tens of hours. Such very slow growth can be explained by a very slow diffusion in a solid phase. Therefore it is possible to assume the existence of a hydrogen intercalation into AC carbon controlled by slow solid-phase hydrogen diffusion. This is also evidenced by a proportionality of the limiting current to the square root of the potential sweep rate and also by a number of other experimental data obtained in [25] (memory effect under potential sweep in the positive direction, absence of correlation between the specific surface area value and specific capacitance etc.) It was assumed in [25] that during a deep cathodic charging of AC, as a result of hydrogen intercalation a compound C_xH and in he limiting case C_6H is formed. On the basis of the shape of the potentiodynamic curves it is also possible to assume that two processes take place: a fast hydrogen chemosorption at the interface of carbon and electrolyte in the pores and a process of hydrogen intercalation into AC with a slow hydrogen solid–phase diffusion.

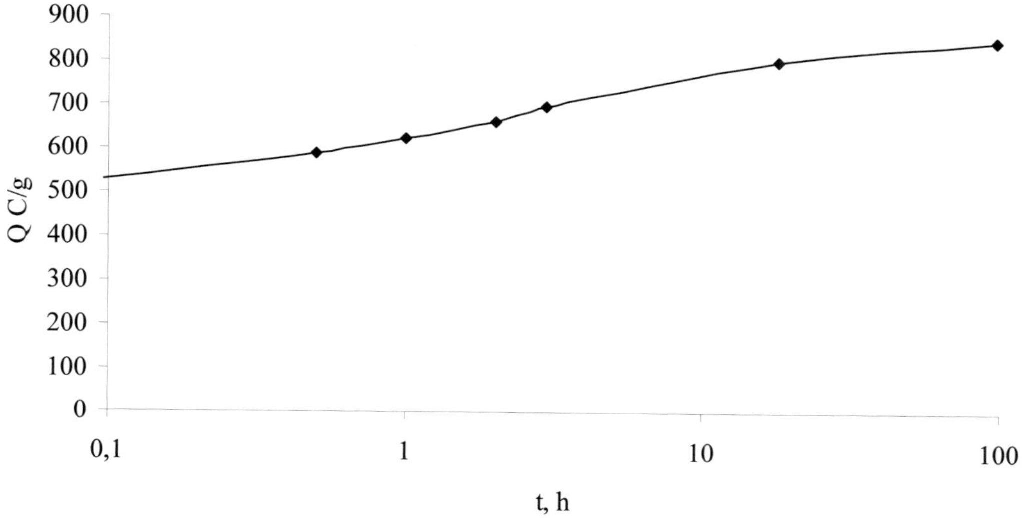

Fig. 7. Dependence of specific amount of electricity under discharge of the CH900-20 electrode in 40.3% sulfuric acid on the charging time under the charging potential of - 250 mV.

Fig. 8 presents the dependence of the active resistance of the electrode on charging time at E = -400 mV measured using the impedance technique in 40% sulfuric acid. As can be seen, the active resistance of the electrode grows in the course of the charging process. Both the solid phase and electrolyte in pores contribute to the resistance. The dependence of electric resistance on the charging time measured under constant current according to the above mentioned technique is more significant. Fig. 9 presents such a dependence of the dc resistance of a CH900-20 electrode on the charging time in 40.3 % sulfuric acid at E = -0.34 V. As seen from this figure, the electrode resistance grows significantly with an increase in the AC charging time. One should keep in mind that the above parameters were obtained on activated carbon cloth (CH900-20), which practically eliminates the effect of contact resistance between its separate fibers, as the fibers represent practically parallel transport paths for electrons. Therefore, the above data may be explained by a change in the bulk phase chemical composition in the course of charging. Thus, both resistance measurement techniques point to its increase as dependent on the charging time. This can also be explained by a change in the solid phase composition: from C to C_xH and, in the limit, to C_6H.

The maximum specific charge of 1560 C/g was obtained after charging for 22 h at the potential of E = -0.31 V RHE in 56.4% H_2SO_4. No close value was described in the literature

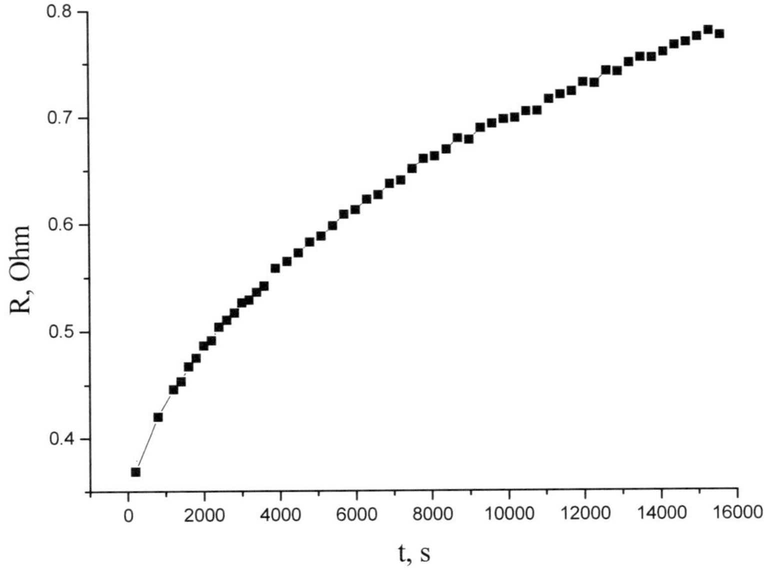

Fig. 8. Dependence of active resistance of the electrode on the charging time at E = –400 mV measured using the impedance technique in 40% sulfuric acid.

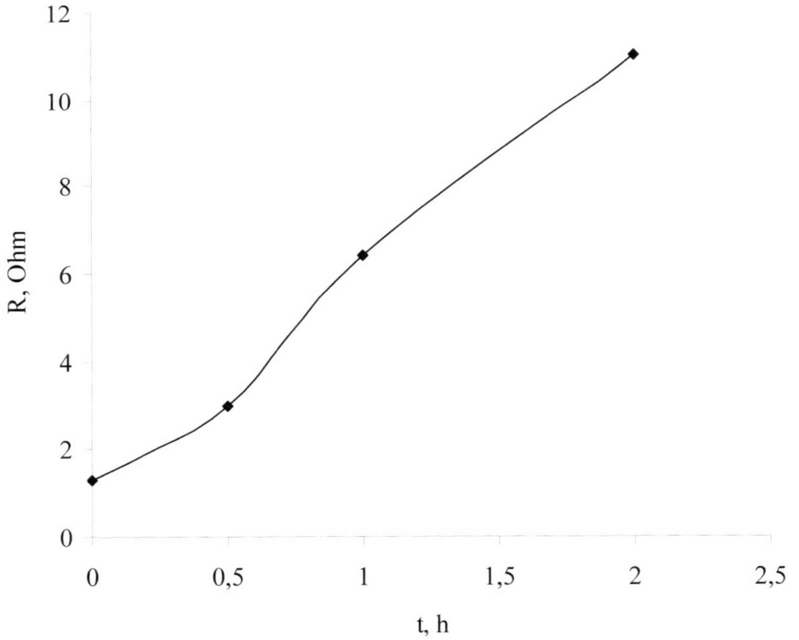

Fig. 9. Dependence of electron resistance of the CH900-20 electrode on the charging time in 40.3 % sulfuric acid at E = –0.34 V measured under constant current.

Fig. 10 presents the dependence of the maximum specific charge of CH900-20 on the sulfuric acid concentration (in the range from 34 to 56%) for the discharge time of 18 h. As follows from this figure, the specific charge grows with an increase in the sulfuric acid concentration. The following explanation is possible: It is known from the literature that: 1) in concentrated sulfuric acid solutions a sulfuric acid intercalation occurs into graphite and graphite-like materials, that is enhanced at an increase in the concentration [35-37]. During this intercalation the gap between graphene layers where the acid penetrates grows. 2) Activated carbons to a certain degree are swelling during adsorption of different adsorptives [38]. As graphite-like impurities are contained in AC [39, 40]. Taking intp account these facts, it is possible to assume that in the described conditions a **double intercalation** occurs. Sulfuric acid is intercalated into AC expanding the interlayer (intergraphene) space. Hydrogen atoms are then directed into this space under deep cathodic charging of AC. This interlayer space serves as a transport route for hydrogen. Then hydrogen interacts with the graphene layers with the ultimate formation of compound **C_6H (carbon hydride or hydrogen carbide)** (see Fig. 11).

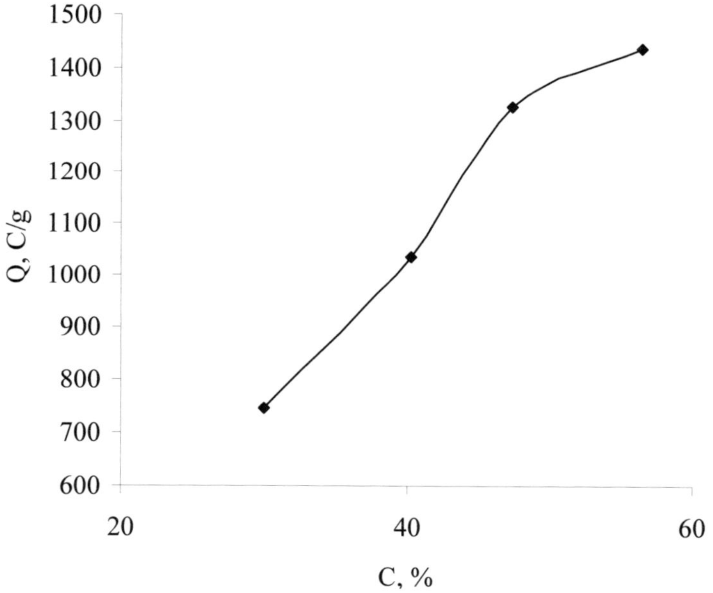

Fig. 10. Dependence of the maximum specific charge of CH900-20 on the sulfuric acid concentration for the charging time of 18 h.

According to Faraday's law, formation of compound C_6H requires 1320 C/g. The maximum value of Q_{max} = 1560 C/g was obtained in our investigation. Therefore, other rechargeable processes require 240 C/g. This value includes in the first instance the value of $Q_{EDL} = C_{EDL} \times \Delta E$, where ΔE is the range of potentials. In this case, ΔE = 1.4 V, so to the first approximation, C_{EDL} = 240/1.4 = 170 F/g. This is approximately the same value as the one obtained from curve 4 in Fig. 3. It must be noted that , this is an approximate estimate, as the C_{EDL} value somewhat depends on the potential and also as a small contribution into the Q_{max} value is made by the pseudocapacitance of redox reactions of surface groups. Nevertheless, this approximate estimate shows the correctness of the assumed mechanism of deep AC charging. Though the process of electrochemical chemosorption on the carbon/electrolyte

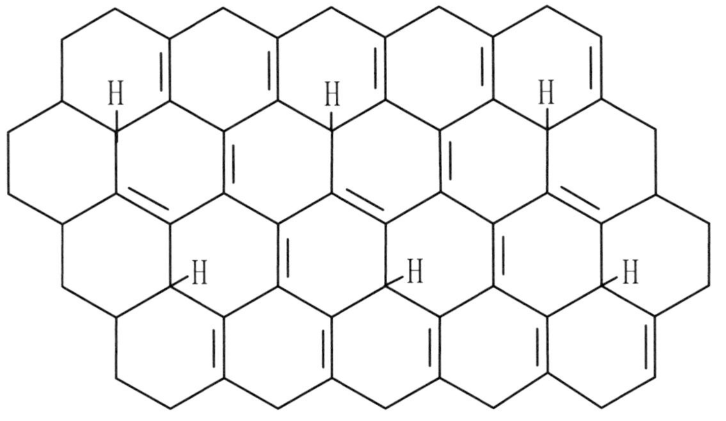

Fig. 11. Structural formula of compound C_6H.

interface and the bulk process of hydrogen intercalation occur at different rates, both of them eventually lead to formation of the single bulk compound C_6H.

Fig. 12 presents the obtained dependence of Q on the charging potential. It can be seen that the curve reaches a plateau at E < -250 mV. The very fact of reaching a plateau points to a saturation of the carbon bulk and surface by hydrogen atoms, i.e., evidences indirectly the saturation of the C_x-H chemical bonds at the given potential and H_2SO_4 concentration.

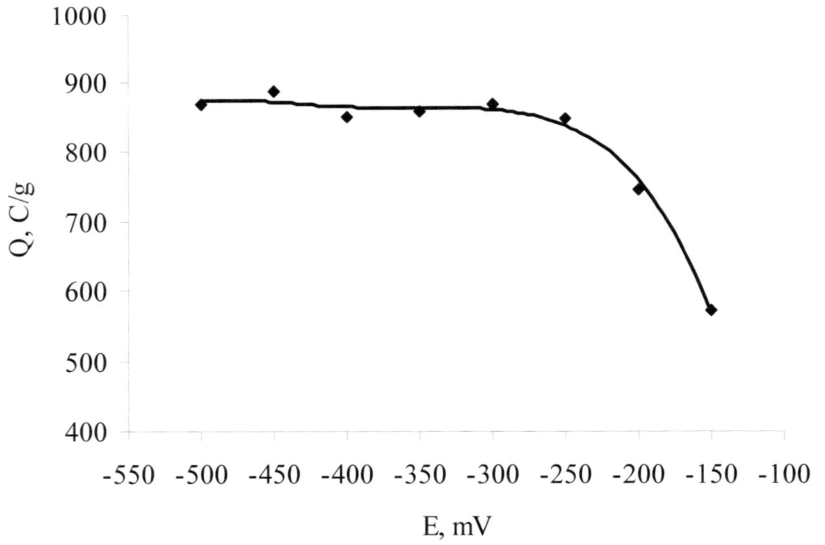

Fig. 12. Dependence of specific capacitance on the potential of charging for 18 h in 40.3% H_2SO_4.

Thus, neither the further increase in the concentration of H_2SO_4 above 56.4%, nor the further decrease in the charging potential below -250 mV, nor any increase in the charging time result in an increase in the maximum specific charge Q_{max} = 1560 C/g. As pointed out above, the contribution into this DEL capacitance value is 240 C/g, so the dominating contribution to the Q_{max} value is made by the specific charge Q = 1320 C/g corresponding to formation of the C_6H compound (pseudocapacitance charge). In principle, if one uses other activated carbons and cloths, the DEL capacitance value may somewhat increase due to an increase in the AC specific surface area that in our case was 1520 m^2/g. However, this must not result in any significant increase in the Q_{max} value.

Fig. 13 shows an impedance Cole-Cole plot (capacitive impedance component vs. resistitive component) for an activated carbon electrode in a 30% sulfuric acid solution at E=0.430 V measured in the frequency range from 100 kHz to 0.01 GHz. In this figure also shown is a fitting plot calculated from the staircase-type equivalent circuit shown in Fig. 14. Such a circuit-type is often used for EDLC electrodes [1]. We interprete this circuit by the simultaneous proceeding of different events: DL-charging, redox-reactions of different surface groups, hydrogen chemosorption and hydrogen intercalation. As the size of micropores is comparable to the DL- thickness, the DL capacity and the kinetic parameters of surface-group reactions can vary for pores of different size. Each of these processes has its own "stairstep" in the equivalent circuit. The absence in this circuit of a Warburg impedance, corresponding to a hydrogen diffusion in the solid phase, can be explained by the time constants of such a diffussion being much higher than the lowest frequencies (0.01 Hz) which can be used for such measurements.

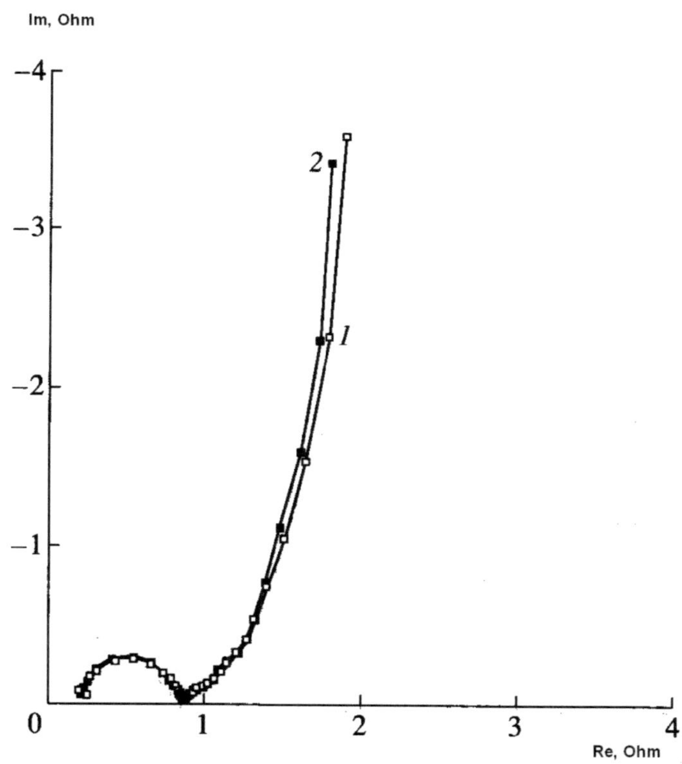

Fig. 13. Impedance plot for an activated carbon electrode in a 30% sulfuric acid solution at E=0.430 V

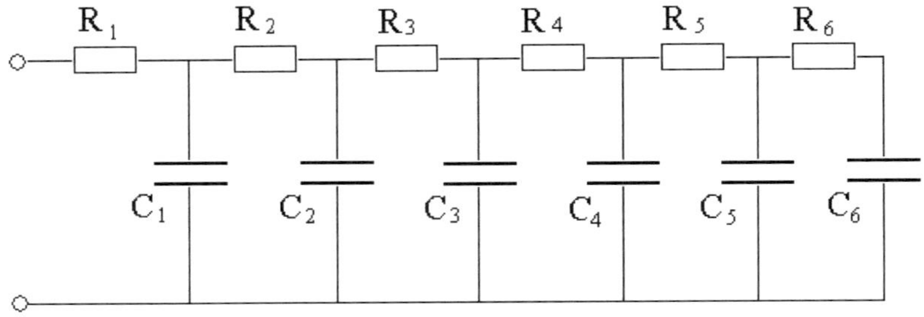

N	1	2	3	4	5	6
R, Ohm	0.29	1.99	1.2	0.57	2.22	2.86
C, F	$3.2 \cdot 10^{-5}$	1.41	3.43	$8.2 \cdot 10^{-2}$	$4.0 \cdot 10^{-3}$	$4.4 \cdot 10^{-5}$

Fig. 14. Equivalent circuit fitting the plot on Fig. 13, the values of the circuit's components are presented in the table below the figure

Galvanostatic cycling of the CH900-20 electrode was carried out at the current density of 1.2 mA/cm^2 in the range of potentials from -0.32 V to 1.03 V. Fig. 15 presents the dependence of specific discharge capacitance Q on number of cycles N. One may see that the Q value changed little in 100 cycles. Then the cycling was stopped, as it had already taken a lot of time: 588 h. The mean Q value was 940 C/g. At N = 100, the obtained overall discharge capacitance was 26.3 A h/g. Fig. 16 presents cyclic charge-discharge curves for 8 cycles. The difference between these curves and the ideal "sawtooth" is related to occurrence of the above pseudocapacitance processes.

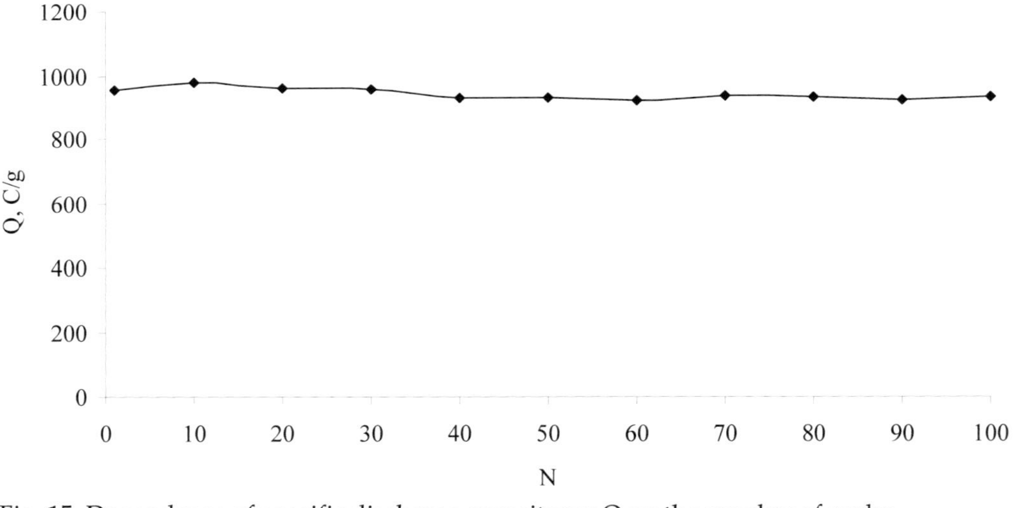

Fig. 15. Dependence of specific discharge capacitance Q on the number of cycles.

Besides the masurements of deep cathodic charging of CH900-20 ACC in H_2SO_4 solutions, similar measurements were carried out in a 90% aqueous H_3PO_4 solution. The specific charge value calculated using this plot was 1200 C/g. As according to [37], phosphoric acid is also intercalated into graphite and graphite-like compounds, these results agree with the above mechanism of double intercalation of H_3PO_4 and hydrogen into AC carbon during deep cathodic charging.

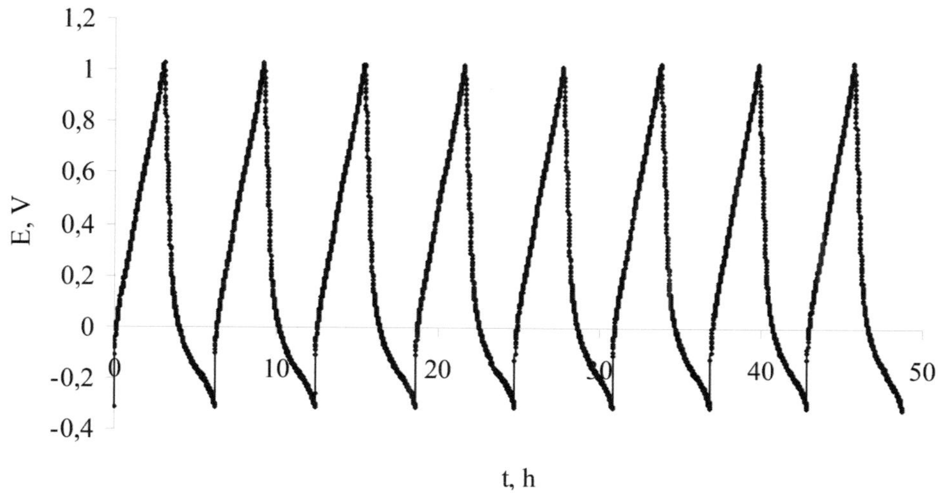

Fig. 16. Cyclic charging-discharge curves for 8 cycles.

The above data point to high prospects of using ACs under the conditions of deep cathodic charging for energy type supercapacitors.

4. Mathematic charging–discharge model of electrode based on activated carbon

We developed a two-dimensional mathematical model of charging-discharge for highly dispersed carbon accounting for the obtained experimental results.

Hydrogen solid-phase diffusion may be assumed to be one-dimensional due to strong anisotropy of carbon crystallites forming the porous structure, as there is practically no diffusion in transverse direction to graphene layers [40]. Then the equation for hydrogen diffusion for crystallites may be written in a one-dimensional form:

$$\frac{\partial c}{\partial t} = D\frac{\partial^2 c}{\partial x^2}, \tag{3}$$

where c is the hydrogen concentration in the AC carbon, t is the time, x is the shortest axial coordinate of the crystallite, D is the longitudinal solid-phase diffusion coefficient of hydrogen in the crystallite. The boundary condition at the interface of carbon/electrolyte in pores is:

$$i_p = nFD\left.\frac{\partial c}{\partial x}\right|_{x=0}, \tag{4}$$

where n is the charge transferred, F is Faraday's number, i_r is the current density determined by the kinetic dependence of the reaction. Let us also assume that there is a certain length H corresponding to half the crystallite length, where there is no hydrogen diffusion flux due to the symmetry:

$$0 = \left.\frac{\partial c}{\partial x}\right|_{x=H} \quad (5)$$

The intercalation current density is determined by the Butler–Volmer kinetics:

$$i_i(\eta) = i_0\left(\frac{c}{c_0} e^{\frac{\alpha\eta F}{RT}} - \frac{c - c_H}{c_0} e^{-\frac{(1-\alpha)\eta F}{RT}}\right) \quad (6)$$

where i_0 is the exchange current density, c_0 is the hydrogen concentration at the interface, α is the electrochemical reaction transfer coefficient. Let us assume that the initial state of the discharge process is saturation of carbon by hydrogen; then the initial condition for equation (2) is:

$$c = c_H\big|_{t=0} \quad (7)$$

Apart from the current density described in (4), there is also hydrogen adsorption current and double electric layer (DEL) charging current in the system. Let us assume that the hydrogen adsorption current is determined by Temkin's adsorption equation [40]:

$$i_A(\eta,\theta) = i_{0A}\left(e^{-\frac{\alpha_1 s_T(\theta-0.5)}{2}} e^{\frac{\alpha_1\eta F}{RT}} - e^{-\frac{(1-\alpha_1)\eta F}{RT}} e^{\frac{s_T(1-\alpha_1)(\theta-0.5)}{2}}\right), \quad (8)$$

where $i_{0,A}$ is the adsorption exchange current density, θ is the surface coverage, α_T is the adsorption transfer coefficient, s_T is the adsorption heat decrease coefficient, K_0 is the adsorption constant, c_H is the concentration of hydrogen ions in electrolyte. Coverage and adsorption current are connected through the following relationship:

$$\frac{\partial\theta}{\partial t} = \frac{i_A}{q_1} \quad (9)$$

where q_1 is the full potential adsorption capacitance. The DEL capacitance current may be written as:

$$i_{EDL} = C_{EDL}\frac{\partial\eta}{\partial\tau} \quad (10)$$

where η is the potential. Intercalation, adsorption, and EDL charging currents are included into the equation determining the polarization distribution across the porous electrode thickness that may be presented as:

$$\frac{\partial}{\partial y}\left(\kappa\frac{\partial\eta}{\partial y}\right) = \gamma s i_i + s i_{EDL} + s i_A \quad (11)$$

where κ is the conductivity of electrolyte in the pores determined according to the Archie relationship: $\kappa = \kappa_0\varepsilon^2$ [41], s is the specific surface area, ε is the porosity, γ is the ratio of half the mean thickness of walls between pores l and value H. It was assumed in equation (9) that specific conductivity of the carbon material is much higher than conductivity of electrolyte. The H value may be provisionally determined on the basis of the literature on activated carbons

[38]: $l \sim 1/ s\rho$, where ρ is the carbon density in activated carbons. $\rho \sim 2$ g/cm³. Assuming that s = 10^7 cm²/g, $l \sim 0.5$ nm. The H value is approximately 50 nm. Therefore, $\gamma \sim 0.01$.

If the adsorption rate is much higher than the rate of hydrogen solid-phase diffusion (as in our case), then the nonsteady-state charging process is described by equation (12):

$$\frac{\partial}{\partial y}\left(\kappa \frac{\partial \eta}{\partial y}\right) = \gamma s i_i + \left(sC_{EDL} + sC_A\right)\frac{\partial \eta}{\partial \tau} \quad (12)$$

where C_A is the hydrogen concentration at the interface. In the case of the galvanostatic mode, the boundary conditions for equation (10) are written as:

$$\kappa \cdot \left.\frac{\partial \eta}{\partial y}\right|_{y=0} = 0, -\kappa \cdot \left.\frac{\partial \eta}{\partial y}\right|_{y=L} = I, \quad (13)$$

where L is the porous electrode thickness. I is the overall current density per electrode unit visible surface.

The system of equations (3)-(13) is the system with parameters twice distributed along the x and y axes; it describes hydrogen intercalation and adsorption in the porous carbon structure and also EDL charging.

This system of equations was solved numerically using the COMSOL Multiphysics FemLab3.5 software package.

The fitting yielded better convergence of these curves for the set of system parameters presented in Table 1 as followed from the comparison of calculated and experimental galvanostatic discharge curves.

Parameters, [dimension]	Values
α, α_1 are the transfer coefficients	0.5
D is the hydrogen solid-phase diffusion coefficient, [cm²/s]	10^{-14}
H is the crystallite half-length, [cm]	4×10^{-6}
L is the electrode thickness, [cm]	0.05
i_0 is the exchange current density of hydrogen intercalation, [A/cm²]	$0.5*10^{-8}$
i_{0A} is the exchange current density of hydrogen adsorption, [A/cm²]	$>10^{-6}$
γ is the crystallite shape factor	0.01
C_{EDL} is the DEL specific capacitance, [F/cm²]	1.84×10^{-5}
S_{phi} is the specific hydrophilic surface, [cm⁻¹]	0.87×10^{7}
ε is the porosity	0.86
Q_1 is the maximum adsorption charge capacitance, [C/cm]	$1.3\ 10^{-5}$
s_T is the adsorption heat decrease coefficient	40
c_H is the limiting concentration of protons in carbon, [m/cm³]	1 /74
T is the system temperature, [K⁰]	298
κ is the specific conductivity of electrolyte, [S/cm]	0.1
I is the overall current density per unit visible surface, [A/cm²]	0.0020

Table 1. System parameters used for calculation

Herewith, most of the parameters were taken from the experiment. The parameters obtained as a result of simulation are as follows: D, i_0; q_1, i_{0A}. It was assumed in the calculations that the slowest process is hydrogen solid-phase diffusion (fast adsorption).

Fig. 17 presents the results of simulation of the galvanostatic discharge curve (dependence of the potential on the discharge time) at the current density of 2 mA/cm^2.

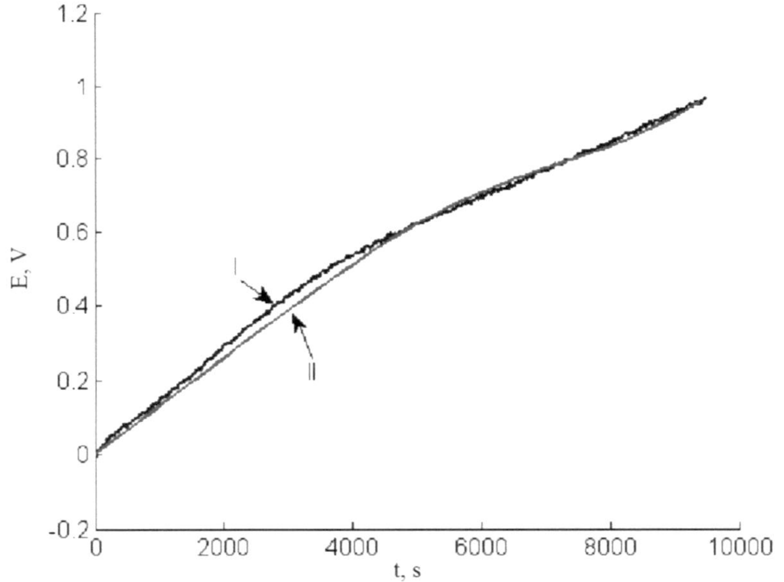

Fig. 17. Fitting of the discharge curve for the current density of 2 mA/cm^2. (I) The experimental curve, (II) the calculated curve.

This figure shows good agreement between the calculated and experimental curves, which evidences the correction of the model assumed.

The fitting yielded an approximate value of hydrogen solid-phase diffusion coefficient into AC carbon $D \sim 10^{-14}$ cm^2/s.

Fig. 18 presents the distribution (profile) of dimensionless concentration at the end of discharge by two coordinates, x and y. One may see thence that nonuniform hydrogen concentration distribution, by both coordinates, x and y, occurs at the given system parameters.

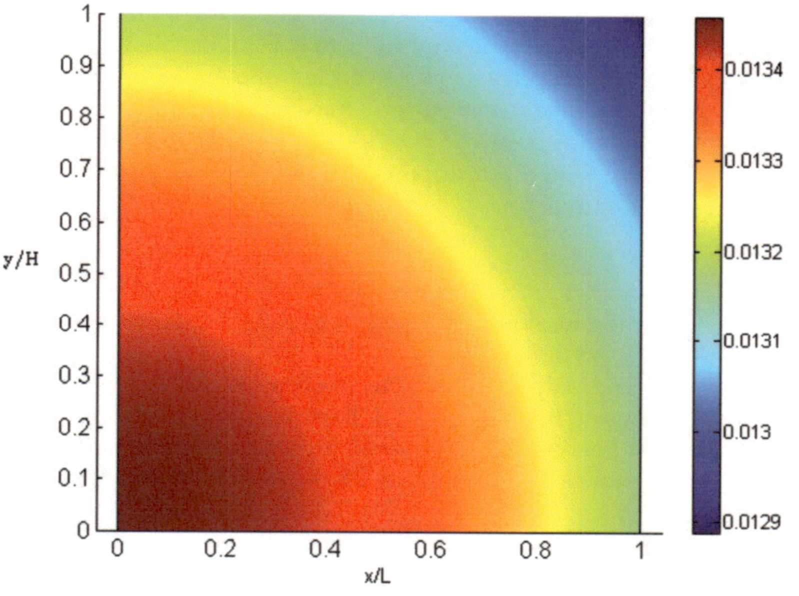

Fig. 18. Dimensionless concentration profile at the end of discharge.

5. Conclusion

Electrochemical properties of electrodes based on CH900-20 activated carbon (AC) cloth were studied in concentrated H_2SO_4 solutions in a wide range of potentials from -0.8 to +1 V RHE. Cyclic voltammetric curves measured in two ranges of potential were studied: in the reversibility range (from 0.1 to 0.9 V) and in the deep cathodic charging range (from -0.8 to +1 V). Electric double electric layer (EDL) charging occurs in the reversibility range, while faradaic processes of hydrogen chemosorption (at the interface of carbon and electrolyte in pores) and its intercalation into AC carbon takes place in the range of negative potentials (< -0.1 V). The intercalation process is controlled by slow solid-phase hydrogen diffusion into AC carbon. For the first time, the maximum value of specific discharge capacitance of 1560 C/g was obtained, which is much higher than the values known from the literature for carbon electrodes. On the basis of this value and Faraday's law, it was assumed that the compound of C_6H is formed in the limiting case of AC deep cathodic charging. The specific charge value grows at an increase in the concentration of H_2SO_4 and also at an increase in the charging time.

The obtained experimental data were interpreted by a mechanism of double inrercalation under AC deep cathodic charging. Sulfuric acid is intercalated into AC expanding the interlayer (intergraphene) space. Hydrogen atoms are then directed into this space. This interlayer space serves as a transport route for hydrogen. Then hydrogen interacts with graphene layers with formation in the limit of compound C_6H. The data obtained were used to develop a mathematical charging-discharge model for an AC electrode taking into account the EDL charging, chemosorption, and hydrogen intercalation.

The data obtained in this work according to which at deep cathodic charging of activated carbon based electrodes to potential values from –0.1 to –0.5 V RHE very high discharge capacities can be achieved point to high prospects of such a use of these electrodes in energy type supercapacitors. At the same time they allow to explain the high specific energy values (up to 20 Wh/kg) observed ten years ago for hybride $(+)PbO_2/H_2S)_4/AC(-)$ supercapacitors [17,18]. These high values were achieved at a maximal charging voltage U_{max} of 2.2 V. Taking into account that for the positive PbO_2 electrode E^{+}_{max} =1.85 V RHE and therefore the value of E^{-}_{min} was –0.35 V RHE corresponding to the region of maximal capacity for activated carbon based electrodes.

6. References

[1] B.E. Conway. Electrochemical supercapacitors. Kluwer Academic / Plenum Publishers, New York. 1999. 698 p.
[2] Yu.M. Volfkovich, T.M. Serdyuk. Russ. J. Electrochem. 38 (2002) 935.
[3] I.V. Barsukov, C. Johnson, E. Doninger, V.Z. Barsukov, New Carbon Based Materials for Electrochemical Energy Storage Systems: Batteries, Supercapacitors and Fuel Cells (NATO Science Series II: Mathematics, Physics and Chemistry), Springer, NY, 2006. 297 p.
[4] R. Kotz, M. Carlen. Electrochimica Acta 45 (2000) 2483.
[5] A.G. Pandolfo, A.F. Hollenkamp. J. Power Sources 157 (2006) 11.
[6] P. Simon, Y. Gogotsi. Nature Materials 7 (2008) 845.
[7] Y. Chen, X. Zhang, P. Yu, Y. Ma. J. Power Sources 195 (2010) 3031.
[8] Y. Chen, X. Zhang, D. Zhang, P. Yu, Y. Ma. Carbon 49 (2011) 573.
[9] W. Lu, L. Qu, K. Henry, L. Dai. J. Power Sources 189 (2009) 1270.
[10] M.D. Stoller, S. Park, Z. Yanwu, J. An, R.S. Ruoff. Nano Letters 8 (2008) 3498.
[11] R.C. Vivekchand, C.S. Rout, K.S. Subrahmanyam, A. Govindaraj, C.N.R. Rao. J. Chem. Sci. 120 (2008) 9.
[12] H. Zhang, G. Cao, Y. Yang, Z. Gu. J. Electrochem. Soc. 155 (2008) K.19.
[13] B.P. Bakhmatyuk, B.Y. Venhryn, I.I. Grygorchak, M.M. Micov, Y. Kulyk. Electrochimica Acta 52 (2007) 6604.
[14] S. Lidorenko. Doklady Akademii Nauk USSR . 216 (1974) 1261.
[15] A.I. Beliakov., A.M. Brintsev. Proc. 7th Int. Seminar on Double Layer Capacitors and Similar Energy Storage Devices. Deerfield Beach. Florida. 1997. V. 7.
[16] A.I. Beliakov. Proc. 8th Int. Seminar on Double Layer Capacitors and Similar Energy Storage Devices. Deerfield Beach. Florida. 1998. V. 8.
[17] Belyakov, Yu.M. Volfkovich, P.A. Shmatko at al. US Patent 6,195,252 B1 (2001).
[18] Yu.M. Volfkovich, P.A. Shmatko. US Patent 6,628,504 (2003).

[19] J. P.Zheng, S. P.Ding, T. R. Jow. Proc. 7th Int. Seminar on Double Layer Capacitors and Similar Energy Storage Devices. Deerfield Beach. Florida. 1997.
[20] B. Fang, L. Binder. J. Power Sources, 163 (2006) 616.
[21] T. Centeno, F. Stoeckli, Electrochimica Acta 52 (2006) 560.
[22] M.J. Bleda-Martinez, J.A. I. Agull, D. Lozano-Caste, E. Morall, D. Cazorla-Amor, A. Linares-Solano. Carbon 43 (2005) 2677.
[23] Yu.M. Volfkovich, A.Yu. Rychagov, V.E.Sosenkin, A.V. Krestinin. Elektrokhimicheskaya Energetika. 8 (2008) 106.
[24] M.Yu. Izmailova, A.Yu. Rychagov, K.K. Den'shchikov, Yu.M. Volfkovich, Ya.S. Vygodskii, E.I. Lozinskaya. Russ. J. Electrochem., 45 (2009) 949.
[25] A.Yu. Rychagov, Yu.M. Volfkovich. Russ. J. Electrochem., 45 (2009) 304.
[26] Yu.M. Volfkovich, V.M. Mazin, N.A. Urisson. Russ. J. Electrochem., 34 (1998) 740.
[27] Yu.M. Volfkovich, O.A. Petrii, A.A. Zaitsev, I.V. Kovrigina, Vestnik MGU. Ser. 2, Khimiya. 29 (1988) 173.
[28] Yu. M. Volfkovich, V. S. Bagotzky, T. K. Zolotova and E. Yu. Pisarevskaya. Electrochimica Acta 41 (1996) 1905.
[29] Yu. M. Volfkovich, A. G. Sergeev, T. K. Zolotova, S.D. Afanasiev, O.N. Efimov, E.P. Krinichnaya. Electrochimica Acta, 44 (1999) 1543.
[30] S. Fernandez, E.B. Cartro, S.G. Real, M.E. Martines. Internat. J. Hydrogen Energy, 34 (2009) 8115.
[31] B.P. Bakhmatyuk, B.Ya. Venhryn, I.I. Grigorchak, m.M. Micov, Yu.O.Kulik. Electrochimica Acta, 52 (2007) 6604.
[32] Yu.M. Volfkovich, V.S. Bagotzky, V.E. Sosenkin, I.A. Blinov. In: Colloid and Surfaces A: Physicochemical and Engineering Aspects. 187-188 (2001) 349.
[33] A.N. Frumkin, V.S. Bagotskii, Z.A. Iofa, B.N. Kabanov. Kinetika elektrodnykh protsessv (Kinetics of Electrode Processes). Moscow: Izd. MSU, 1952.
[34] M.R. Tarasevich. Elektrokhimiya uglerodnykh materialov (Electrochemistry of Carbon Materials). Moscow.: Nauka. 1984. P. 251.
[35] A.R. Ubellode, F.A. L'yuis. Grafit I ego kristallicheskie soedineniya (Graphite and Its Crystalline Compounds). Moscow: Mir, 1965. 256 pp.
[36] A.S. Fialkov. Uglerod, mezhsloevye soedineniya i kompozity na ego osnove (Carbon, Interlayer Compounds and Composites on Its Basis). Moscow: Aspect press, 1997.
[37] N.E. Sorokina, I.V. Nikol'skaya, S.G. Ionov, V.V. Avdeev, Izv RAS, Ser. Khim. 54 (2005) 1.
[38] V. Yu. Yakovlev, A.A.Fomkin, A.V. Tvardovski. J. Colloid Interface Sci. 280 (2004) 305.
[39] A.Yu. Rychagov, N.A. Urisson, Yu.M. Volfkovich. Russ. J. Electrochem. 37 (2001) 1172.
[40] A.Yu. Rychagov, Yu.M. Volfkovich, Russ.J. Electrochem., 43 (2007) 1343.
[41] V.B. Fenelonov. Poristyi uglerod (Porous Carbon). Novosibirsk. 1995.
[42] P. Delahay. Double layer and electrode kinetics. John Wiley & sons, Inc., New York. 1965. 340 p.

[43] Yu.A. Chizmadzhev, V.S. Markin, M.R. Tarasevich, Yu.G. Chirkov. Makrokinetika protsessov v poristykh sredakh (Macrokinetics of Processes in Porous Media). Moscow: Nauka, 1971. 364 p.

8

Water Management and Experimental Diagnostics in Polymer Electrolyte Fuel Cell

Kosuke Nishida[1], Shohji Tsushima[2] and Shuichiro Hirai[2]
[1]*Department of Mechanical and System Engineering, Kyoto Institute of Technology*
[2]*Department of Mechanical and Control Engineering, Tokyo Institute of Technology*
Japan

1. Introduction

Polymer electrolyte fuel cell (PEFC) is a promising candidate for mobile and vehicle applications and distributed power systems because of its high power density and low operating temperature. However, there are several technical problems to be solved in order to achieve practicability and popularization. Especially, water management inside a PEFC is essential for high performance operation. At high current densities, excessive water generated by the electrode reaction is rapidly condensed in the cathode electrode. When the open pores in the catalyst layer (CL) and gas diffusion layer (GDL) are filled with liquid water, oxygen cannot be supplied to the reaction sites. Furthermore, water migrates significantly through the electrolyte membrane from the anode to cathode owing to electro-osmotic drag. Thus, the membrane dehydration occurs mainly on the anode side and causes the low proton conductivity during low-humidity operation. These phenomena known as "water flooding" and "dryout" are a critical barrier for high efficiency and high power density. To alleviate these issues, it is necessary to develop various diagnostic tools for understanding the fundamental phenomena of water transport between cathode and anode in PEFC.

Experimental approaches to probe water transport in PEFCs have been attempted in the previous studies. Liquid water formation, transport and removal in cathode flow channel and GDL were investigated by neutron radiography (Bellows et al., 1999; Satija et al., 2004; Kramer et al., 2005; J. Zhang et al., 2006; Turhan et al., 2006; Hickner et al., 2006; Yoshizawa et al., 2008), soft X-ray radiography (Sasabe et al., 2010), X-ray computed tomography (Lee et al., 2008), and optical visualization using transparent fuel cell (Tüber et al., 2003; Yang et al., 2004; F.Y. Zhang et al., 2006; Nishida et al., 2010a). Water content distribution in polymer electrolyte membrane (PEM) was measured by using magnetic resonance imaging (MRI) (Tsushima et al., 2004). Although various diagnostic techniques were developed as mentioned above, there are few experimental efforts to measure water distribution inside cathode GDL because of the difficulty in observing internal microstructure of opaque porous layer. Boillat et al. (Boillat et al., 2008) resolved the water distribution between the different layers of the membrane electrode assembly (MEA) in an operating PEFC using high-resolution neutron radiography. Sinha et al. (Sinha et al., 2006) have explored the possibility of using X-ray micro-tomography to quantify liquid water distribution along the

GDL thickness of a PEFC. Litster et al. (Litster et al., 2006) developed the fluorescence microscopy technique for visualizing liquid water in hydrophobic fibrous media, and applied to ex-situ measurement of water transport in a GDL. To predict two-phase flow across cathode GDL in PEFCs, numerical simulations have been also performed by many researchers. Wang et al. (Wang et al., 2001) applied a two-phase flow model based on computational fluid dynamics (CFD) to the air cathode of PEFC with a hydrophilic GDL. He et al. (He et al., 2000) and Natarajan and Nguyen (Natarajan & Nguyen, 2001) proposed two-dimensional two-phase models for PEFCs with interdigitated and conventional flow fields, respectively. Subsequently, Pasaogullari and Wang (Pasaogullari & Wang, 2004a) developed a theory describing liquid water transport in hydrophobic GDL, and explored the effect of GDL wettability on liquid water transport. Recently, Sinha and Wang (Sinha & Wang, 2007, 2008), Gostick et al. (Gostick et al., 2007) and Rebai and Prat (Rebai & Prat, 2009) have developed a pore-network model to understand the liquid water transport in a hydrophobic GDL with the GDL morphology taken into account.

This chapter introduces several novel measurement techniques for evaluating the water transport inside a PEFC (Nishida et al., 2009, 2010a, 2010b). In section 3, the experimental method for quantitatively estimating the liquid water content in the cathode gas diffusion electrode (GDE) is presented based on the weight measurement (Nishida et al., 2010a). Furthermore, the visualization tool to probe the liquid water behavior at the cathode is provided by using an optical diagnostic, and the influences of operating condition and GDL properties on the water transport through the porous electrode are discussed. Under high current density conditions, water flooding occurs significantly at the interface between cathode CL and GDL. Section 4 presents the ex-situ measurement method for evaluating the amount of liquid water accumulated at the cathode CL|GDL interface using near-infrared reflectance spectroscopy (NIRS) (Nishida et al., 2010b). NIRS is a non-invasive optical technique for quantitatively estimating the amount and concentration of water, and make it possible to determine the thickness of the liquid water film attached to the cathode CL surface after fuel cell operation. In this section, the effects of GDL hydrophobicity and microporous layer (MPL) addition on the water accumulation at the cathode CL|GDL interface are investigated. During low-humidity operation, water management on anode side is essential for achieving sufficient membrane hydration and high proton conductivity. In section 5, the imaging technique to observe the water distribution in the anode flow field of an operating PEFC is provided using water sensitive paper (WSP) (Nishida et al., 2009). WSP is a test paper for detecting water droplets, fog and high humidity. This paper is inserted into the transparent fuel cell, and makes it possible to visualize the water condensation process in the anode flow channel under low-humidity PEFC operation. To achieve better water management and alleviate membrane dryout in PEFC, the optimum operating condition is explored based on the WSP measurement. Finally, in section 6, the main conclusions derived from the present work are summarized.

2. Water transport in PEFC

Water transport in PEFC is extremely complex, and has an important impact on cell performance. Figure 1 schematically shows a cross-sectional view of PEFC and its water transport processes. A PEM film coated with CLs on both sides is sandwiched between two hydrophobic GDLs. Hydrogen and oxygen as fuel and oxidant are supplied to the anode and cathode sides, respectively. Hydrogen gas diffuses through the anode GDL to the active reaction sites inside the CL. At the anode CL, hydrogen dissociates into protons (H^+) and electrons. Protons migrate through the electrolyte membrane to the cathode electrode. On the cathode side, protons combine with electrons and oxygen, and produce water. The external flow of electrons can be utilized for electric power. At high current operations, excessive water is generated in the cathode CL. A small part of the product water is reversely diffused through the membrane from the cathode to anode, and the electrolyte membrane is hydrated. On the other hand, most of the product water is condensed and accumulated inside the cathode CL and GDL. If open pores in the CL and GDL are filled with liquid water, or if the gas channels are clogged by liquid water, oxygen transport to the reaction sites is hindered. These phenomena known as "water flooding" and "plugging" are an important limiting factor for PEFC performance. Furthermore, water is transported from the anode to cathode in the PEM by electro-osmotic effect. Thus, under low-humidity conditions, the membrane dehydration proceeds mainly on the anode side and the proton conductivity declines. This anode dryout causes a substantial drop in cell voltage, resulting in not only temporary power loss but also cell degradation. In operating PEFC, transversal water distribution in MEA is complicatedly determined as a result of coupled processes including water generation, evaporation, condensation, back-diffusion, electro-osmotic and interfacial mass transfer. To achieve proper water management and improve cell performance, it is necessary to obtain the fundamental understandings of water transport inside fuel cell.

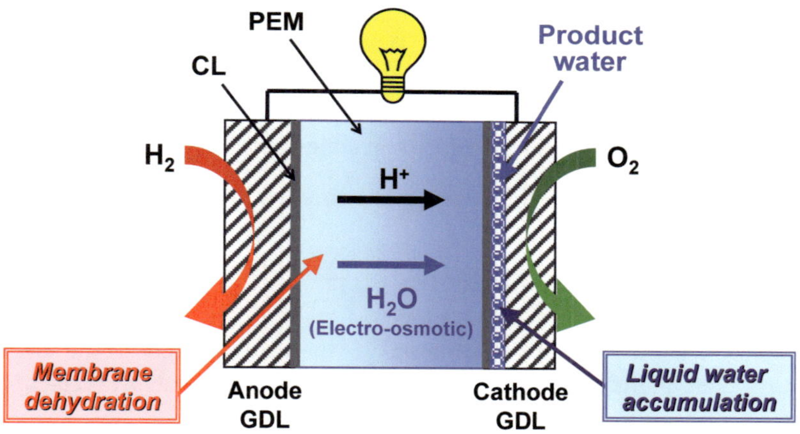

Fig. 1. Cross-sectional view of PEFC and its water transport processes

3. Measurement of liquid water content in cathode gas diffusion electrode

This section presents a novel method for quantitatively estimating the average liquid water content inside the cathode gas diffusion electrode (GDE) of a PEFC based on the weight measurement (Nishida et al., 2010a). In addition, the liquid water behavior at the cathode during cell operation is visualized using an optical diagnostic, and the influences of current density and GDL thickness on the water transport through the cathode GDE are also discussed.

3.1 Estimation method for liquid water content in cathode GDE

By measuring the weights of liquid water accumulated in the MEA and PEM, the average liquid water content in the cathode GDE of an operating cell can be predicted. The cathode GDE structurally consists of a cathode GDL and CL. In this experiment, the time-series data of cell voltage is also monitored during cell operation to investigate the relationship between the water accumulation in the cathode GDE and the voltage change. However, the evaluation of liquid water content in the GDE and the sequential monitor of cell voltage are conducted at difference times, because the assembled cell must be composed in measuring the weight of liquid water in the MEA.

The average liquid water content in the cathode GDE, X_W, is defined as the averaged volume fraction of liquid water in the porous media, and given by

$$X_W = \frac{\left(\Delta m_{MEA} - \Delta m_{PEM}\right) \cdot v_W}{V_{GDL}} \times 100 \tag{1}$$

where Δm_{MEA} and Δm_{PEM} are the weight increases of the MEA and PEM due to the liquid water generation, respectively. v_W denotes the specific volume of liquid water, and V_{GDL} the pore volume inside the cathode GDL. The MEA used in this experiment is constructed of a PEM film and two GDEs including CL. In this estimation, the liquid water volume in the CLs is neglected because the thickness of CL is very thin and the pore volume is extremely small. Furthermore, since dry hydrogen is supplied to the anode side without humidification, the water condensation in the anode GDE hardly occurs. The water influx to the anode is only due to the back diffusion through the membrane. Therefore, the liquid water accumulation in the anode GDE can be also ignored. Under these assumption, the average liquid water content in the cathode GDE including the CL is described by Equation (1). Δm_{MEA} is given by measuring the weights of the MEA experimentally before and after operation test.

The weight of the PEM, m_{PEM}, in Equation (1) is estimated by

$$m_{PEM} = m_{dry}\left(\frac{18\lambda}{EW} + 1\right) \tag{2}$$

where m_{dry} is the weight of the dry membrane, λ the water content in the membrane, and EW the equivalent weight of the dry membrane. The water content, λ, is calculated by

$$\lambda = \frac{\sigma}{0.005139} \exp\left[1268\left(\frac{1}{273+T} - \frac{1}{303}\right)\right] + 0.63436 \tag{3}$$

where σ is the membrane conductivity and T is the cell temperature. This equation was empirically obtained from measuring the membrane water content and conductivity under a range of water vapor activities at 30°C (Springer et al., 2005). The membrane conductivity, σ, is also given by Equation (4)

$$\sigma = \frac{t_{PEM}}{R_{PEM} \cdot A} \tag{4}$$

where t_{PEM} is the membrane thickness, R_{PEM} the membrane resistance, and A the electrode reaction area. R_{PEM} is measured by using AC impedance method.

3.2 Experimental

Figure 2 shows the experimental setup used for PEFC operation test, which consists of a constant temperature chamber, a gas supply unit, a high-resolution digital CCD camera, a transparent fuel cell, an electronic load, a data logger, and a personal computer. To directly observe the liquid water behavior at the cathode electrode, a quartz glass is inserted into the experimental fuel cell as a window. The experimental cell equipped with the transparent window is operated in the constant temperature chamber in order to maintain the cell temperature. The digital CCD camera with the zoom lens for optical visualization is set outside of the constant temperature chamber, and the working distance from the cathode electrode of the transparent fuel cell is adjusted. The cathode flow field is illuminated by a halogen light source and the close-up images of the GDL surface can be clearly captured. The time-series output voltage and temperature of the operating fuel cell are recorded by the data logger. The cell temperature is measured using a thermocouple. The high frequency resistance (HFR) of the PEM is also measured by the LCR meter.

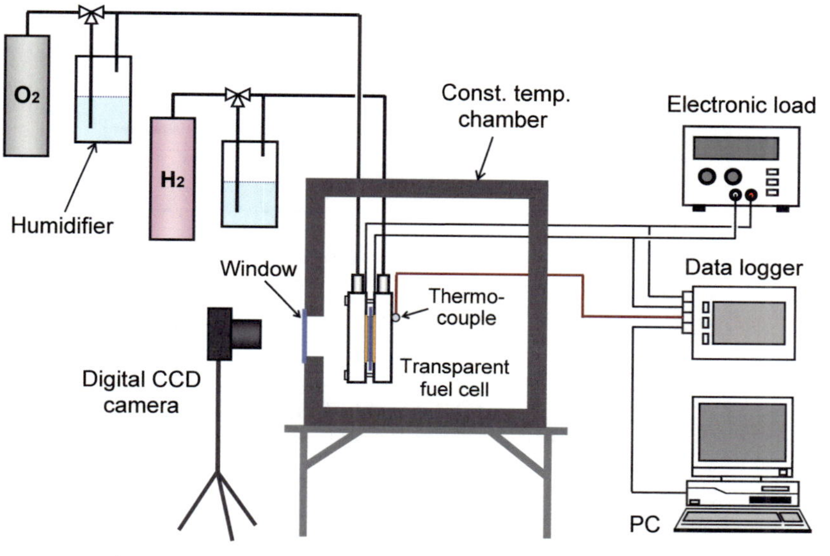

Fig. 2. Experimental setup used for PEFC operation test

The schematic diagram and photograph of the transparent fuel cell are shown in Figure 3. The catalyst coated membrane (CCM) on which Pt particles are loaded is sandwiched between two PTFE-proofed GDLs. In addition, the MEA constructed of the PEM, two CLs and two GDLs is sandwiched between two copper current collector plates with gold coating. The active area of the experimental cell used in this study is 5 cm^2. Two stainless steel separators which have a single-pass serpentine flow channel are placed outside the current collectors and held together by four M6 bolts. The width, depth and overall length of the serpentine channel are 2 mm, 2 mm, and 10.5 cm, respectively. In this experiment, dry hydrogen and oxygen are fed into the anode and cathode channels at constant flow rates without humidification.

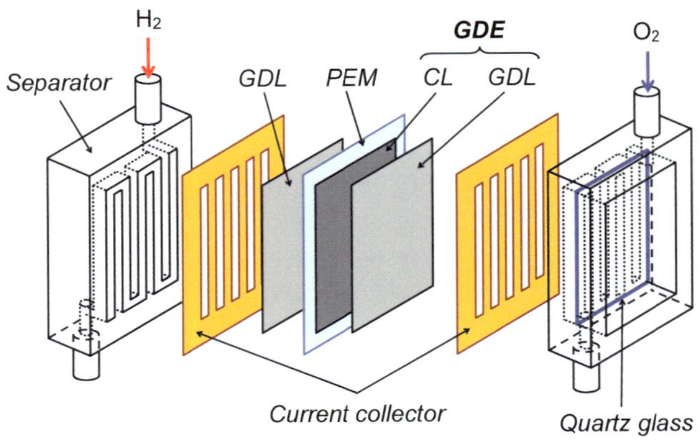

(a) Cell structure

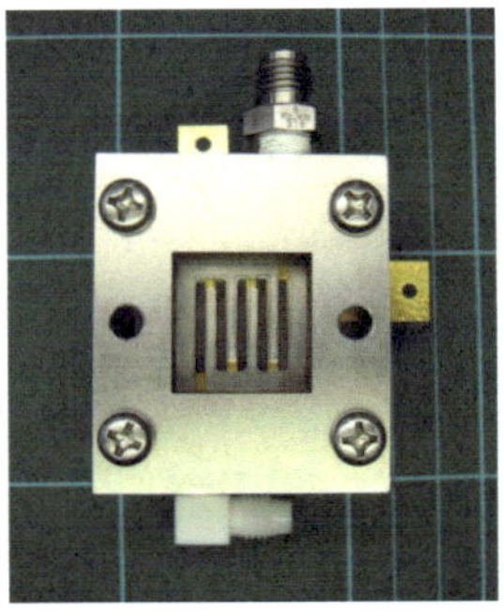

(b) Photograph

Fig. 3. Shematic diagram and photograph of the transparent fuel cell

Figure 4 shows the experimental procedure for estimating the liquid water content in the cathode GDE. Beforehand, the weight of the dry MEA is measured by an electronic balance. Furthermore, the HFR of the electrolyte membrane is also measured by the LCR meter, and the water content in the PEM is predicted. Subsequently, the pre-operation of the experimental fuel cell is carried out at 0.16 A/cm^2 for 2 hours in order to hydrate the

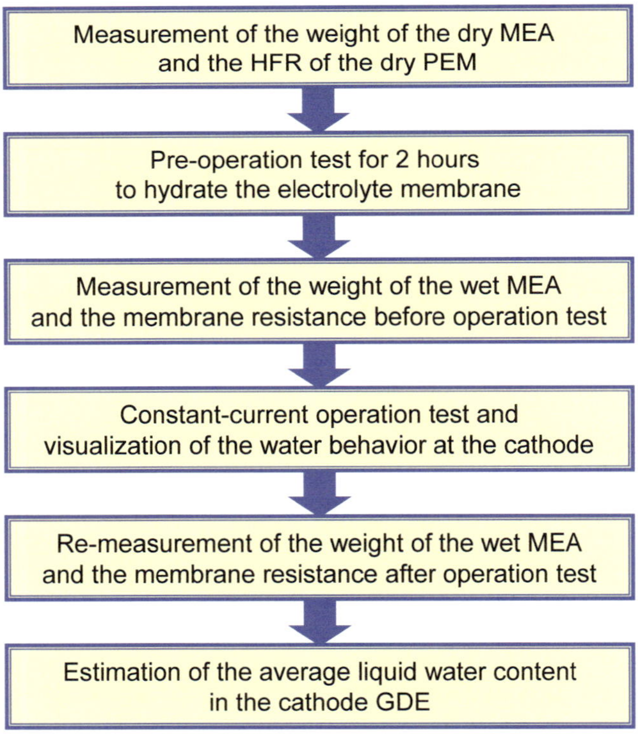

Fig. 4. Experimental procedure for estimating the liquid water content in the cathode GDE

electrolyte membrane. After the pre-operation, the assembled fuel cell is decomposed into the MEA, current collectors and separators, and the weight of the wet MEA and the membrane resistance are measured. The adherent liquid droplets on the GDL surface are wiped away before the measurement. Then the MEA is slowly dried until the liquid water weight in the cathode GDE is adjusted to the initial state. The liquid water weight in the cathode GDE is given by subtracting Δm_{PEM} from Δm_{MEA}. Following the membrane hydration, the constant-current operation test is conducted, and the liquid water behavior at the cathode is directly visualized by using the digital CCD camera. The cell voltage during operation is also monitored. Finally, after the operation test, the experimental cell is decomposed and the weight of the wiped MEA and the membrane resistance are measured again. The average liquid water content in the cathode GDE is quantitatively evaluated by Equation (1).

In order to investigate the relationship between the water content in the cathode GDE and the operation time, the operation test and the estimation of water content need to be repeated again and again because the assembled cell must be decomposed in measuring the weight of water in the MEA.

3.3 Cell voltage vs. liquid water content in cathode GDE

PEFC performance is largely influenced by liquid water accumulation in cathode GDL and CL. The characteristics of the cell voltage and the average liquid water content in the cathode GDE are shown in Figure 5. The GDL used for this experiment is Toray TGP-H-120 carbon paper (360 μm thick). The current density as an operating parameter is set to 0.16, 0.24 and 0.3 A/cm^2. In all experiments, the fuel cell is held under the open circuit condition for the first 50 s. Figure 5(a) presents the voltage change during startup operation at 20°C. At the low current density of 0.16 A/cm^2, the fuel cell operates stably for 2000 s though the cell voltage decreases a little. When the current density increases up to 0.3 A/cm^2, the sudden voltage drop occurs immediately after starting the operation.

The relationships between the average liquid water content in the cathode GDE and the operation time are plotted in Figure 5(b). Since the assembled cell must be decomposed in measuring the weight of liquid water in the MEA, these plots were obtained from many different operation tests. In the case of 0.16 A/cm^2, the liquid water content in the cathode GDE increases rapidly up to approximately 15% for 500 s after starting the operation. After t=500 s, the rate of increase of the water content slows down because the liquid water accumulated in the GDL is drained to the flow channel. When the current density increases, the rate of increase of the water content in the GDE increases due to the production of much water. In the case of 0.24 and 0.3 A/cm^2, the cell voltages reduce to zero at t=400 and 200 s, though the water contents reach only 16 and 12%, respectively. It can be considered that the oxygen transport through the GDL is not limited because of low water content. Therefore, these sudden voltage drops are probably due to the mass transfer limitation within the CL. If most of the cathode CL is covered with the condensed water, oxygen cannot be sufficiently supplied to the reaction sites and the concentration overpotential is remarkably increased. The amount of liquid water accumulated in the CL tends to increase with an increase in current density.

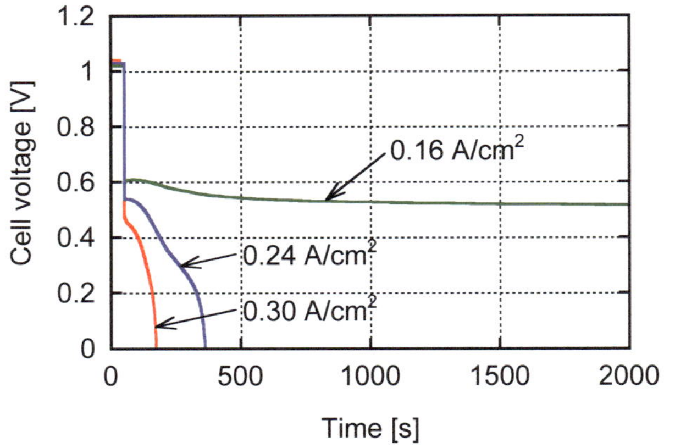

(a) Cell voltage

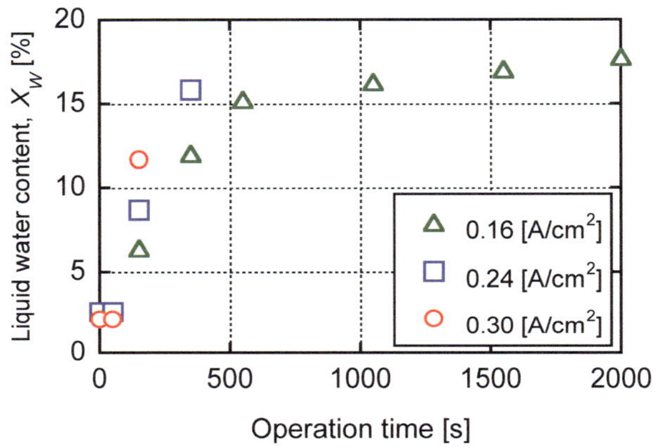

(b) Water content in cathode GDE

Fig. 5. Comparison of the cell voltage and average liquid water content in the cathode GDE at 0.16, 0.24 and 0.3 A/cm2

3.4 Visualization image of liquid water behavior at cathode

Figure 6 shows the sequential images of liquid water behavior on the cathode GDL at the current density of 0.16 A/cm2. The cathode gas flows from the upper right to the lower left in the serpentine flow field which has five straight channels. Liquid water is hardly drained from the cathode GDE for 300 s after starting the operation. However, a few liquid droplets accumulated in the cathode GDE appear on the electrode surface at t=500 s. These droplets on the GDL surface grow and the number of water droplets increases after 500 s of operation. It is noted that the gradual increase of the water content in the cathode GDE after t=500 s shown in Figure 5(b) is due to the liquid water removal from the GDE.

3.5 Effect of GDL thickness on liquid water accumulation in cathode GDE

Liquid water transport in cathode GDE is also affected by thickness and porous structure of GDL. Figure 7 presents the effect of GDL thickness on the cell voltage and liquid water

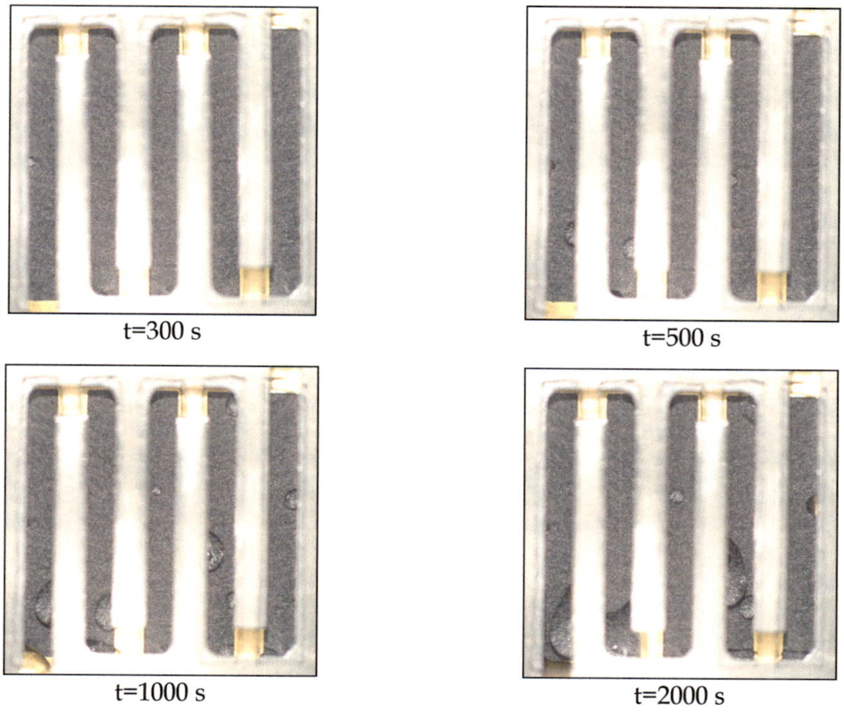

Fig. 6. Visualization images of liquid water behavior on the cathode GDL at 0.16 A/cm^2

accumulation in the cathode GDE at 0.24 A/cm^2. In this experiment, two different thickness paper-type GDLs of Toray TGP-H-060 and TGP-H-120 are used. The properties of these GDLs are described in Table 1. The thickness of TGP-H-060 is approximately half of that of TGP-H-120. Figure 7(a) shows the voltage changes for two different GDLs during startup at 20°C. In the case of thick GDL (TGP-H-120), the cell voltage drops suddenly after starting the fuel cell operation. On the other hand, the cell voltage for the thin GDL (TGP-H-060) case remains constant for 8000 s, because liquid water in the cathode CL and GDL is smoothly removed and oxygen is stably supplied to the reaction sites.

	TGP-H-060	TGP-H-120
Type	Carbon paper	Carbon paper
Thickness	190 μm	360 μm
Porosity	0.78	0.78

Table 1. Properties of GDLs

The weights of liquid water accumulated in the cathode GDE in both cases are plotted in Figure 7(b). The rate of the water accumulation for the thin GDL (TGP-H-060) is slower than that for the thick GDL (TGP-H-120). This is because the liquid water inside cathode GDE is quickly drained to the flow channel after starting the operation in the case of thin GDL. In the case of thick GDL, the cell voltage decreases to zero at t=400 s though the weight of liquid water in the GDE reaches only 21 mg. The weight of water of 21 mg is equivalent to the low water volume fraction of 15.8% in the GDE. This result indicates that most of the product water tends to remain inside the cathode CL in the case of thick GDL.

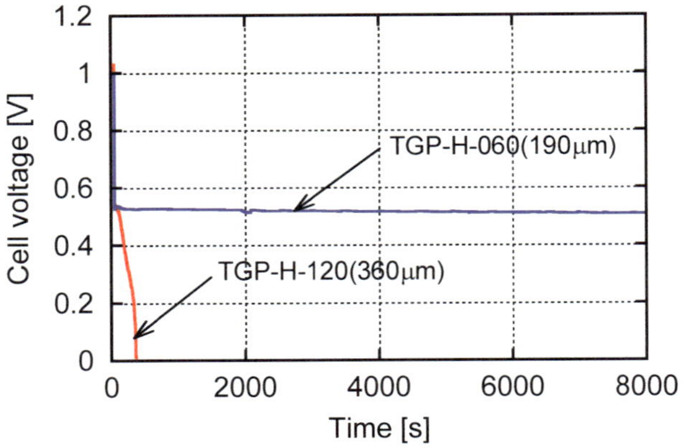

(a) Cell voltage

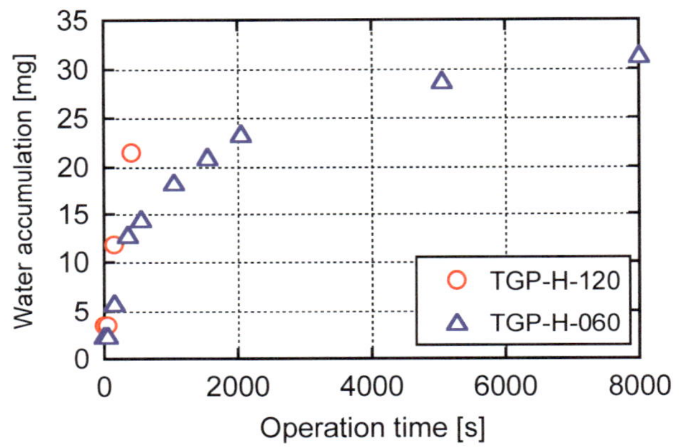

(b) Water accumulation in cathode GDE

Fig. 7. Effect of GDL thickness on the cell voltage and liquid water accumulation in the cathode GDE at 0.24 A/cm^2

4. Quantitative evaluation of liquid water at cathode CL|GDL interface using near-infrared reflectance spectroscopy

The significant performance loss at high current densities is attribute to severe oxygen transport limitation inccured by water flooding at the interface between cathode CL and GDL. This section presents a diagnostic method for quantitatively evaluating the amount of liquid water accumulated at cathode CL|GDL interface of a PEFC using near-infrared reflectance spectroscopy (NIRS) (Nishida et al., 2010b). In this measurement, the effects of GDL hydrophobicity and MPL addition on the water accumulation at the CL|GDL interface are investigated.

4.1 Near-infrared reflectance spectroscopy (NIRS)

Near-infrared reflectance spectroscopy (NIRS) is an absorption spectroscopy technique for quantitatively estimating the amount of water based on Lambert-Beer's law. Figure 8 shows the fiber-optic NIRS measurement system used in this study. This NIRS system consists of a tungsten light source, optical filter, optical fiber, PbS detector, sampling circuit and analog computing circuit. The incident light emitted from the tungsten light source is separated into the measuring light (wavelength: λ_1=1.94 μm) and two reference lights (wavelength: λ_2, λ_3=1.8, 2.1 μm) by the optical filter. Each light passes through the optical fiber and illuminates the cathode CL surface of the catalyst coated membrane (CCM) removed from a PEFC. The reflected lights from the CL surface are conducted along the optic cable to the PbS detector and converted the electrical signals. The output value, α, from the NIRS system is expressed by Equation (5)

$$\alpha = K \cdot \log \frac{E_2 + E_3}{2E_1} \tag{5}$$

where K is gain constant. E_1, E_2 and E_3 are the energies of the measuring light and two reference lights, respectively. Since the reference lights are also received by the detector, the impacts of light collection efficiency and environmental changes on the measurement results can be neglected. When the measuring light travels through the liquid water film attached to the CL surface, the light intensity decreases exponentially. The thickness of the water film at the cathode CL|GDL interface is quantitatively obtained from the absorbance of the measuring light.

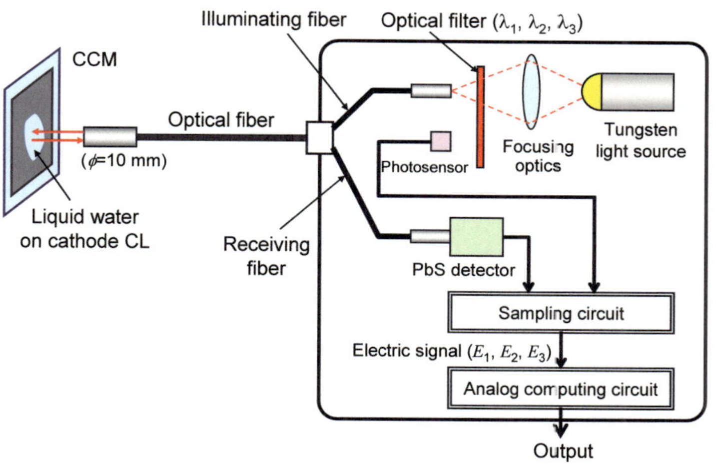

Fig. 8. Fiber-optic NIRS measurement system

For near-infrared (NIR) absorption spectroscopy of liquid water, it is essential to select a strong absorption band. Figure 9 presents the absorption coefficients of liquid water at 20°C in the wavelength range from 0.7 to 2.5 μm (Curcio & Petty, 1951). There are five prominent water absorption bands in the NIR which occur at 0.76, 0.97, 1.19, 1.45 and 1.94 μm. In the NIRS system, the maximum absorption band at 1.94 μm is used for the measuring light.

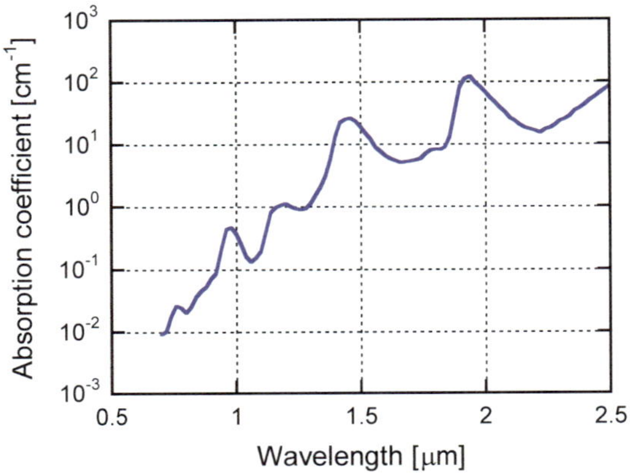

Fig. 9. Absorption coefficients of liquid water at 20°C in the wavelength range of 0.7-2.5 μm (Curcio & Petty, 1951)

4.2 Experimental

The calibration must be carried out beforehand to obtain the amount of liquid water attached to the cathode CL from the absorbance of the NIR light. To establish the calibration curve, the liquid water accumulation on the cathode CL surface is quantitatively estimate by the weight measurement. The average thickness of liquid water film on the CL surface, t (μm), is calculated by Equation (6)

$$t = \frac{w_{wet} - w_{dry} - w_{liq}}{\rho A} \times 10^4 \quad (6)$$

where w_{wet} is the weight (g) of the CCM with a wet cathode surface, w_{dry} is the weight (g) of the dry CCM, w_{liq} the weight (g) of liquid water that penetrates into the CCM, ρ the density of liquid water (g/cm^3), and A the electrode area (cm^2). Figure 10 presents the calibration curve for the conversion of the measurement value, α to the average thickness of liquid water on the CL surface. The calibration curve is formulated as

$$\alpha = -0.095216t^2 + 8.0483t + 871.27 \quad (7)$$

The structure of the experimental fuel cell used in this measurement is the same as that in Figure 3. The CCM on which platinum particles (0.5 mg/cm^2) are loaded as a catalyst layer is sandwiched between two PTFE-proofed paper-type GDLs (Toray TGP-H-060). After cell operation, the wet CCM is removed from the assembled cell, and the amount of liquid water on the cathode CL surface is measured using the fiber-optic NIRS system. Figure 11 shows four measurement positions of NIRS on the cathode CL surface of the CCM. The cathode gas flows from the upper right to the lower left through the serpentine channel. Thus, position (1) and (2) are located in the upstream section of the cathode flow field, and (3) and (4) are located in the downstream.

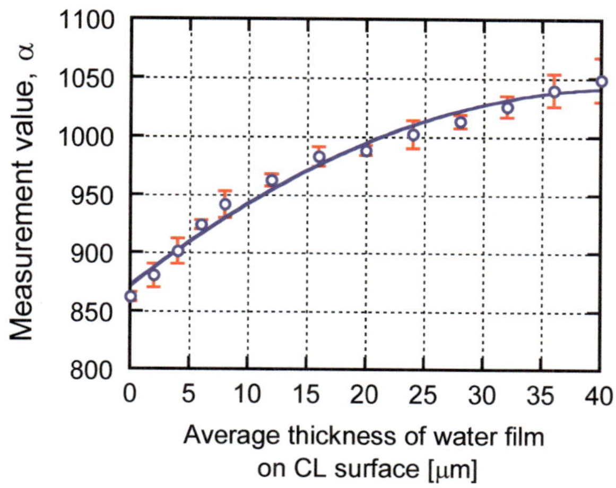

Fig. 10. Calibration curve for the conversion of the measurement value, α to the average thickness of liquid water on the CL surface

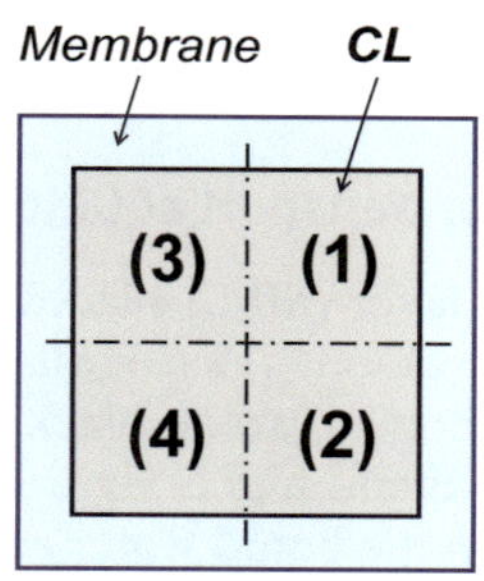

Fig. 11. Measurement positions of the fiber-optic NIRS system on the cathode CL surface

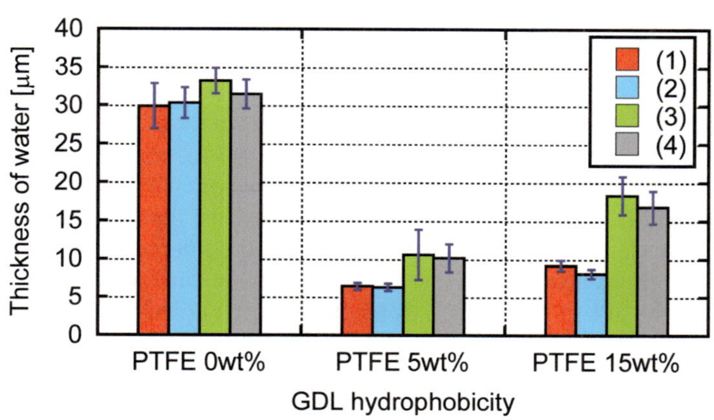

Fig. 12. Effect of GDL hydrophobicity on liquid water accumulation on the cathode CL surface

4.3 Effect of GDL hydrophobicity on water transport at cathode CL|GDL interface

Liquid water accumulation at cathode CL|GDL interface is largely affected by GDL hydrophobicity. In this subsection, the effect of PTFE content of GDL on the water accumulation at the cathode interface is investigated. Figure 12 presents the thickness of liquid water film attached to the cathode CL surface for hydrophilic (PTFE content: 0 wt%) and hydrophobic (5 and 15 wt%) GDLs. The water measurements in both cases were carried out using the fiber-optic NIRS system shown in Figure 8 after fuel cell operation. The experimental fuel cell is operated at 20°C and 0.3 A/cm^2 for 60 min. In each case, many operation tests were repeated to confirm the reproducibility of the NIRS results. Each bar graph was obtained by averaging several experimental data, and the error bars were provided. (1), (2), (3) and (4) denote the NIRS measurement positions shown in Figure 11. In the case of hydrophilic GDL (0 wt% PTFE), the thickness of water film reaches more than 30 μm in all positions in spite of the short operation time. When the thickness of liquid film accumulated at the cathode CL|GDL interface increases beyond 30 μm, oxygen gas cannot be sufficiently supplied to the reaction sites, and the fuel cell operates unstably. The amount of adherent water for the hydrophobic GDLs (5 and 15 wt% PTFE) is much less than that for the hydrophilic GDL. The hydrophobic treatment of GDL is effective in alleviating liquid water accumulation at cathode CL|GDL interface. However, the thickness of water film on the CL surface at 15 wt% PTFE is thicker than that at 5 wt%. This is probably because the highly hydrophobic treatment of GDL blocks the inflow of liquid water into the porous GDL.

4.4 Effect of MPL addition on water transport at CL|GDL interface

It is well known that microporous layer (MPL) placed between CL and GDL reduces the negative effect of water flooding at cathode (Pasaogullari & Wang, 2004b; Nam & Kaviany, 2003; Nam et al., 2003). MPL consisting of carbon black and PTFE is coated on one side of a coarse GDL, and has micro-structural and highly hydrophobic characteristics. In subsection 4.4, the influence of MPL addition on the liquid water accumulation at the interface between the cathode CL and MPL is investigated using the NIRS measurement. Table 2 presents the thickness of water film on the cathode CL in the cases without and with MPL. The amounts of carbon black and PTFE in the MPL are 2.0 mg/cm^2 and 2.0 mg/cm^2, respectively. The NIRS measurements in both cases were conducted after 60 min of operation. The current density and PTFE content for GDL are 0.3 A/cm^2 and 15 wt%. It is noted that the liquid water film on the cathode CL is not formed at all by the MPL addition.

Figure 13 shows the schematic diagram of water transport mechanisms at cathode CL|GDL interface without and with MPL. In the case without MPL, micro-scale liquid droplets attached to the cathode CL grow up and fill the large pores of GDL. The formation of liquid water film at the cathode CL|GDL interface is governed by the coarse pore structure of GDL. On the other hand, small pores in the MPL prevent the interfacial water droplets at the CL|MPL interface from growing large, and reduce the water saturation level. The fine pore structure of MPL is effective in preventing water flooding at cathode CL|MPL interface.

	(1)	(2)	(3)	(4)
Without MPL	9.2 μm	8.1 μm	18.4 μm	16.8 μm
With MPL	0.1 μm	0.0 μm	0.1 μm	0.0 μm

Table 2. Effect of MPL addition on the thickness of water film on the cathode CL surface

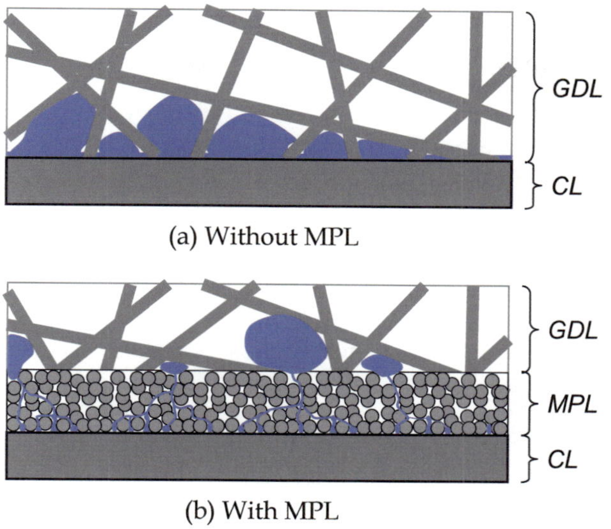

Fig. 13. Schematics of liquid water transport at the cathode CL | GDL interface without and with MPL

5. Measurement of water distribution in anode flow field under low-humidity conditions

In operating PEFCs, water migrates through electrolyte membrane from anode side to cathode side due to electro-osmotic effect. Thus, membrane dehydration occurs mainly near anode inlet under low-humidity conditions. In this section, the water vapor condensation in a PEFC is optically visualized by using water sensitive paper (WSP), and the water distribution in the anode flow field is investigated during low-humidity operation (Nishida et al., 2009).

5.1 Water sensitive paper

In order to visualize the water condensation in the anode flow field of an operating fuel cell, water sensitive paper (WSP), which is a test paper for water detection manufactured by Syngenta, was used in this experiment. Figure 14 shows the photograph of WSP sheet. The thickness of WSP is approximately 100 μm. WSP is coated with a yellow surface, which is changed into dark blue when exposed to water droplets, fog and high humidity. Figure 15

Fig. 14. Photograph of water sensitive paper (WSP)

presents the discoloration images of WSP when exposed to three different high humidity environments (RH=60, 70 and 80%) at 70°C. At the relative humidity of 60%, the surface color of WSP is hardly changed for 1000 s. On the other hand, when the relative humidity increases up to 70%, the yellow surface of WSP discolors to blue at t=500 s. The discoloration of WSP proceeds quickly when the environmental relative humidity reaches more than 80%.

5.2 Experimental

To directly observe the discoloration image of WSP in the anode flow channel, the experimental equipment and transparent fuel cell shown in Figure 2 and 3 are used. For water measurement, three sheets of WSP (1.0 mm x 30 mm) are inserted between the anode GDL and current collector, and a quartz glass is installed in the anode end plate as a window. The photo image of the anode flow field with three WSP sheets in the transparent cell is shown in Figure 16. Pure hydrogen and oxygen as the fuel and oxidant are fed into the anode and cathode channels at RH=0 or 30%. The utilization of hydrogen and oxygen are 0.4 and 0.2, respectively. The cell temperature in this experiment is set to 70°C.

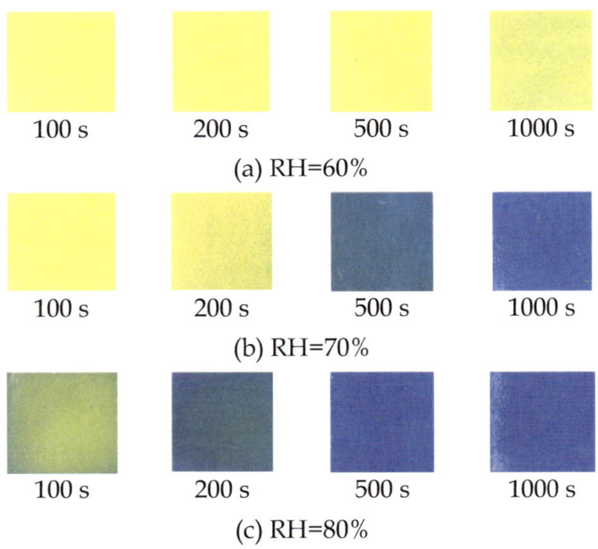

Fig. 15. Discoloration images of WSP when exposed to high humidity environments at 70°C

Fig. 16. Photo image of the anode flow field with three WSP sheets in the transparent fuel cell

5.3 Water distribution in anode flow field for operation without humidification

Figure 17 shows the time-sequential images of the WSP discoloration in the anode flow field under non-humidity condition. The operation test was conducted at 70°C and 0.1 A/cm^2. The gas streams along the anode and cathode channels are arranged in co-flow. The anode gas (dry H_2) flows from the upper right to the lower left in the serpentine flow channel. The WSP surface begins to discolor from yellow to blue in the downstream section of the anode channel at t=100 s. This result suggests that the product water on the cathode side is reversely diffused toward the anode through the electrolyte membrane. Water concentration profile in anode channel during non-humidity operation is dominated by water generation at cathode and back-diffusion of water in electrolyte membrane. As seen in the photograph, the anode water condensation occurs gradually from the downstream section after startup, because of the strong back-diffusion of water from the cathode downstream.

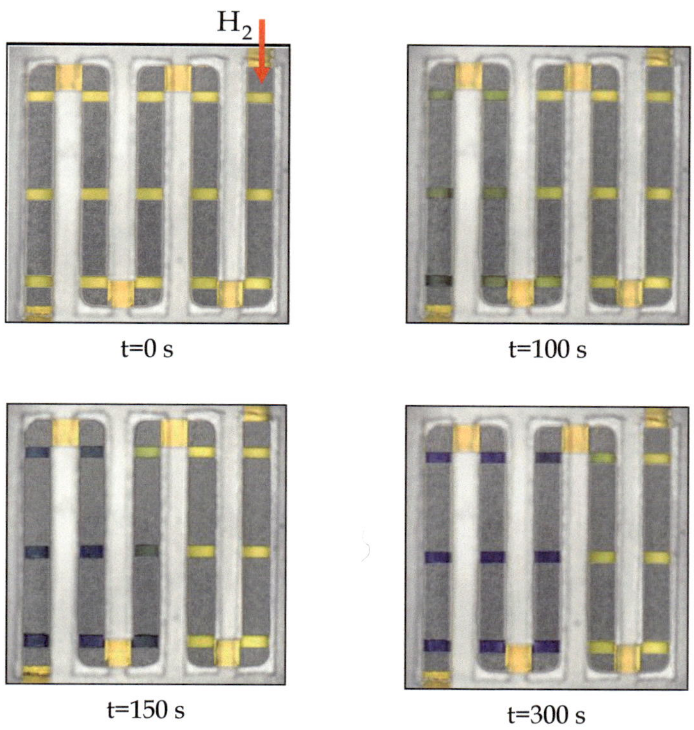

Fig. 17. Visualization images of the WSP discoloration in the anode channel under non-humidity condition at 0.1 A/cm^2

5.4 Effect of inlet gas humidification on anode water distribution and cell performance

In this subsection, the influence of inlet gas humidification on the anode water distribution and the proton conductivity of the electrolyte membrane is discussed under low and high current densities.

5.4.1 Low current density operation

Figure 18 presents the discoloration images of WSP in the anode flow channel taken after 1000 s of operation at two different humidification conditions. In this experiment, either anode or cathode inlet gas is humidified to 30% RH. The experimental fuel cell is operated at the low current density of 0.1 A/cm^2. The anode gas flows from the upper right to the lower left in both images. In both cases of A/C=30/0 and 0/30%, the WSP sheets positioned from the second to fifth channel are changed into dark blue by the inlet humidification and water back-diffusion. Under low current conditions, the water distribution in the anode flow field for the anode humidification case (A/C=30/0%) is almost similar to that for the cathode humidification (A/C=0/30%).

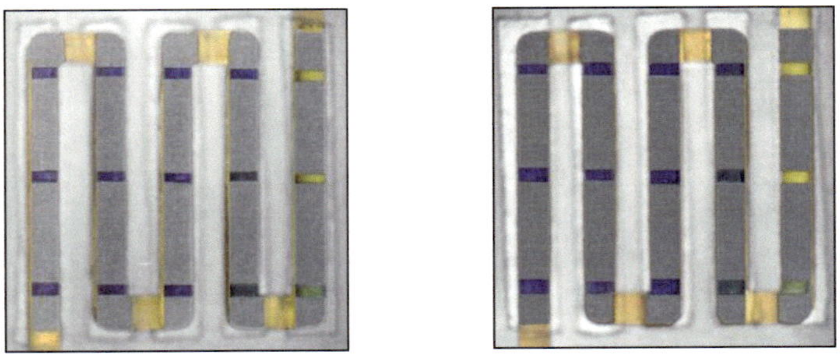

(a) A/C=30/0% RH (b) A/C=0/30% RH

Fig. 18. Discoloration images of WSP in the anode channel for two different humidification conditions (Anode/Cathode(A/C)=30/0%, 0/30% RH) at low current density of 0.1 A/cm^2

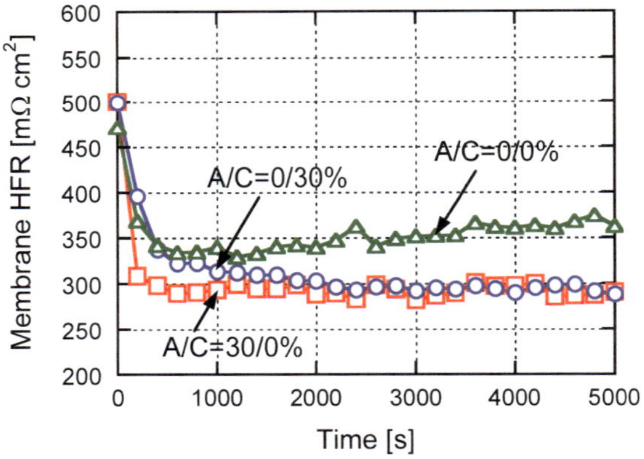

Fig. 19. Membrane HFRs for different humidification conditions (A/C=0/0%, 30/0%, 0/30% RH) at low current density of 0.1 A/cm^2

Figure 19 shows the high frequency resistance (HFR) of the electrolyte membrane during cell operation at three humidification conditions (A/C=0/0, 30/0 and 0/30% RH). The humidification of anode or cathode inlet is effective in reducing the membrane resistance and improving the cell performance because of encouraging the membrane hydration. The membrane HFR in the anode humidification case is almost same as that in the cathode humidification.

5.4.2 High current density operation

During low-humidity operation at high current densities, much water is significantly transported through electrolyte membrane from anode to cathode by strong electro-osmotic effect. Figure 20 presetns the discoloration images of WSP in the anode flow channel at the high current density of 0.3 A/cm². The inlet humidification condition is A/C=30/0 and 0/30% RH. Each picture was taken after 1000 s of operation. As seen in the photograph, the anode water concentration at 0.3 A/cm² is lower than that at 0.1 A/cm² shown in Figure 18. The decrease of the water concentration on the anode side is due to the strong water transport to the cathode driven by electro-osmotic drag. In the case of cathode humidification (A/C=0/30%), the WSP sheets positioned in only the fifth channel are changed into blue by the back-diffusion of water. On the other hand, when the anode inlet is humidified to 30%, the discoloration area of WSP in the anode flow field is expanded from the fifth to fourth channel. Under high current density condition, water shortage occurs mainly on anode side owing to electro-osmotic effect. Therefore, anode inlet humidification is effective in increasing water concentration in anode flow channel.

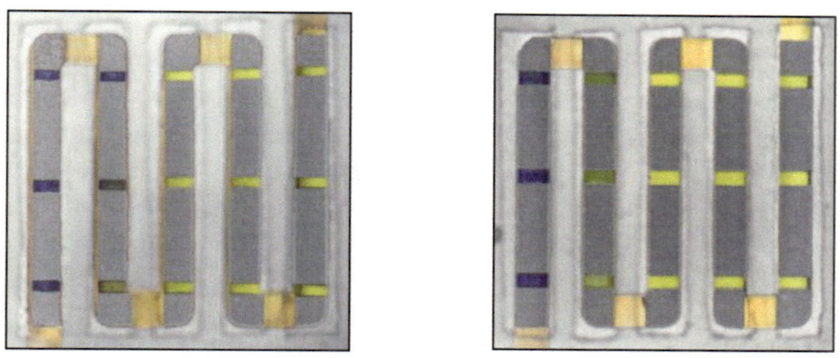

(a) A/C=30/0% RH (b) A/C=0/30% RH

Fig. 20. Discoloration images of WSP in the anode channel for two different humidification conditions (Anode/Cathode(A/C)=30/0%, 0/30% RH) at high current density of 0.3 A/cm²

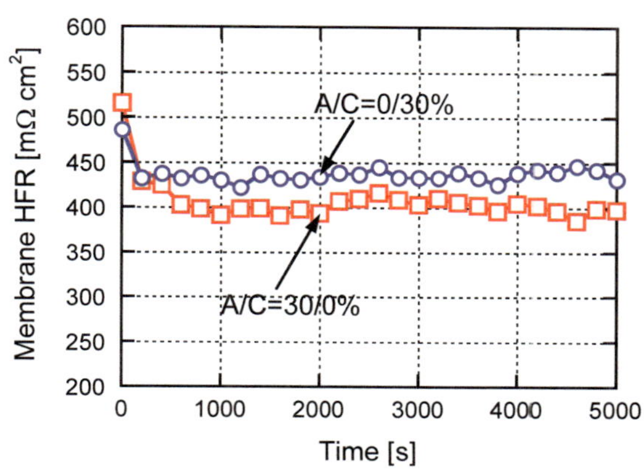

Fig. 21. Membrane HFRs for different humidification conditions (A/C=30/0%, 0/30% RH) at high current density of 0.3 A/cm²

Figure 21 shows the membrane HFR during high current density operation at two humidification conditions. Since the membrane dehydration proceeds with an increase in current density, the membrane resistance at 0.3 A/cm^2 becomes higher than that at 0.1 A/cm^2 provided in Figure 19. In addition, during high current density operation, the membrane HFR for the anode humidification case of A/C=30/0% is lower than that for the cathode humidification of A/C=0/30%. It should be noted that humidification of anode inlet gas alleviates membrane dryout and improves cell performance under low-humidity conditions.

6. Conclusions

In this chapter, three novel diagnostic tools for investigating the water transport phenomena inside a PEFC were developed. These measurement techniques can provide important information about water transport in fuel cell without the use of specialized equipments such as neutron radiography and X-ray computed tomography. In the first part of this chapter, the experimental method for estimating the liquid water content in the cathode GDE of a PEFC was presented based on the weight measurement. Furthermore, the optical visualization tool to explore the liquid water behavior was provided using a transparent fuel cell, and the impacts of current density and GDL thickness on the water transport in the cathode electrode were discussed. In the second part, the liquid water accumulation at the interface between the cathode CL and GDL was quantitatively measured using near-infrared reflectance spectroscopy (NIRS). The results showed that hydrophobic treatment of GDL is effective in alleviating water flooding at cathode CL|GDL interface. In addition, liquid water film at cathode interface is not formed at all by microporous layer (MPL) addition. The third part introduced the unique imaging technique to observe the water distribution in the anode flow field of a low-humidity PEFC using water sensitive paper (WSP). It was found that humidification of anode inlet gas prevents membrane dryout and enhances fuel cell power effectively under low-humidity conditions.

7. Acknowledgements

This study was supported by Grant-in-Aid for Young Scientists (B) (No.17760157, 20760134) of Japan Society for the Promotion of Science (JSPS), and New Energy and Industrial Technology Development Organization (NEDO) of Japan. The authors wish to thank a number of current and former students in Thermal Energy Engineering Laboratory at Kyoto Institute of Technology for technical assistance and useful discussions.

8. References

Bellows, R.J.; Lin, M.Y.; Arif, M.; Thompson, A.K. & Jacobson, D. (1999). Neutron Imaging Technique for In Situ Measurement of Water Transport Gradients within Nafion in Polymer Electrolyte Fuel Cells, *Journal of the Electrochemical Society*, Vol.146, No.3, (March 1999), pp. 1099-1103, ISSN 0013-4651

Boillat, P.; Kramer, D.; Seyfang, B.C.; Frei, G.; Lehmann, E.; Scherer, G.G.; Wokaun, A.; Ichikawa, Y.; Tasaki, Y. & Shinohara, K. (2008). In situ observation of the water distribution across a PEFC using high resolution neutron radiography, *Electrochemistry Communications*, Vol.10, (April 2008), pp. 546-550, ISSN 1388-2481

Curcio, J.A. & Petty, C.C. (1951). The Near Infrared Absorption Spectrum of Liquid Water, *Journal of The Optical Society of America*, Vol.41, No.5, (February 1951), pp. 302-304, ISSN 1084-7529

Gostick, J.T.; Ioannidis, M.A.; Fowler, M.W. & Pritzker, M.D. (2007). Pore network modeling of fibrous gas diffusion layers for polymer electrolyte membrane fuel cells, *Journal of Power Sources*, Vol.173, (November 2007), pp. 277-290, ISSN 0378-7753

He, W.; Yi, J.S. & Nguyen, T.V. (2000). Two-Phase Flow Model of the Cathode of PEM Fuel Cells Using Interdigitated Flow Fields, *AIChE Journal*, Vol.46, No.10, (October 2000), pp. 2053-2064, ISSN 1547-5905

Hickner, M.A.; Siegel, N.P.; Chen, K.S.; McBrayer, D.N.; Hussey, D.S.; Jacobson, D.L. & Arif, M. (2006). Real-Time Imaging of Liquid Water in an Operating Proton Exchange Membrane Fuel Cell, *Journal of the Electrochemical Society*, Vol.153, No.5, (March 2006), pp. A902-A908, ISSN 0013-4651

Kramer, D.; Zhang, J.; Shimoi, R.; Lehmann, E.; Wokaun, A.; Shinohara, K. & Scherer, G.G. (2005). In situ diagnostic of two-phase flow phenomena in polymer electrolyte fuel cells by neutron imaging, Part A. Experimental, data treatment, and quantification, *Electrochimica Acta*, Vol.50, (April 2005), pp. 2603-2614, ISSN 0013-4686

Lee, S.J.; Lim, N.Y.; Kim, S.; Park, G.G. & Kim, C.S. (2008). X-ray imaging of water distribution in a polymer electrolyte fuel cell, *Journal of Power Sources*, Vol.185, (December 2008), pp. 867-870, ISSN 0378-7753

Litster, S.; Sinton, D. & Djilali, N. (2006). Ex situ visualization of liquid water transport in PEM fuel cell gas diffusion layers, *Journal of Power Sources*, Vol.154, (March 2006), pp. 95-105, ISSN 0378-7753

Nam, J.H. & Kaviany, M. (2003). Effective diffusivity and water-saturation distribution in single- and two-layer PEMFC diffusion medium, *International Journal of Heat and Mass Transfer*, Vol.46, (November 2003), pp. 4595-4611, ISSN 0017-9310

Nam, J.H.; Lee, K.J.; Hwang, G.S.; Kim, C.J. & Kaviany, M. (2009). Microporous layer for water morphology control in PEMFC, *International Journal of Heat and Mass Transfer*, Vol.52, (May 2009), pp. 2779-2791, ISSN 0017-9310

Natarajan, D. & Nguyen, T.V. (2001). A Two-Dimensional, Two-Phase, Multicomponent, Transient Model for the Cathode of a Proton Exchange Membrane Fuel Cell Using Conventional Gas Distributors, *Journal of the Electrochemical Society*, Vol.148, No.12, (November 2001), pp. A1324-A1335, ISSN 0013-4651

Nishida, K.; Yokoi, Y.; Tsushima, S. & Hirai, S. (2009). Measurement of Water Distribution in Anode of Polymer Electrolyte Fuel Cell Under Low Humidity Conditions, *Proceedings of the ASME 2009 Seventh International Fuel Cell Science, Engineering and Technology Conference*, Paper No. FUELCELL2009-85128, ISBN 978-0-7918-4881-4, California, USA, June 8-10, 2009

Nishida, K.; Murakami, T.; Tsushima, S. & Hirai, S. (2010a). Measurement of liquid water content in cathode gas diffusion electrode of polymer electrolyte fuel cell, *Journal of Power Sources*, Vol.195, (June 2010), pp. 3365-3373, ISSN 0378-7753

Nishida, K.; Ishii, M.; Taniguchi, R.; Tsushima, S. & Hirai, S. (2010b). Quantitative Evaluation of Liquid Water at Cathode Interface of Polymer Electrolyte Fuel Cell using Near-Infrared Reflectance Spectroscopy, *Proceedings of the ASME 2010 Eighth International Fuel Cell Science, Engineering and Technology Conference*, Paper No. FUELCELL2010-33226, ISBN 978-0-7918-4404-5, New York, USA, June 14-16, 2010

Pasaogullari, U. & Wang, C.Y. (2004a). Liquid Water Transport in Gas Diffusion Layer of Polymer Electrolyte Fuel Cells, *Journal of the Electrochemical Society*, Vol.151, No.3, (February 2004), pp. A399-A406, ISSN 0013-4651

Pasaogullari, U. & Wang, C.Y. (2004b). Two-phase transport and the role of micro-porous layer in polymer electrolyte fuel cells, *Electrochimica Acta*, Vol.49, (October 2004), pp. 4359-4369, ISSN 0013-4686

Rebai, M. & Prat, M. (2009). Scale effect and two-phase flow in a thin hydrophobic porous layer. Application to water transport in gas diffusion layers of proton exchange membrane fuel cells, *Journal of Power Sources*, Vol.192, (July 2009), pp. 534-543, ISSN 0378-7753

Sasabe, T.; Tsushima, S. & Hirai, S. (2010). In-situ visualization of liquid water in an operating PEMFC by soft X-ray radiography, *International Journal of Hydrogen Energy*, Vol.35, (October 2010), pp. 11119-11128, ISSN 0360-3199

Satija, R.; Jacobson, D.L.; Arif, M. & Werner, S.A. (2004). In situ neutron imaging technique for evaluation of water management systems in operating PEM fuel cells, *Journal of Power Sources*, Vol.129, (April 2004), pp. 238-245, ISSN 0378-7753

Sinha, P.K.; Halleck, P. & Wang, C.Y. (2006). Quantification of Liquid Water Saturation in a PEM Fuel Cell Diffusion Medium Using X-ray Microtomography, *Electrochemical and Solid-State Letters*, Vol.9, No.7, (May 2006), pp. A344-A348, ISSN 1099-0062

Sinha, P.K. & Wang, C.Y. (2007). Pore-network modeling of liquid water transport in gas diffusion layer of a polymer electrolyte fuel cell, *Electrochimica Acta*, Vol.52, (November 2007), pp. 7936-7945, ISSN 0013-4686

Sinha, P.K. & Wang, C.Y. (2008). Liquid water transport in a mixed-wet gas diffusion layer of a polymer electrolyte fuel cell, *Chemical Engineering Science*, Vol.63, (February 2008), pp. 1081-1091, ISSN 0009-2509

Springer, T.E.; Zawodzinski, T.A. & Gottesfeld, S. (1991). Polymer Electrolyte Fuel Cell Model, *Journal of the Electrochemical Society*, Vol.138, No.8, (August 1991), pp. 2334-2342, ISSN 0013-4651

Tsushima, S.; Teranishi, K. & Hirai, S. (2004). Magnetic Resonance Imaging of the Water Distribution within a Polymer Electrolyte Membrane in Fuel Cells, *Electrochemical and Solid-State Letters*, Vol.7, No.9, (July 2004), pp. A269-A272, ISSN 1099-0062

Tüber, K.; Pócza, D. & Hebling, C. (2003). Visualization of water buildup in the cathode of a transparent PEM fuel cell, *Journal of Power Sources*, Vol.124, (November 2003), pp. 403-414, ISSN 0378-7753

Turhan, A.; Heller, K.; Brenizer, J.S. & Mench, M.M. (2006). Quantification of liquid water accumulation and distribution in a polymer electrolyte fuel cell using neutron imaging, *Journal of Power Sources*, Vol.160, (October 2006), pp. 1195-1203, ISSN 0378-7753

Wang, Z.H.; Wang, C.Y. & Chen, K.S. (2001). Two-phase flow and transport in the air cathode of proton exchange membrane fuel cells, *Journal of Power Sources*, Vol.94, (February 2001), pp. 40-50, ISSN 0378-7753

Yang, X.G.; Zhang, F.Y.; Lubawy, A.L. & Wang, C.Y. (2004). Visualization of Liquid Water Transport in a PEFC, *Electrochemical and Solid-State Letters*, Vol.7, No.11, (October 2004), pp. A408-A411, ISSN 1099-0062

Yoshizawa, K.; Ikezoe, K.; Tasaki, Y.; Kramer, D.; Lehmann, E. & Scherer, G.G. (2008). Analysis of Gas Diffusion Layer and Flow-Field Design in a PEMFC Using Neutron Radiography, *Journal of the Electrochemical Society*, Vol.155, No.3, (January 2008), pp. B223-B227, ISSN 0013-4651

Zhang, F.Y.; Yang, X.G. & Wang, C.Y. (2006). Liquid Water Removal from a Polymer Electrolyte Fuel Cell, *Journal of the Electrochemical Society*, Vol.153, No.2, (December 2005), pp. A225-A232, ISSN 0013-4651

Zhang, J.; Kramer, D.; Shimoi, R.; Ono, Y.; Lehmann, E.; Wokaun, A.; Shinohara, K. & Scherer, G.G. (2006). In situ diagnostic of two-phase flow phenomena in polymer electrolyte fuel cells by neutron imaging, Part B. Material variations, *Electrochimica Acta*, Vol.51, (March 2006), pp. 2715-2727, ISSN 0013-4686

Part 4

Bioelectrochemistry

9

Spectroelectrochemical Investigation on Biological Electron Transfer Associated with Anode Performance in Microbial Fuel Cells

Okamoto Akihiro[1], Hashimoto Kazuhito[1,2] and Nakamura Ryuhei[1]
[1]*The University of Tokyo, School of Engineering, Department of Applied Chemistry,*
[2]*JST-ERATO Project,*
Japan

1. Introduction

Microbial fuel cells (MFCs) are considered to be an attractive future option for the treatment of organic wastes and recovery of bioenergy from renewable biomass resources (Kim et al., 1999; Schroder et al., 2003; Lovley 2006; Bretschger et al., 2007; Logan et al., 2010). MFCs are devices that generate electricity from organic matter by exploiting the catabolic activities of microbes. Although MFCs have operational and functional advantages over the current technologies used for generating energy from organic matter, the full potential and application of these devices remains unclear, primarily because it is an emerging technology that requires further technical development. Another limitation of MFCs is their relatively low process performance in comparison with competing technologies. Therefore, improvement of the efficiency of electricity generation in MFCs will require both biological and engineering research approaches (Schroder et al., 2003; Zhao et al., 2010).

Recently, electrochemists have focused on the mechanisms of microbial current generation in an attempt to improve the anode performance in MFCs (Reguera et al., 2006; Torres et al., 2010). In particular, a metal-reducing bacterial species of the genus *Shewanella, S. oneidensis* MR-1, is one of the most extensively studied organisms for microbial current generation (Heidelberg et al., 2002; Bretschger et al., 2007; Newton et al., 2009). This microorganism has a significant quantity of redox proteins, namely *c*-type cytochromes (*c*-Cyts), in the outer membrane (OM) (Myers & Myers 1992; Heidelberg et al., 2002) that are proposed to transport electrons generated by the intracellular metabolic oxidation of organic matter to extracellular electrodes as a terminal process for anaerobic respiration (Myers & Myers 1997; Shi et al., 2009; Gralnick et al., 2010). To date, however, the basic electrochemical analyses on the extracellular electron transfer (EET) process has been strictly limited to studies with purified OM *c*-Cyts or synthesized model compounds immobilized on graphite electrodes (Hartshorne et al., 2007; Wigginton et al., 2007; Marsili et al., 2008). Thus, the processes by which living microorganisms deliver electrons to extracellular solid electrodes remain largely unknown. The main obstacle to determining the underlying mechanisms of electron transfer (ET) is the shortage of *in-vivo* spectroscopic and electrochemical techniques to directly differentiate OM *c*-Cyts from the numerous uncharacterized biological molecules, such as self-secreted redox shuttles, present in bacterial cells. Furthermore, the highly dynamic nature of *in-vivo* ET resulting from protein-protein and protein-membrane interactions impedes our ability to utilize knowledge and methodologies developed in studies with purified proteins.

Herein, we report the *in-vivo* spectroscopic and electrochemical techniques recently developed in our laboratory for the study of microbial EET. The following three techniques are introduced in this chapter:

- Whole-microorganism electrochemistry with a three-electrode system to examine EET pathways and dynamics under physiological conditions.
- UV-vis evanescent wave (EW) spectroscopy for monitoring the electronic state of *c*-Cyts located at the cellular membrane/electrode interface in the course of microbial current generation.
- In-frame deletion mutants of OM and periplasmic *c*-Cyts to determine the specific location of proteins responsible for the respiratory EET chain.

The application of these techniques with intact *Shewanella* cells allowed us to reveal the thermodynamic properties of OM *c*-Cyts under *in-vivo* conditions, which significantly differed from those previously found *in vitro*. Together with previous biochemical investigations of purified *c*-Cyts, the techniques introduced in this chapter will aid basic electrochemists to perform experiments using living microorganisms as the subject of electrochemical and spectroscopic studies. We speculate that such studies will be an important precondition to cultivate and advance the emerging technology of MFCs as a feasible option for the treatment and recovery of bioenergy from organic wastes and renewable biomass.

2. Whole-microorganism electrochemistry

2.1 The role of OM *c*-Cyts for EET processes under *in-vivo* conditions

Using protein film voltammetry (PFV), several purified decaheme OM *c*-Cyts of *S. oneidensis* MR-1, particularly OmcA and MtrC, have been demonstrated to mediate direct electron transfer (DET) to graphite electrodes (Hartshorne et al., 2007; Firer-Sherwood et al., 2008). Kinetic analyses revealed that ET at the protein/electrode interface occurs within the order of milliseconds (Firer-Sherwood et al., 2008). Moreover, the importance of protein-protein interactions to facilitate DET to a graphite electrode has been demonstrated using the MtrCAB protein complex (see Fig. 1) (Hartshorne et al., 2009). Although the fundamental properties of protein-mediated EET have become clearer through studies involving PFV, the mechanisms by which living organisms deliver electrons to extracellular electrodes under physiological conditions are largely unknown, and a number of discrepant results exist concerning the kinetics and energetics of bacterial EET.

To confirm the existence of EET pathways mediated by OM *c*-Cyts under physiological conditions (Fig. 1), we have developed a highly sensitive, whole-cell voltammetry technique. Electrochemical measurements with *S. oneidensis* MR-1 cells were conducted using an in-house constructed, vacuum-tight, single-chamber, three-electrode reactor (Fig. 2). A flat, uniform layer of tin-doped In_2O_3 (ITO) grown on a glass substrate by r.f. magnetron sputtering (8 Ω/square resistance, 1.0-mm glass thickness, 3.1-cm^2 surface area; SPD

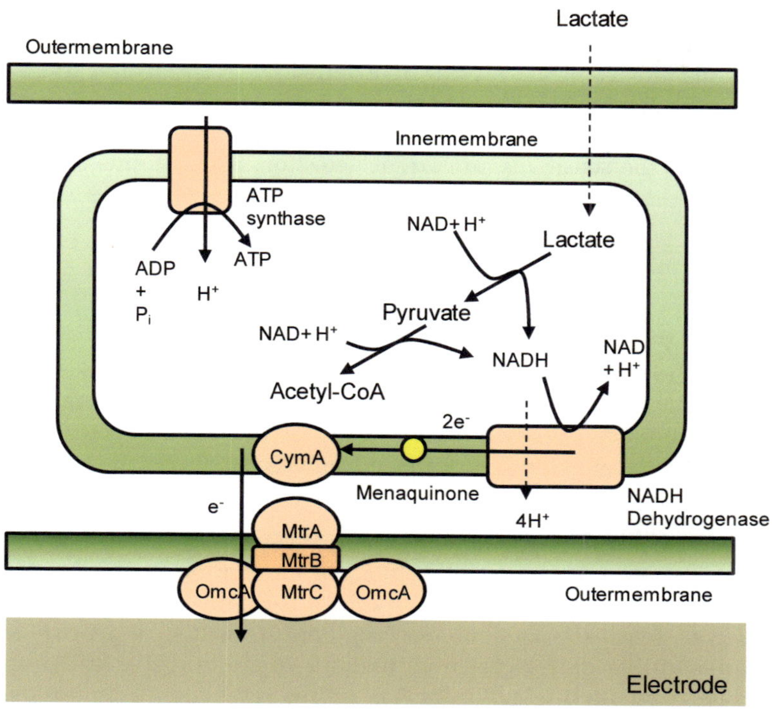

Fig. 1. A schematic illustration of the EET process of *S. oneidensis* MR-1 cells under anaerobic conditions. Respiratory electrons are continuously generated by the metabolic oxidation of lactate to pyruvate and acetyl-CoA, and are transferred to the extracellular electrode via the electron transport system consisting of the NADH dehydrogenase and menaquinone redox couple, followed by the protein network of *c*-Cyts (CymA and OmcA-MtrCAB protein complexes) spanning the inner and outer membranes.

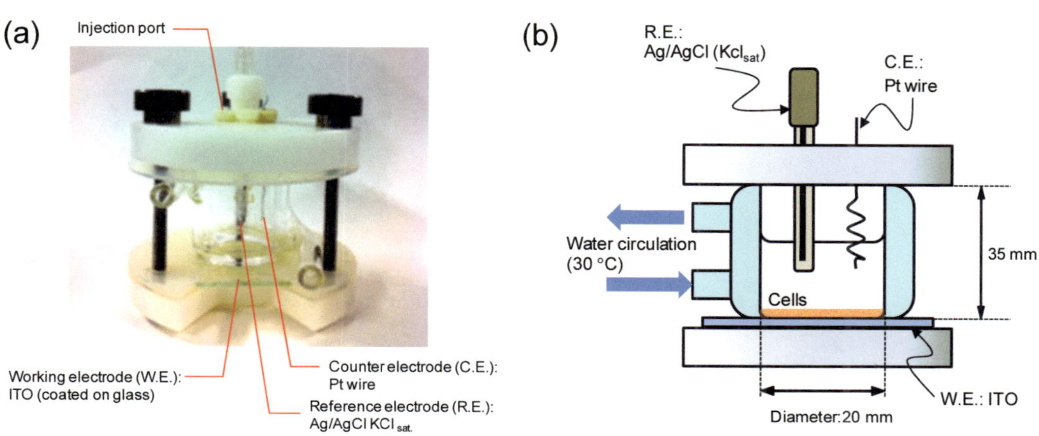

Fig. 2. (a) Photograph and (b) schematic illustration of the electrochemical reactor used for whole-microorganism electrochemistry.

Laboratory, Inc.,) was used as the working electrode, while Ag/AgCl ($KCl_{sat.}$) and a platinum wire were used as the reference and counter electrodes, respectively. The reactor temperature was maintained at 303 K using an external water circulation system (Fig. 2 (b)). It is important to note that the ITO electrode was placed on the bottom surface of the reactor, allowing the injected cells to statically settle on the electrode surface within a few minutes. This design improves the sensitivity of current detection, as it permits the cells to form a uniform electroactive biofilm on the electrode with high surface coverage.

2.2 EET from living microbes to ITO electrodes

To investigate EET *in vivo*, electrochemical experiments were performed under conditions of static potential at 0.4 V (vs SHE). As an electrolyte solution, 4.7-mL of defined medium (DM) containing lactate (10 mM) and yeast extract (0.5 g per liter) was added into the electrochemical reactor (Roh et al. 2006), which was subsequently deaerated by bubbling with dry N_2 (99.999% purity). After a dissolved O_2 concentration of 0.1 ppm was reached (PreSens, Microx TX3 trace), 0.3 mL of a freshly prepared cell suspension with an optical density at 600 nm (OD_{600}) of 1.6 was injected into the electrochemical reactor to give a final OD_{600} inside the reactor of 0.1.

As shown in Fig. 3 trace (a), upon adding the cell suspension to the reactor, an anodic current was immediately generated and reached 1.0 μA cm^{-2} after 2 h of cultivation. The current continued to increase with time, reaching approximately 2.0 μA cm^{-2} after 12 h. In contrast, when medium lacking lactate was used as an electrolyte solution, the microbial current was strongly impaired (Fig. 3 trace (b)). In this system, a maximum density of less than 0.3 μA cm^{-2} was attained, with the small amount of current generation likely attributable to the trace amounts of organic compounds present in the yeast extract. These results demonstrate that the observed current is due to the metabolic oxidation of lactate, followed by EET from the cell surfaces to the ITO electrode. Given that the anaerobic lactate oxidation pathway in strain MR-1 has been elucidated (Scott & Nealson 1994), as schematically shown in Fig. 1, we can conclude that the generated current is associated with the production of NADH via the conversion of lactate to pyruvate and acetyl-CoA. Susequently, menaquinone and *c*-Cyts located in the inner and outer membranes, respectively, sequentially transport the electrons to the electrode surface.

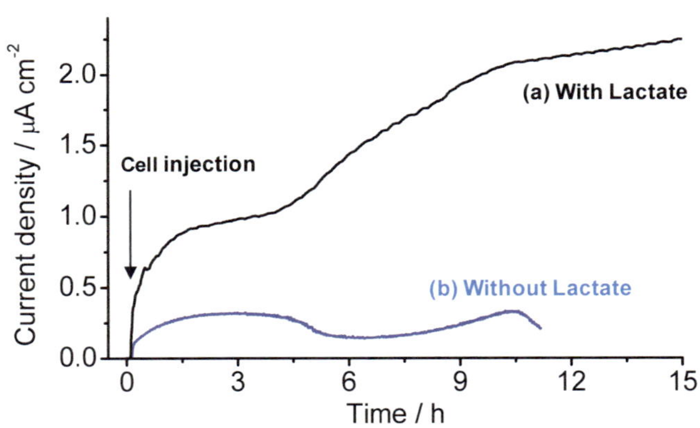

Fig. 3. Microbial current generation versus time for *S. oneidensis* MR-1 cells inoculated in electrochemical reactors containing medium with (a) and without (b) 10 mM lactate.

2.3 Whole-microorganism voltammetry

To identify the redox molecules responsible for the EET, we conducted whole-cell voltammetry measurements of the cell suspension of MR-1 strain. As shown in the cyclic voltammogram (CV) in Fig. 4(a), the electrode with attached MR-1 cells exhibited a single redox wave with a midpoint potential (E_m) of 50 mV (vs SHE). The waveform remained stable even after hundreds of repeated potential scans from 0.7 to -0.8 V at scan rates ranging from 0.01 to 200 V s^{-1}. As the plot of the peak current as a function of scan rate gave a linear relationship, the redox wave at 50 mV was attributed to an electron-exchange process mediated by adsorbed species. In other words, the bacterial cells localized abundant and robust redox species at the cell membrane/electrode interface.

Following the CV measurements, the reactor electrode was washed with standard HEPES buffer and subjected to scanning electron microscopy (SEM). The SEM image shown in Fig. 4 (b) revealed that rod-shaped cells of approximately 2 μm in length and 0.5 μm in width were immobilized on the ITO electrode surface, with most cells orientated horizontally to the electrode surface, rather than via the apex of their long axis. As MR-1 cells possess a high quantity of *c*-Cyts on the OM, we can reasonably assign the observed redox signal with an E_m of 50 mV in Fig. 4 (a) to the redox reaction of the major OmcA-MtrCAB protein complex with the ITO electrode (Fig. 1).

The kinetic analysis of the electron exchange reaction at the cell-electrode interface also supports the assignment of the redox species as the OmcA-MtrCAB complex. The heterogeneous standard rate constant (k_0) for intact cells was estimated to be 300 ± 10 s^{-1} at 303 K by Trumpet plot analysis (Laviron 1979). This value is the same order of magnitude with the k_0 values previously determined for purified MtrC and the MtrCAB protein complex immobilized on a basal-plane graphite electrode of 220 and 195 s^{-1}, respectively (Hartshorne et al., 2009). Note that these k_0 values are two orders of magnitude higher than that for riboflavin immobilized on an electrode surface ($k_0 < 0.7$ s^{-1}) (Okamoto et al., 2009), which is reported to be a major electron shuttle secreted by *Shewanella* cells for EET (Marsili et al., 2008).

(a)

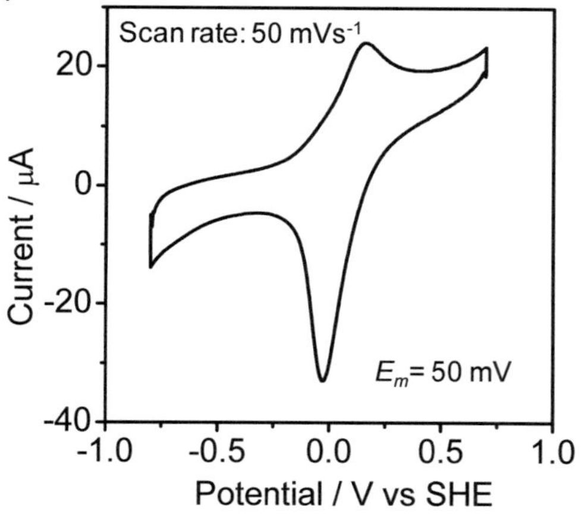

(b)

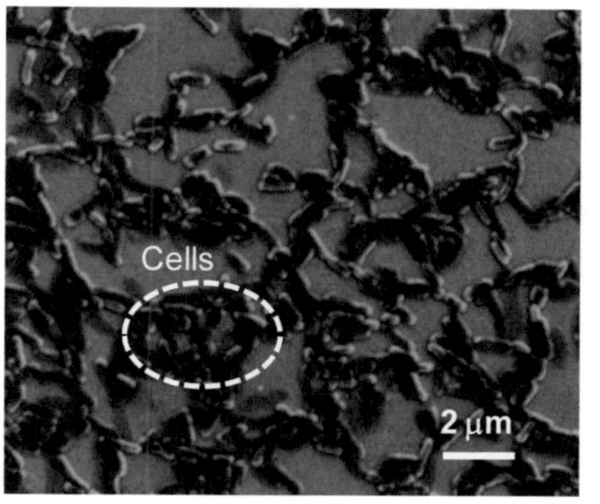

Fig. 4. (a) Cyclic voltammogram of *S. oneidnesis* MR-1 cells at a scan rate of 50 mV s^{-1}. (b) SEM image of the ITO electrode surface (top view) after 25 h of current generation in the presence of lactate.

It should be noted that the E_m of 50 mV detected for intact MR-1 cells is a markedly more positive potential compared to those reported for purified OM *c*-Cyts (–170 mV for MtrC and –210 mV for OmcA) (Eggleston et al., 2008; Firer-Sherwood et al., 2008; Hartshorne et al., 2009). In addition to OM *c*-Cyts, MR-1 cells are known to synthesize several flavin and quinone derivatives that function as electron shuttles, a few of which strongly bind to the outer cell surface and provide redox signals characteristic of immobilized electroactive species (Saffarini et al., 2002; Okamoto et al., 2009; Bouhenni et al., 2010). Therefore, the large difference in E_m between purified OM *c*-Cyts and intact cells has led many researchers to speculate that the observed redox wave at 50 mV is attributable to cell-attached menaquinone, whose midpoint potential is approximately 80 mV (Li et al., 2010; Zhao et al., 2010), rather than OM *c*-Cyts.

In the following sections, we will demonstrate based on the results of spectroscopic and molecular biological techniques that the redox wave at +50 mV is indeed assignable to the OmcA-MtrCAB protein complex of *S. oneidensis* MR-1. The definitive assignment of this peak will also highlight the importance of *in-vivo* electrochemistry to verify the true energetics and kinetics of EET in living microbes.

3. *In-vivo* EW spectroscopy

3.1 EW spectroscopy for selective monitoring of the electronic state of OM *c*-Cyts

UV-vis EW spectroscopy is a powerful technique that allows the selective monitoring of molecules adsorbed on an electrode surface (Fig. 5) and has been used to investigate the EET mechanisms of *Shewanella* cells under physiological conditions. The *c*-Cyts located in the cytoplasmic (inner) membrane can be spectroscopically detected owing to both their large quantity and the intense visible absorption band of heme groups present in the proteins. To obtain conclusive evidence for the assignment of the redox signal at +50 mV to OM *c*-Cyt, we subjected an ITO electrode with attached *Shewanella* cells to EW spectroscopy and monitored changes in the electronic state of heme in OM *c*-Cyts by measuring electrode potentials.

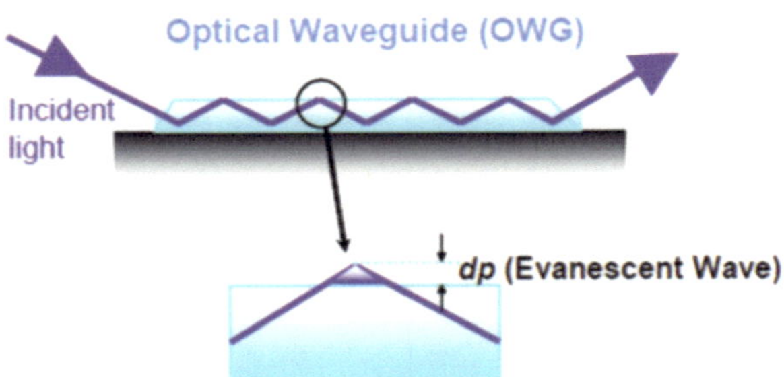

Fig. 5. Schematic illustration of EW spectroscopy with an optical waveguide.

As a potential approach to differentiate between the redox signal derived from OM *c*-Cyts and that of menaquinone or other uncharacterized redox shuttles, we noted that the hemes of *c*-Cyts have the ability to bind small molecules such as O_2, NO, CO, and imidazole, because of the existence of an open coordination site on the centered heme iron (Andrew et al., 2002; Fornarini et al., 2008). The ligation reaction of heme is an established method for analyzing the redox and electronic states of purified proteins (Hoshino et al., 1993). Previous *in-vitro* studies showed that the NO ligation of heme caused a large positive shift (~600 mV) in the E_m due to the strong back donation of NO to the centered iron (Chiavarino et al., 2008). More importantly, NO has a high binding affinity to metal centers, particularly for the hemes of *c*-Cyts (the equilibrium binding constant of NO to a ferrous heme is ~10^5 M^{-1} (Hoshino et al., 1993)).

We first performed NO ligation of the hemes of strain MR-1 by exposing a bacterial suspension to NO for 10 min under anaerobic conditions. The suspension was centrifuged and the supernatant was replaced with fresh medium to remove the excess dissolved NO. The *in-vivo* optical absorption spectra of the bacterial suspension before and after the NO ligation were measured in diffuse-transmission mode. Prior to the NO ligation, strain MR-1 exhibited an intense band at 419 nm and weak bands at 522 and 552 nm which originated from the Soret and Q bands, respectively, of the reduced ferrous hemes in *c*-Cyts (Fig. 6, trace (a)) (Nakamura et al., 2009). The obtained spectral position and shape are consistent with those reported for purified OmcA and MtrC of MR-1 cells (Shi et al., 2006; Hartshorne et al., 2007; Wigginton et al., 2009). Following NO ligation, however, the spectrum showed a distinct blue shift of the Soret band to 408 nm, concomitantly with the loss of Q band absorption (Fig. 6, trace (b)). This change in the spectrum agreed well with those reported for nitrosyl adduct formation in purified proteins and synthetic hemes (de Groot et al., 2007).

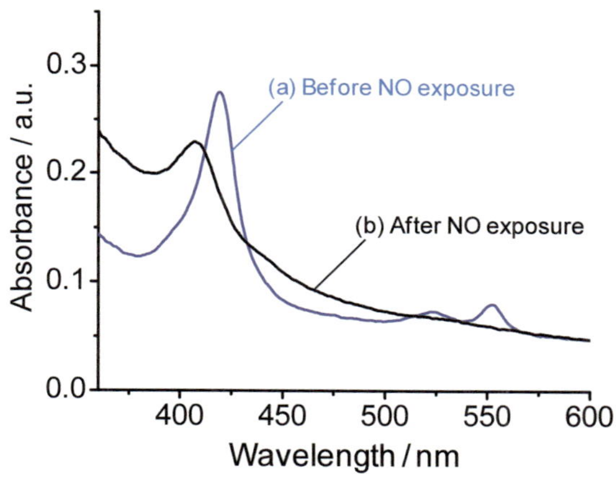

Fig. 6. Diffuse-transmission UV-vis absorption spectra of a cell suspension of *S. oneidensis* MR-1 (a) before and (b) after NO exposure.

Second, to confirm if NO-ligated hemes were present in the *c*-Cyts located in both the OM and periplasmic regions, UV-vis EW experiments were conducted (Fig. 7 (a)). The UV-vis EW spectra of intact MR-1 cells were measured using a SIS-5000 Surface and Interface Spectrometer System (System Instruments) equipped with a 150-W xenon lamp. The angle of incident light was 8° relative to the surface of the optical waveguide (OWG) substrate (reflective index of 1.89). The bacterial suspension casted on a quartz OWG showed a blue shift of the Soret band from ~419 to ~408 nm after the NO ligation of ferrous hemes, consistent with the results obtained by diffuse-transmission UV-Vis spectroscopy (Fig. 7 (b)). Because EW is only generated in the vicinity of the substrate surface and has a penetration depth of less than 30 nm (Nakamura et al., 2009), these results confirm that the nitrosyl adducts are located at the cell-substrate interface, as the thicknesses of the OM, periplasm, and inner membrane of *Shewanella* are ~15, ~10, and ~8 nm, respectively.

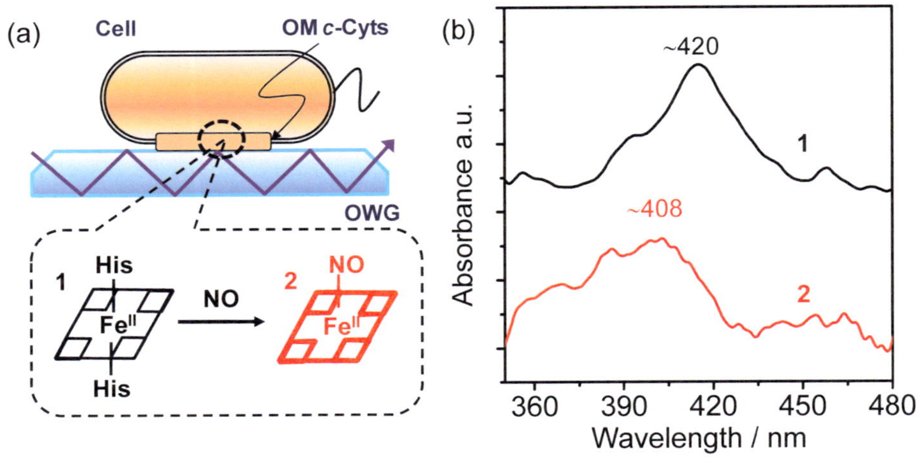

Fig. 7. (a) Schematic illustration of the NO coordination reaction to heme molecules in OM *c*-Cyts at the cell/electrode interface detected by UV-vis EW spectroscopy. (b) UV-vis EW spectra of whole cells of *S. oneidensis* MR-1 before (1) and after NO exposure (2).

The NO-ligated cells were next subjected to CV experiments to examine the effects of NO ligation of hemes on the redox properties of MR-1 cells. As expected, the NO-ligated cells were found to have an open circuit potential of 270 mV, which was 470 mV more positive than that for cells not exposed to NO. Furthermore, in the CV of NO-MR-1 cells obtained at a scan rate of 10 mV s^{-1}, the redox wave observed at 50 mV for untreated cells completely disappeared, and a new redox wave appeared at 650 mV (Okamoto et al., 2010). A similarly large positive shift in potential was also reported following the NO ligation of purified *c*-Cyts and is consistent with the strong back-donation effect of NO ligands (de Groot et al., 2007). Accordingly, we can conclude that the bacterial redox wave at 50 mV is attributable to *c*-Cyts localized at the cell membrane/electrode interface. In addition, it should be noted that menaquinone possesses no specific affinity for NO, excluding the contribution of this mediator to the redox signal at 50 mV.

3.3 EW spectroscopy combined with whole-microorganism electrochemistry

Based on the aforementioned UV-vis EW experiments, we further investigated the role of OM *c*-Cyts in the movement of metabolically generated electrons by monitoring the redox state of hemes under potential-controlled conditions. In this experiment, a quartz OWG coated with an ITO film was placed on the bottom surface of the electrochemical reactor to serve as an internal reflection element for the generation of an EW at the interface between the electrode and electrolyte solution (Fig. 8). A single-chamber electrochemical reactor consisting of Ag/AgCl ($KCl_{sat.}$) and platinum wire as the counter and reference electrodes, respectively, was mounted onto the ITO substrate and then sealed with a silicon rubber O-ring (Fig. 8).

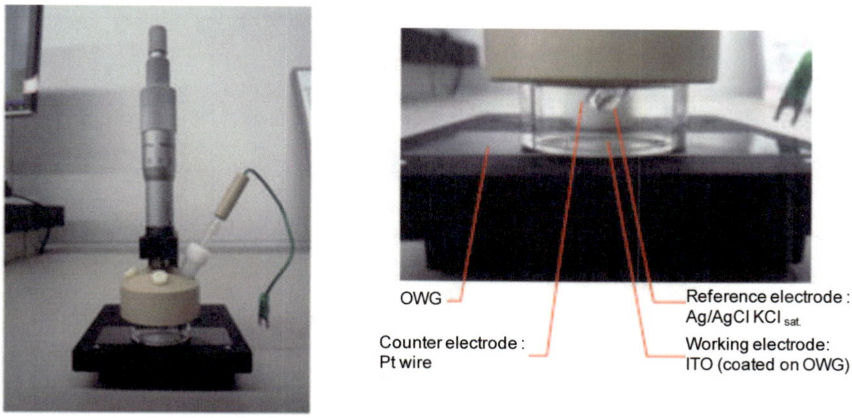

Fig. 8. Photographs of the evanescent wave electroabsorption spectroscopy system used for *in-situ* electrochemistry with an ITO-coated optical waveguide (OWG).

In the course of microbial current generation with an electrode potential of 320 mV (vs SHE), the UV-vis EW spectrum exhibited a Soret absorption band at ~410 nm, characteristic of the oxidized form of *c*-Cyts (Fe^{3+}) (Richardson et al., 2000; Hartshorne et al., 2007). The penetration depth at which the EW field (reflective index of 1.47) decayed to $1/e$ was estimated to be ~110 nm at a wavelength of 400 nm. Accordingly, the appearance of the Soret peak at ~410 nm demonstrates that the oxidized *c*-Cyts are adjacent to the electrode surface during the respiratory ET reaction. However, upon lowering the electrode potential from 320 to −180 mV, a rapid red shift (419 nm) of the Soret band was observed, which was due to the reduced form of *c*-Cyts (Fe^{2+}). The spectral changes were fully reversible for both the positive- and negative-direction scans. The plot of the peak intensity (at 419 nm) against electrode potential shows that the *c*-Cyts of strain MR-1 have a potential distribution ranging approximately 300 mV, from fully oxidized (270 mV) to fully reduced (20 mV) (Fig. 9). The E_{m} of the *c*-Cyts under operating conditions of the reactor was spectroscopically estimated to be 145 ± 50 mV, which is shifted by +100 mV from the E_m electrochemically determined for purified OM *c*-Cyts. In contrast to purified proteins, electrons in living cells are continuously supplied to the heme groups of OM *c*-Cyts through the oxidation reaction of lactate. The constant influx of electrons results in a displacement of the equilibrium and production of the reduced form of OM *c*-Cyts, which is likely reflected in the positive shift of the E_{m} in the living system. The occurrence of the E_m shift further demonstrates that hemes localized at the cellular membrane are reduced by metabolically generated electrons. Namely, the microbial current generation originates from the electrons supplied from NADH, which is continuously generated by microbial metabolic processes, such as the conversion of lactate → pyruvate → acetyl-CoA, or an incomplete TCA cycle, to the electrode via the electron transport system consisting of quinol derivatives and *c*-Cyts in the inner and outer membranes, respectively (see Fig. 1).

4. In-frame deletion mutants of OM *c*-Cyts

The experimental data presented in the above sections clearly demonstrates that the redox wave at 50 mV is attributable to the OM *c*-Cyts of strain MR-1. To identify the specific location of the heme molecules involved in the observed redox reaction, we employed

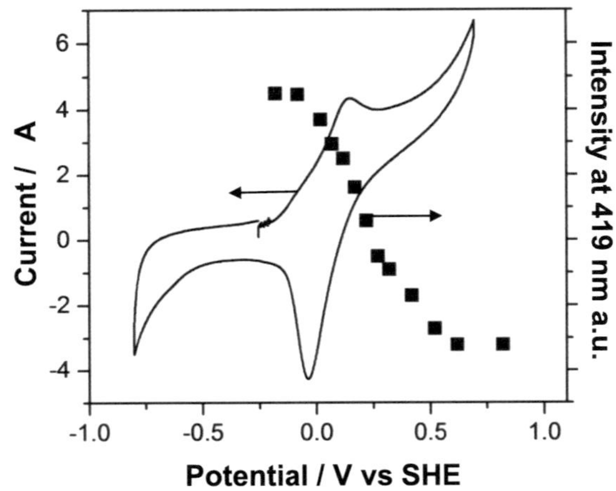

Fig. 9. Plot of absorbance at 419 nm as a function of electrode potential (black squares) and a cyclic voltammogramof a whole cell of strain MR-1 (scan rate, 10 mV s^{-1}).

several mutant MR-1 strains of membrane-bound *c*-Cyt proteins (OmcA, MtrA, MtrB, MtrC, and CymA). The location and function of these five proteins in MR-1 have been identified using genome sequence analyses (Heidelberg et al., 2002) in conjunction with the biochemical analyses of purified membrane fractions (Myers & Myers 1992). As schematically shown in Fig. 1, OmcA and MtrC are integral OM proteins that are exposed to the extracellular space, and are proposed to mediate DET to solid substrates. These two proteins are a part of the OmcA-MtrCAB a protein complex, which delivers respiratory electrons over a distance of approximately 10 nm, spanning the periplasm to the outer cell surface (Myers & Myers 1997; Pitts et al., 2003). MtrB is a β-barrel porin of the MtrCAB complex and serves as an electron conduit between MtrA to MtrC (Hartshorne et al., 2009), while MtrA is predicted to be located on the periplasmic side of the OM. CymA is an inner-membrane-bound tetraheme *c*-Cyt that acts as a menaquinone oxidase (Firer-Sherwood et al., 2008; Hartshorne et al., 2009). To confirm the involvement of heme molecules in the OM *c*-Cyts proteins of MR-1 cells in the observed EET reaction between bacteria and electrodes, we examined mutant strains with in-frame deletions of the genes encoding MtrA, MtrB, MtrC, OmcA, and CymA. Each mutant was constructed by allele replacement using a two-step homologous recombination method, as previously reported (Saltikov & Newman 2003; Kouzuma et al., 2010).

We first introduced a double-deletion mutant lacking the genes encoding OmcA and MtrC (ΔomcA/ΔmtrC) into the electrochemical reactor and found that the respiratory current generation ability of the mutant was impaired by 25% relative to wild-type (WT) cells at 400 mV (vs SHE). After the mutant cells were immobilized onto the electrode surface, whole-microorganism voltammetry was conducted. The deletion of the omcA and mtrC genes decreased the anodic (I_{pa}) and cathodic peak currents (I_{pc}) by 68% and 67%, respectively, relative to those observed in WT, with a corresponding shift of E_m from 50 to 135 mV. When the I_{PA} was normalized by total protein content in the biofilms formed by WT and ΔomcA/ΔmtrC cells, the coulomb value of ΔomcA/ΔmtrC displayed a 78% decrease compared to WT (Fig. 10). We confirmed that deletion of the gene encoding PilD (ΔpilD), a predicted prepilin peptidase, also resulted in a comparable reduction with ΔomcA/ΔmtrC (Fig. 10). Western blot analysis of cell extracts of strain ΔpilD has demonstrated that PilD is involved in the processing of type IV, Msh, and T2S prepilin proteins, and this strain is also reported to lack OMCs, including OmcA and MtrC (Bouhenni et al., 2010). Our electrochemical comparisons among WT, ΔomcA/ΔmtrC, and ΔpilD cells clearly indicate that OmcA and MtrC function as the major reductases for EET from attached cells to electrode surfaces (*i.e.*, DET reaction).

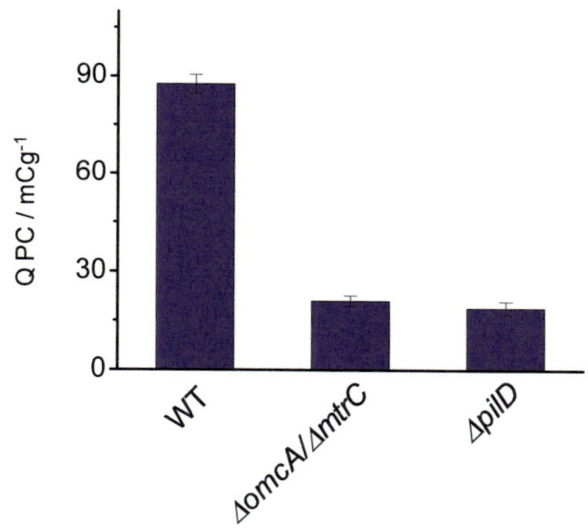

Fig. 10. Coulomb area of the redox peak signals for wild-type (WT) *S. oneidensis* MR-1 and the ΔomcA/ΔmtrC and ΔpilD mutatns normalized by the total protein content of cells attached on the electrode surface (QPC). Error bars indicate the standard error of the means calculated with data obtained from greater than three individual experiments.

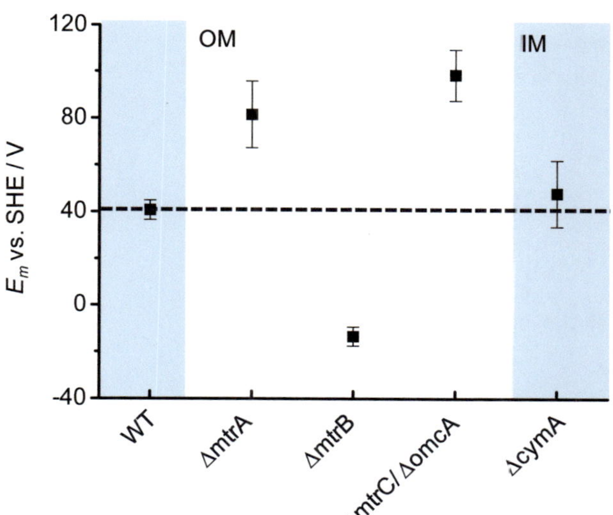

Fig. 11. Midpoint potentials (E_m) of the redox signal for wild-type (WT) *S. oneidensis* MR-1 and the ΔmtrA, ΔmtrB, ΔmtrC/ΔomcA, and ΔcymA mutant strains. Error bars indicate the standard error of the means calculated with data obtained from greater than three individual experiments.

To confirm that the observed redox reaction involves the OmcA-MtrCAB protein complex, the in-frame deletion mutants of MtrA, MtrB, or CymA were further subjected to CV measurements. The effects of each gene deletion on the redox wave of whole cells were compared using E_m values, as changes in the E_m would sensitively reflect conformational and functional changes in the protein complex. The E_m of the ΔmtrA and ΔmtrB mutants exhibited a significant E_m shift relative to WT, whereas deletion of the CymA gene caused only a subtle change in the E_m (Fig. 11). As only MtrA and MtrB are directly associated with OmcA and MtrC, comparison of the E_m values among the mutant strains indicates that the most plausible assignment for the redox signal at 50 mV is the OM-bound complex composed of OmcA-MtrCAB.

5. *In-vivo* vs *in-vitro* study of bacterial EET

As can be seen from the energy diagram of bacterial EET presented in Fig. 12, the midpoint potential of OM *c*-Cyts (50 mV) is close to that of menaquinone (80 mV), which is localized to the inner membrane. This result is thermodynamically consistent with the reported ET pathway for *S. oneidensis* MR-1 (Fig. 12) (Shi et al., 2007). However, our studies have revealed that the redox properties of OM *c*-Cyts under *in-vivo* conditions significantly differ from those under *in-vitro* conditions. Specificaly, purified OM *c*-Cyts have an E_m that is 200 mV more negative than that observed under *in-vivo* conditions. This finding indicates that the heme environment *in vivo* significantly differs from that *in vitro*, as the redox potential of heme molecules is strongly affected by ligation and hydration (Sola et al., 2002). Therefore, the large potential difference for OM *c*-Cyts under these two conditions highlights the limitation of *in-vitro* studies for understanding the *in-vivo* energetics and dynamics of bacterial EET.

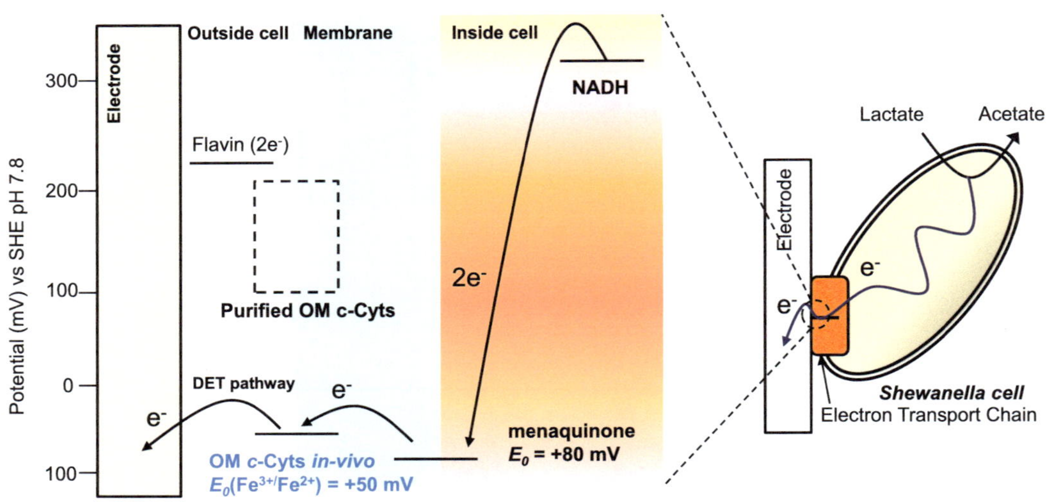

Fig. 12. Energy diagram for the DET pathway from inside to outside cells. Electrons generated by bacterial metabolic processes, such as the conversion of lactate to acetate, are transferred to an extracellular electrode through the redox reaction of menaquinone in the inner membrane and OM *c*-Cyts. The E_m region of purified OM *c*-Cyts is represented as a dashed-line box.

The redox potential of OM *c*-Cyts determined under *in-vivo* conditions severely complicates the current understanding of the flavin-mediated EET pathway, which is considered the primary EET path for microbial current production by *Shewanella* species (Marsili et al., 2008). In this pathway, flavin is proposed to function as a redox mediator that receives two electrons from the reduced form of OM *c*-Cyts (Fe^{2+}) and subsequently delivers the electrons to an extracelluar acceptor, such as an electrode surface ($FMN + 2e^- + 2H^+ \rightarrow FMNH_2$, E_m = −210 mV) (Marsili et al., 2008; Gralnick et al., 2010). However, the E_m of $FMNH_2$ (−210 mV) is approximately 260 mV more negative than that of OM *c*-Cyts determined under *in-vivo* conditions (Fig. 12). As a 260-mV energy barrier would strongly inhibit ET from OM *c*-Cyts to flavin, a different mechanism is likely operating in cells that accounts for the flavin reduction reaction by OM *c*-Cyts. We are currently conducting investigations on the *in-vivo* EET process to resolve this apparent conflict in energetics. Recently, our group has elucidated a novel mechanism underlying the flavin-mediated EET pathway, which will be described in the near future, by noting the specific affinity of flavin for OM *c*-Cyts.

6. Conclusion

We have introduced three techniques to tackle the study of *in-vivo* EET mediated by OM *c*-Cyts: whole-microorganism electrochemistry, UV-vis electrochemical EW spectroscopy, and the in-frame deletion of OM *c*-Cyts. The EET process of *S. oneidensis* MR-1 to an ITO electrode was confirmed by monitoring metabolic current using a single-chamber, three-electrode system. The application of an NO chemical labeling technique indicated that the redox wave detected at +50 mV was assignable to *c*-Cyts localized at the cellular membrane, and EW electrospectroscopy confirmed that the ET process was coupled with metabolic lactate oxidation. Furthermore, using the mutant MR-1 strains, the specific *c*-Cyts responsible for EET were determined to be members of the OmcA-MtrCAB protein complex. Together, these findings represent the first *in-vivo* evidence for the involvement of OM *c*-Cyts in direct EET by *S. oneidensis* MR-1, in which the metabolic electrons generated by intracellular lactate oxidation are delivered to extracellular electrode surfaces as a terminal step for anaerobic respiration.

The results of the new spectroscopic and electrochemical techniques presented in this chapter emphasize the importance of *in-vivo* bioelectrochemistry for the chemical and physical elucidation of the EET process.

7. Acknowledgements

We thank Prof. K. H. Nealson for providing the ΔmtrC/omcA, ΔmtrB, ΔmtrA, ΔcymA, and ΔpilD mutants. This work was financially supported by the Exploratory Research for Advanced Technology (ERATO) program of the Japan Science and Technology Agency (JST), and partially by Research on Priority Areas from the Ministry of Education, Culture, Sports, Science, Technology (MEXT) of the Japanese Government (21750186), The Canon Foundation, and Research Fellowships of the Japan Society for Promotion of Science (JSPS) for Young Scientists (00218864).

8. References

Andrew, C. R., George, S. J., Lawson, D. M. & Eady, R. R. (2002). Six- to five-coordinate heme-nitrosyl conversion in cytochrome *c*' and its relevance to guanylate cyclase. *Biochemistry*, Vol. 41, No. 7, pp. 2353-2360, ISSN 0006-2960.

Bouhenni, R. A., Vora, G. J., Biffinger, J. C., Shirodkar, S., Brockman, K., Ray, R., Wu, P., Johnson, B. J., Biddle, E. M., Marshall, M. J., Fitzgerald, L. A., Little, B. J., Fredrickson, J. K., Beliaev, A. S., Ringeisen, B. R. & Saffarini, D. A. (2010). The Role of Shewanella oneidensis MR-1 Outer Surface Structures in Extracellular Electron Transfer. *Electroanalysis*, Vol. 22, No. 7-8, pp. 856-864, ISSN 1040-0397.

Bretschger, O., Obraztsova, A., Sturm, C. A., Chang, I. S., Gorby, Y. A., Reed, S. B., Culley, D. E., Reardon, C. L., Barua, S., Romine, M. F., Zhou, J., Beliaev, A. S., Bouhenni, R., Saffarini, D., Mansfeld, F., Kim, B. H., Fredrickson, J. K. & Nealson, K. H. (2007). Current production and metal oxide reduction by *Shewanella oneidensis* MR-1 wild type and mutants. *Appl Environ Microbiol*, Vol. 73, No. 21, pp. 7003-7012, ISSN 0099-2240.

Chiavarino, B., Crestoni, M. E., Fornarini, S., Lanucara, F., Lemaire, J., Maitre, P. & Scuderi, D. (2008). Direct probe of NO vibration in the naked ferric heme nitrosyl complex. *Chemphyschem*, Vol. 9, No. 6, pp. 826-828, ISSN 1439-4235.

de Groot, M. T., Evers, T. H., Merkx, M. & Koper, M. T. M. (2007). Electron transfer and ligand binding to cytochrome c' immobilized on self-assembled monolayers. *Langmuir*, Vol. 23, No. 2, pp. 729-736, ISSN 0743-7463.

Eggleston, C. M., Voros, J., Shi, L., Lower, B. H., Droubay, T. C. & Colberg, P. J. S. (2008). Binding and direct electrochemistry of OmcA, an outer-membrane cytochrome from an iron reducing bacterium, with oxide electrodes: A candidate biofuel cell system. *Inorganica Chimica Acta*, Vol. 361, No. 3, pp. 769-777, ISSN 0020-1693.

Firer-Sherwood, M., Pulcu, G. S. & Elliott, S. J. (2008). Electrochemical interrogations of the Mtr cytochromes from *Shewanella*: opening a potential window. *J Biol Inorg Chem*, Vol. 13, No. 6, pp. 849-854, ISSN 0949-8257.

Fornarini, S., Chiavarino, B., Crestoni, M. E. & Rovira, C. (2008). Unravelling the intrinsic features of NO binding to iron(II)- and iron(III)-hemes. *Inorganic Chemistry*, Vol. 47, No. 17, pp. 7792-7801, ISSN 0020-1669.

Gralnick, J. A., Coursolle, D., Baron, D. B. & Bond, D. R. (2010). The Mtr Respiratory Pathway Is Essential for Reducing Flavins and Electrodes in *Shewanella oneidensis*. *Journal of Bacteriology*, Vol. 192, No. 2, pp. 467-474, ISSN 0021-9193.

Hartshorne, R. S., Jepson, B. N., Clarke, T. A., Field, S. J., Fredrickson, J., Zachara, J., Shi, L., Butt, J. N. & Richardson, D. J. (2007). Characterization of *Shewanella oneidensis* MtrC: a cell-surface decaheme cytochrome involved in respiratory electron transport to extracellular electron acceptors. *J Biol Inorg Chem*, Vol. 12, No. 7, pp. 1083-1094, ISSN 0949-8257.

Hartshorne, R. S., Reardon, C. L., Ross, D., Nuester, J., Clarke, T. A., Gates, A. J., Mills, P. C., Fredrickson, J. K., Zachara, J. M., Shi, L., Beliaev, A. S., Marshall, M. J., Tien, M., Brantley, S., Butt, J. N. & Richardson, D. J. (2009). Characterization of an electron conduit between bacteria and the extracellular environment. *Proc. Natl. Acad. Sci. U S A*, Vol. 106, No. 52, pp. 22169-22174, ISSN 1091-6490.

Heidelberg, J. F., Paulsen, I. T., Nelson, K. E., Gaidos, E. J., Nelson, W. C., Read, T. D., Eisen, J. A., Seshadri, R., Ward, N., Methe, B., Clayton, R. A., Meyer, T., Tsapin, A., Scott, J., Beanan, M., Brinkac, L., Daugherty, S., DeBoy, R. T., Dodson, R. J., Durkin, A. S., Haft, D. H., Kolonay, J. F., Madupu, R., Peterson, J. D., Umayam, L. A., White, O., Wolf, A. M., Vamathevan, J., Weidman, J., Impraim, M., Lee, K., Berry, K., Lee, C., Mueller, J., Khouri, H., Gill, J., Utterback, T. R., McDonald, L. A., Feldblyum, T. V., Smith, H. O., Venter, J. C., Nealson, K. H. & Fraser, C. M. (2002). Genome sequence of the dissimilatory metal ion-reducing bacterium *Shewanella oneidensis*. *Nature Biotechnology*, Vol. 20, No. 11, pp. 1118-1123, ISSN 1087-0156.

Hoshino, M., Ozawa, K., Seki, H. & Ford, P. C. (1993). Photochemistry of Nitric-Oxide Adducts of Water-Soluble Iron(Iii) Porphyrin and Ferrihemoproteins Studied by

Nanosecond Laser Photolysis. *Journal of the American Chemical Society*, Vol. 115, No. 21, pp. 9568-9575, ISSN 0002-7863.

Kim, B. H., Kim, H. J., Hyun, M. S. & Park, D. H. (1999). Direct electrode reaction of Fe(III)-reducing bacterium, *Shewanella putrefaciens*. *Journal of Microbiology and Biotechnology*, Vol. 9, No. 2, pp. 127-131, ISSN 1017-7825.

Kouzuma, A., Meng, X. Y., Kimura, N., Hashimoto, K. & Watanabe, K. (2010). Disruption of the Putative Cell Surface Polysaccharide Biosynthesis Gene SO3177 in *Shewanella oneidensis* MR-1 Enhances Adhesion to Electrodes and Current Generation in Microbial Fuel Cells. *Appl. Environ. Microbiol.*, Vol. 76, No. 13, pp. 4151-4157, ISSN 0099-2240.

Laviron, E. (1979). General Expression of the Linear Potential Sweep Voltammogram in the Case of Diffusionless Electrochemical Systems. *Journal of Electroanalytical Chemistry*, Vol. 101, No. 1, pp. 19-28, ISSN 0022-0728.

Li, S. L., Freguia, S., Liu, S. M., Cheng, S. S., Tsujimura, S., Shirai, O. & Kano, K. (2010). Effects of oxygen on Shewanella decolorationis NTOU1 electron transfer to carbon-felt electrodes. *Biosens. Bioelectron.*, Vol. 25, No. 12, pp. 2651-2656, ISSN 1873-4235.

Logan, B. E., Kiely, P. D., Call, D. F., Yates, M. D. & Regan, J. M. (2010). Anodic biofilms in microbial fuel cells harbor low numbers of higher-power-producing bacteria than abundant genera. *Applied Microbiology and Biotechnology*, Vol. 88, No. 1, pp. 371-380, ISSN 0175-7598.

Lovley, D. R. (2006). Bug juice: harvesting electricity with microorganisms. *Nat Rev Microbiol*, Vol. 4, No. 7, pp. 497-508, ISSN 1740-1526.

Marsili, E., Baron, D. B., Shikhare, I. D., Coursolle, D., Gralnick, J. A. & Bond, D. R. (2008). *Shewanella* Secretes flavins that mediate extracellular electron transfer. *Proc Natl Acad Sci U S A*, Vol. 105, No. 10, pp. 3968-3973, ISSN 0027-8424.

Myers, C. R. & Myers, J. M. (1992). Localization of Cytochromes to the Outer-Membrane of Anaerobically Grown Shewanella-Putrefaciens MR-1. *Journal of Bacteriology*, Vol. 174, No. 11, pp. 3429-3438, ISSN 0021-9193.

Myers, C. R. & Myers, J. M. (1997). Outer membrane cytochromes of *Shewanella putrefaciens* MR-1: spectral analysis, and purification of the 83-kDa *c*-type cytochrome. *Biochim Biophys Acta*, Vol. 1326, No. 2, pp. 307-318, ISSN 0006-3002.

Nakamura, R., Ishii, K. & Hashimoto, K. (2009). Electronic absorption spectra and redox properties of C type cytochromes in living microbes. *Angew Chem Int Ed*, Vol. 48, No. 9, pp. 1606-1608, ISSN 1521-3773.

Newton, G. J., Mori, S., Nakamura, R., Hashimoto, K. & Watanabe, K. (2009). Analyses of current-generating mechanisms of *Shewanella loihica* PV-4 and *Shewanella oneidensis* MR-1 in microbial fuel cells. *Appl Environ Microbiol*, Vol. 75, No. 24, pp. 7674-7681, ISSN 1098-5336 (Electronic) 0099-2240.

Okamoto, A., Nakamura, R., Ishii, K. & Hashimoto, K. (2009). In vivo electrochemistry of C-type cytochrome-mediated electron-transfer with chemical marking. *Chembiochem*, Vol. 10, No. 14, pp. 2329-2332, ISSN 1439-7633.

Pitts, K. E., Dobbin, P. S., Reyes-Ramirez, F., Thomson, A. J., Richardson, D. J. & Seward, H. E. (2003). Characterization of the *Shewanella oneidensis* MR-1 decaheme cytochrome MtrA: expression in Escherichia coli confers the ability to reduce soluble Fe(III) chelates. *J. Biol. Chem.*, Vol. 278, No. 30, pp. 27758-27765, ISSN 0021-9258.

Reguera, G., Nevin, K. P., Nicoll, J. S., Covalla, S. F., Woodard, T. L. & Lovley, D. R. (2006). Biofilm and nanowire production leads to increased current in *Geobacter sulfurreducens* fuel cells. *Appl. Environ. Microbiol.*, Vol. 72, No. 11, pp. 7345-7348, ISSN 0099-2240.

Richardson, D. J., Field, S. J., Dobbin, P. S., Cheesman, M. R., Watmough, N. J. & Thomson, A. J. (2000). Purification and magneto-optical spectroscopic characterization of cytoplasmic membrane and outer membrane multiheme *c*-type cytochromes from *Shewanella frigidimarina* NCIMB400. *Journal of Biological Chemistry*, Vol. 275, No. 12, pp. 8515-8522, ISSN 0021-9258.

Roh, Y., Gao, H. C., Vali, H., Kennedy, D. W., Yang, Z. K., Gao, W. M., Dohnalkova, A. C., Stapleton, R. D., Moon, J. W., Phelps, T. J., Fredrickson, J. K. & Zhou, J. Z. (2006). Metal reduction and iron biomineralization by a psychrotolerant Fe(III)-reducing bacterium, *Shewanella* sp strain PV-4. *Applied and Environmental Microbiology*, Vol. 72, No. 5, pp. 3236-3244, ISSN 0099-2240.

Saffarini, D. A., Blumerman, S. L. & Mansoorabadi, K. J. (2002). Role of menaquinones in Fe(III) reduction by membrane fractions of *Shewanella putrefaciens*. *Journal of Bacteriology*, Vol. 184, No. 3, pp. 846-848, ISSN 0021-9193.

Saltikov, C. W. & Newman, D. K. (2003). Genetic identification of a respiratory arsenate reductase. *Proc. Natl. Acad. Sci. U S A*, Vol. 100, No. 19, pp. 10983-10988, ISSN 0027-8424.

Schroder, U., Niessen, J. & Scholz, F. (2003). A generation of microbial fuel cells with current outputs boosted by more than one order of magnitude. *Angew Chem Int Ed*, Vol. 42, No. 25, pp. 2880-2883, ISSN 1433-7851.

Scott, J. H. & Nealson, K. H. (1994). A Biochemical-Study of the Intermediary Carbon Metabolism of *Shewanella-Putrefaciens*. *Journal of Bacteriology*, Vol. 176, No. 11, pp. 3408-3411, ISSN 0021-9193.

Shi, L., Chen, B., Wang, Z., Elias, D. A., Mayer, M. U., Gorby, Y. A., Ni, S., Lower, B. H., Kennedy, D. W., Wunschel, D. S., Mottaz, H. M., Marshall, M. J., Hill, E. A., Beliaev, A. S., Zachara, J. M., Fredrickson, J. K. & Squier, T. C. (2006). Isolation of a high-affinity functional protein complex between OmcA and MtrC: Two outer membrane decaheme *c*-type cytochromes of *Shewanella oneidensis* MR-1. *J Bacteriol*, Vol. 188, No. 13, pp. 4705-4714, ISSN 0021-9193.

Shi, L., Squier, T. C., Zachara, J. M. & Fredrickson, J. K. (2007). Respiration of metal (hydr)oxides by *Shewanella* and *Geobacter*: a key role for multihaem *c*-type cytochromes. *Mol Microbiol*, Vol. 65, No. 1, pp. 12-20, ISSN 0950-382X.

Shi, L. A., Richardson, D. J., Wang, Z. M., Kerisit, S. N., Rosso, K. M., Zachara, J. M. & Fredrickson, J. K. (2009). The roles of outer membrane cytochromes of *Shewanella* and *Geobacter* in extracellular electron transfer. *Env. Microbiol. Rep.*, Vol. 1, No. 4, pp. 220-227, ISSN 1758-2229.

Sola, M., Battistuzzi, G., Borsari, M., Cowan, J. A. & Ranieri, A. (2002). Control of cytochrome c redox potential: Axial ligation and protein environment effects. *Journal of the American Chemical Society*, Vol. 124, No. 19, pp. 5315-5324, ISSN 0002-7863.

Torres, C. I., Marcus, A. K., Lee, H. S., Parameswaran, P., Krajmalnik-Brown, R. & Rittmann, B. E. (2010). A kinetic perspective on extracellular electron transfer by anode-respiring bacteria. *FEMS Microbiol Rev*, Vol. 34, No. 1, pp. 3-17, ISSN 1574-6976.

Wigginton, N. S., Rosso, K. M., Shi, L., Lower, B. H. & Hochella, M. F. (2007). Insights into enzymatic reduction of metal-oxides from single-molecule tunneling studies of multiheme cytochromes. *Geochimica Et Cosmochimica Acta*, Vol. 71, No. 15, pp. A1112-A1112, ISSN 0016-7037.

Wigginton, N. S., Rosso, K. M., Stack, A. G. & Hochella, M. F. (2009). Long-Range Electron Transfer across Cytochrome-Hematite (alpha-Fe2O3) Interfaces. *Journal of Physical Chemistry C*, Vol. 113, No. 6, pp. 2096-2103, ISSN 1932-7447.

Zhao, Y., Watanabe, K., Nakamura, R., Mori, S., Liu, H., Ishii, K. & Hashimoto, K. (2010). Three-dimensional conductive nanowire networks for maximizing anode performance in microbial fuel cells. *Chemistry*, Vol. 16, No. 17, pp. 4982-4985, ISSN 1521-3765.

10

The Inflammatory Response of Respiratory System to Metal Nanoparticle Exposure and Its Suppression by Redox Active Agent and Cytokine Therapy

B.P. Nikolaev, L.Yu.Yakovleva, V.A. Mikhalev, Ya.Yu. Marchenko, M.V. Gepetskaya, A.M. Ischenko, S.I. Slonimskaya and A.S. Simbirtsev
Research Institute of Highly Pure Biopreparations
Russian Federation

1. Introduction

Engineered nanoparticles (NPs) are increasingly being developed for needs of electronic, pharmaceutical and chemical industry. Manufactured NPs as a specific subset of ultrafine particles being suspended in air may pose a hazard to human health. Exposures to nanosized airborne metals facilitate respiratory irritation and lung inflammation. The lung is the target portal of environmental and engineered nanopollutants that lead to exacerbation of respiratory diseases, increased risk of infection, allergy and cardiopulmonary mortality. Inflammatory challenge causes hypoxia of tissues. Potential risks of widespread nanoengineered products should be carefully evaluated. The new nanoproducts for therapeutic and diagnostic applications were emerged in pharmaceutical market. The safety and potential hazards of new nanomaterials are not examined properly. Revelation of indicators of pathophysiological response determinants in the reactivity of respiratory tracts is of a great significance, on one hand, for understanding the mechanisms of nanotoxicity of new ultradispersed materials and, on the other hand, for improvement of clinical-diagnostic procedures of studying lung cancer by magnetic-resonance imaging technique (MRI) (Rinck, 2001). The diagnostic efficiency of MRI in elucidation of malignant feature of lung lesion can be improved with the use of contrast agents. The variety of contrast agents for MRI involves a series of preparations on the basis of magnetic NPs of iron oxides. The iron-containing contrast agents are widely used in clinical diagnostics of liver and spleen diseases and in angioplastics due to low toxicity of iron NPs, their ability for prolonged existence in blood flow, and high nonspecific ability of reticular-endothelium system of liver to absorb the dispersed particles (Wang et al., 2001). A contrast agent for increasing contrast of lung tissue in the MRI technique can be delivered by inhalation through respiratory airways. Successful clinical trials of gadolinium chelate inhalation showed a possibility of scanning time reduction in the MRI experiment (Haage et al., 2005). Further development of inhalation procedure of contrasting is governed by the knowledge of the contra-indication pattern, doses, and the scheme of preparation administration. A wide application of iron-containing contrast agents in diagnostics of lung pathology by NMR tomography is determined by magnetic characteristics of contrast agent (Wang et al., 2001; Gubin et al., 2005), possibility of direct clinical administration of contrast agent into lung tissue (Limbach, 2007; Oberdörster G., et al 2005), safety of contrast agent, its biological availability, and physiological response of bronchoalveolar system to the influence of dispersion agent (Li et al., 1999; Gupta et al., 2005; Oberdörster et al., 2000; Tran et al., 2000).

Pulmonary inflammation is known to be controlled by a complex network of cellular and humoral mediators that have pro-inflammatory and regulatory anti-inflammatory functions. The main cellular components in NPs injury are neutrophils and macrophages and the vast number of antibody mediators include cytokines, reactive oxygen species, electron-transport proteins and proteases. The pro-inflammatory cytokine interleukin-1 beta (IL-1β) is produced by alveolar macrophages and neutrophils in response to iron NPs challenge (Simbirtsev, 2011). This cytokine is crucial in the cytokine cascade events because it activates other inflammation stages. IL-1β promotes the movement of neutrophils by augmenting interaction between neutrophils and endothelial cells and increasing capillary leak. Another cytokine as IL-1β receptor antagonist may play an essential role in decreasing inflammatory reactions of certain respiratory states associated with acute inflammation induced by NPs. The animals which have got IL-1β intratracheally rapidly develop lung neutrophil accumulation and neutrophil-dependent acute edemateous lung leak which may be decreased by intravenous administration of IL-1β receptor antagonist. Intratracheal instillation of IL-1β receptor antagonist has been shown to lessen acute lung inflammation. Aerosol delivery of IL-1β antagonist may prevent pneumonia by impeding bacterial access and reducing inflammation phenomena in pulmonary tract. For anti-inflammatory treatment caused by impact of NPs the interleukin-1 receptor antagonist (IL-1ra) was applied. Recent studies had shown that IL-1ra may play an essential role in decreasing inflammatory reactions of certain respiratory disorders induced by administration of metal dust powder (Danilov, 2003). IL-1β may decrease inflammation in the hypoxia cases which as a rule typical for NPs aerosol exposures. Hypoxia transforms the endothelium toward a pro-inflammatory phenotype and enhance leukocyte adhesion on airway walls provoking pathological conditions of vascular remodeling and ischemia–reperfusion injury.

Efficient immunotherapy of inflammatory impact by NPs is energy consumptive procedure. Activation of alveolar macrophages and neutrophils induced by phagocytosis of NPs demands high ATP-dependent energy supply (Buttgereit et al., 2000). Specific immune functions mediated by IL-1 as migration diseases associated with infection and acute inflammation. Aerosol administration of IL-1ra may be employed for efficient delivery of IL-1ra to the bronchoalveolar space versus instillation procedure. Direct delivery of IL-1ra to the target site of lung in a high ratio of local to systemic bioavailability results in the decrease of an effective drug dose. However, the clinical efficiency of antagonist IL-1 of leucocytes, cytokinesis, phagocytosis, antigen processing and its presentation and synthesis of antibodies, cytotoxicity, regulatory functions consume large amounts of nucleoside triphosphate (ATP). Immunocompetent cells affiliated to IL-1 and its antagonist require much energy to maintain cellular integrity and basal metabolism (Dziurla et al., 2010). Most of their specific immune functions directly or indirectly use ATP or other high-energy nucleoside phosphates. Active cell respiration provides free energy in the chemical form of ATP and electrical transmembrane potential (Nicholls, 2000). Oxidative phosphorylation and glycolysis in mitochondria ensures proper functioning of immune cells as lymphocytes.

Mitochondria support the energy-dependent regulation of assorted cell functions, including intermediary metabolism, ion regulation and cell motility, cell proliferation through processes of oxidative phosphorylation and glycolysis. In active metabolic state respiratory cells derive up dominant energy through oxidative phosphorylation. But inflammation leads to decrease of main oxidative substrates in tissue, oxygen and glucose and fall down of synthesis of energy rich compounds ATP, NADH. Oxygen deficiency during inflammatory cause of NPs challenge may produce deep changes in cellular metabolism that may lead to tissue failure. Hypoxia and lack of glucose affect on cellular energy metabolism and on cytokine secretion in stimulated human CD4+ T lymphocytes (Dziurla et al., 2010). Therefore, bioenergetic aspect of cell functioning is very essential for treatment of cell injury by cytokine therapy.

The central point of energy conversion in injured cells under hypoxic stress is the transfer of electron via electron-transport chain (ETC) to final oxygen acceptor coupled with generation proton gradient across mitochondrial membrane. Transmembrane electrical potential $\Delta\psi$ is the measure of cell energetic activity (Skulachev, 1996). Study of metabolic adaptation to hypoxia stress displayed the key role of membrane potential $\Delta\psi$ as the indicator and regulator of metabolic cell activity. The absolute value of transmembrane potential $\Delta\psi$ depends on the physiology of cell and can be easily evaluated by electrode method. In respiratory systems of cells quinone can operate as donor-acceptor junction sites for transfer electrons between redox active enzymes. Quinones are membrane-entrapped redox-active entities that carry 2 electrons and 2 protons in the quinol state. Synthesized species of quinones with different electrochemical potential are able to repair the defected sites of ETC after inflammatory influence of metal NPs. The pharmaceutical "Oliphen" belongs to the class quinones with redox properties compatible to ETC. Oliphen is the mixture of poly-(2,5-dehydro-oxy-phenylene-4)-thiosulfonate sodium redox oligomers with median molecular mass 300-500 D. The trade names of medicines containing oliphen are «Oliphen», «Olifen», «Hipoxen». Redox properties of oliphen are linked with quinone and thiosulfate residues in their molecular structure. Oliphen activates ETC of cells shunting the possible molecular disorders of oxidative reactions in mitochondria. Oliphen enhances the resistance to extreme conditions such as oxygen stress and ischemia. The preparation "Oliphen" is a prescribed medicine for correcting defects of mitochondrial electron-transport chain associated with inflammatory state. The application of redox agent "Oliphen" is proposed to favour the therapy efficiency of ethiotropic preparations of receptor antagonist IL-1β. To be sure in benefits of redox curing of mitochondrial defects of ETC by oliphen in treating tissue hypoxia followed after metal NPs exposure the present mouse model study is aimed to determine antihypoxic efficacy of aerosol oliphen exposures.

Here we present the results of studying the effect of magnetic iron NPs on the bronchoalveolar system of mouse at cellular and biochemical levels in inhalation contact with mucosa (Nikolaev, 2009). Intratracheal instillation was used as well-known standard procedure for evaluation of respiratory toxicity of particles. The iron NPs for intratracheal administration were prepared by gas-phase synthesis and precipitation from suspensions. The ability of the samples for contrasting of magnetic resonance images was tested based on the line broadening of ^{1}H NMR spectra of aqueous iron suspensions in vitro.

The response of bronchoalveolar system to iron NPs was studied based on variations in some characteristics of bronchoalveolar lavage (BAL). A special attention was given to anti-inflammatory activity of the preparation as a factor of its inhalation application. The level of anti-inflammatory activity was estimated from the protein concentration, content of neutrophils, and methabolic fingerprints of cellular respiration elucidated using ^{1}H NMR of supernatant obtained after centrifuging of BAL. The metabolic activity of immunecompetent cells in BAL was assayed by measurement of trans membrane potential $\Delta\psi$ and redox potential (ORP). The choice of these parameters for characterization of inflammation response and hypoxic state was based on the previous study of anti-inflammatory activity of intranasal instillations of bacterial lipopolysaccharides (Ischenko et al., 2007). This study showed that mouse pneumonia causes neutrophil recruitment into alveoli, an increase in the total cell content of BAL, generation of active oxygen species, and variation in the metabolic activity of bronchoalveolar lavage fluid (BALF). To study the characteristics of inflammatory response we used microscopy, high performance liquid chromatography (HPLC), gel electrophoresis, ^{1}H NMR spectroscopy, and various methods of estimating size and concentration of finely dispersed preparations.

2. Materials and methods

2.1 Animal handling and study design

The reactivity of bronchoalveolar system was studied using white outbred mice (n=20) and mice of the C57BL/6 line (n=25). As the preparation for the current study, the mice were kept in vivarium with ordinary food and water ad libitum. A single dose of inflammation inducer in the form of iron NPs (1 mg per mouse) suspended in isotonic sodium chloride solution (0.9% w/v) was administered to a group of anesthetized mice through trachea with the use of a DP-4M aerosol microsyringe (Penn Century, the United States). Before using, an aerosol nebulizer was checked for reproducibility of dosage and fractional composition of the aerosol. After intratracheal administration of iron NPs the mice were killed by cervical dislocation and the thorax was dissected. To obtain 4 ml of lavage, four 1-ml portions of 0.15 M NaCl solution were administered into the free trachea. The cellular component of BAL fluid was separated by centrifuging at 1700 rpm for 5 min and analyzed immediately after preparation. The samples for subsequent studies were kept at 4°C. The inflammatory response of bronchopulmonary system of mice was estimated within 1, 3, and 7 days after inhalation contact with iron NPs from the variation of the relative content of cellular component in BAL. The inflammatory response of mouse bronchopulmonary system was monitored after 3 days from the reliable difference in the characteristics of BAL fluids obtained from intact and infected mice.

The anti-inflammatory treatment by human IL-1ra was studied upon delivery of the aerosol formulation of the drug to mice with acute pulmonary inflammation, induced by iron NPs. IL-1ra was produced by recombinant gene technology in State Research Institute of Highly Pure Biopreparations (Federal Medical -Biological Agency, Russia). The protein IL-1ra (99% purity) had been isolated from E.coli BL21 in solution. The structural identity of IL-1ra to native protein was confirmed by gel electrophoresis and HPLC data.

Redox active drugs were delivered into upper and lower parts of respiratory tract in the form of dry and droplike aerosols. Oliphen was used in the form of preparation "Oliphen" (SRR" Oliphen", Saint-Petersburg, № PN000125/01-2000). Its efficacy was compared with reference drugs: sodium succinate (Reachem), sodium oxybutirate (Reachem). The dosage exposure was selected in conformity with therapeutical effective concentrations achieved by intravenous and peroral administration to humans. To get the respirable medicinal liquid aerosol the jet and ultrasonic nebulizers were applied. The particle-size analysis was monitored by photo-electric counter combined with microcomputer PDP-11. A mass median aerodynamic diameter (MMAD) of droplets was no more than 5 mkm. The aerosol concentration was determined by measuring the mass of deposited particles on the filters or microcyclones (impingers).

The acceptance and compliance of dry powder inhalations for redox agent treatments was accessed in original inhalation system. The inhalation system consisted of pneumatic glass generator (unitary dose 50-250 mg and MMAD 2.5-4 mkm). Micronized powder formulation for mice inhalation was prepared by air-jet milling with inert excipients. Aspiration doses for mice were calculated in approach with coefficient of aerosol deposition 0.88 MMAD 3 mkm and polydispersity square deviation 0.65. The mice were exposed to aerosol therapy for prechallenged terms 10, 30, 60 min. The aerosol inhalations occurred once a day during 1, 3, 5 day curse. Aspiration doses were 0,03 -0,06 mkg/mouse.

For aerosol inhalation exposure of mice by antihypoxic drugs the specially constructed dynamic and static chambers were applied. Groups of 4, 10 mice were placed in chambers, containing special cages to prevent any possibility to get drug from external hair surfaces of animals.

Survival in groups of mice after aerosol exposure to antihypoxic agents was studied in model of acute normobaric hypoxia. Outbred white mice (20-25 g) were used. Animals were placed in the transparent plastic box with volume 30 l with an inlet and outlet through which the nitrogen flowed. Hypoxia was induced by continuous flushing N_2 up to partial pressure pO_2 4.2% during 10 min. The O_2 concentration was monitored with an in-line oxygen analyzer (Clark's electrode and registration block). The final CO_2 level was 20% according measurements of IR absorption of hypoxic gas. The whole number of mice that have received an acute hypoxia challenge and aerosol therapy was 100.

2.2 Synthesis of iron magnetic NPs

Iron NPs were synthesized by two ways: continuous gas-phase synthesis (Kim et al., 2007; Vasilieva et al., 2006; Choi et al., 2002; Montagne et al., 2002) and precipitation from a suspension (Vidal-Vidal et al., 2006; Capek et al., 2004). NPs of iron oxides were synthesized by reduction of iron salts in a solution according to the procedure of suspension precipitation. NPs with an inner iron core were synthesized by pyrolysis of iron pentacarbonyl in a flow-type reactor in the medium of anhydrous carbon oxide at atmospheric pressure (Vasilieva et al., 2006; Choi et al., 2002). The fractional composition of the resulting powders was estimated using light and electron microscopy. The average particle size was approximately 30 nm. Iron NPs were sphere-shaped and formed filamentary structures due to the strong magnetic interaction between spheres. The powder sample NPs consisted of NPs with inner iron core 30 nm in diameter covered with outer Fe_3O_4 shell (2-3 nm). The physicochemical characteristics of iron NPs are given in details in (Kim et al., 2007; Vasilieva et al., 2006; Choi et al., 2002). The magnetic characteristics of iron NPs were observed from the abrupt reduction of transverse nuclear magnetic relaxation time T_2 of solvent in NPs suspension.

2.3 NMR experiment

The samples of BAL fluids were transferred to standard NMR tubes 5 mm in diameter. The ^1H NMR spectra of the samples of BAL fluids were recorded at room temperature on Bruker CXP 300 spectrometer operating at 300 MHz. To detect weak signals of metabolites the large signal of water was suppressed by homonuclear irradiation at the frequency of water resonance using one pulse sequence. The number of acquisition was varied from 500 to 2000. No exponential weighting function was applied to the free induction decay before Fourier transformation. Chemical shifts were measured relative to the signal of ammonium formate at 8.50 ppm; the error in determination of chemical shifts did not exceed 0.02 ppm. The 4.5-6.0 ppm spectral region was removed to eliminate water signal. The concentration of a metabolite was determined from the ratio of integral intensities of the corresponding metabolite signal and the signal of ammonium formate.

2.4 The bronchoalveolar lavage analysis by HPLC

The samples of BAL fluids from exposed mice, mouse serum and one sample of IL-1ra were prepared for HPLC. The total protein was measured by Lowry assay. The bronchoalveolar lavage analysis was realized by reversed-phase HPLC with conditions listed below: chromatograph HP-1090 Hewlett–Packard, USA; column "Jupiter" Phenomenex C-18 (5 µm, length 250 mm, internal diameter 4.6 mm), USA; solvent system gradient, 0.1 % trifluoroacetic acid/acetonitrile, 30% to 70% acetonitrile in 15 minutes; temperature 35°C; flow rate 1.5 ml/min; detection: 220, 254, 280 nm; injection volume 50 µl.

2.5 Polyacrylamide gel electrophoresis

The proteins of BALF from exposed and nonexposed mice were analyzed by SDS-polyacrylamide gel electrophoresis (SDS-PAGE) according to Laemmli (1970). We used 5% (w/v) stacking gel in Tris–HCl buffer pH 6.8 and 12.5% (w/v) separating gel in pH 8.8 Tris–HCl buffer. The BALF samples were prepared by dilution 1:1 in Tris–HCl buffer 50 mM pH 6.8 with 70 mM sodium dodecyl sulfate (SDS), 1.4 M glycerol, 100 mM dithiothreitol and 0.01 mM bromophenol blue. After 5 min of heating at 100°C, samples were pipetted on gels. Electrophoresis was carried out in a Tris-glycine buffer, pH 8.3, containing 25 mM Tris, 250 mM glycine, 3.5 mM SDS, at a constant power of 50 W. After electrophoresis, gels were stained with Coomassie blue G-250 0.1% (w/v), for 2 h, and then de-stained with 4% (v/v) perchloric acid and 50% (v/v) ethanol solutions. Protein patterns in the gels were recorded as digitalized images and analyzed by Gel Analyzer (MediaCybernetics). Molecular mass of proteins was determined according to molecular weight marker (PageRuler™ Unstained Protein Ladder, Fermentas).

2.6 Transmembrane ($\Delta\psi$) and redox (ORP) potentials

$\Delta\psi$ of bacterial and red blood cells were assayed by the distribution of the penetrating tetraphenylphosphonium (TPP^+) cation between the cytoplasm and the external solution using an ion-selective electrode. Sorption TPP+ by cells was estimated as decrease of its concentration in external media. After washing in physiological solution cell suspensions were added to Tris-HCl buffer medium containing 10 mkM TPP^+. Electrode readings were calibrated by TPP^+ solutions with variable concentration. Transmembrane potential $\Delta\psi$ was calculated using Nernst equation (1) (Skulachev, 1996).

$$\Delta\psi = -RT/zF \times \ln(TPP_{+in}/TPP_{+ext}) \quad (1)$$

The functioning of electron transport chain (ETC) of E.coli cells and the membrane permeability of mouse blood red cells in the presence of therapeutical doses of oliphen were investigated. The estimation of bacterial respiration activity was made with the Clark oxygen electrode. Anaerobic state of cells was achieved by blowing through of nitrogen gas. In the anaerobic state no inhibition of oxidize-phosphorilation enzymes was registrated with the help of membrane potential by the mean of cation tetraphenylphosphonium absorption. Oliphen (0.01-0.1% w/v) caused the decrease of membrane potential and of the rate of cell respiration.

ORP (oxidation-reduction potential, E) of the reaction medium was measured by a platinum electrode (EPV-01, Gomel' Instrumentation Plant, Republic of Belarus') and Ag/Ag Cl reference electrode (ER-10103, Gomel' Instrumentation Plant, Republic of Belarus'). Molecular hydrogen release from bacteria was estimated by using the Orion pH electrode.

3. Results

Intratracheal iron NPs exposure provoked intense infiltration of lymphatic interstitium and blood-capillary net. Increase of alveolar-capillary permeability enhances the recruitment of immune cells (lymphocytes) to lung tissue. Iron NPs exposure of mice with the help of an aerosol microsyringe evoked a transient inflammatory increase of macrophages, neutrophils and lymphocytes number during one week. Acute inflammatory response assayed as relative percentage cell count in BAL at 1, 3, 7 day after instillation NPs is represented at Fig.1.

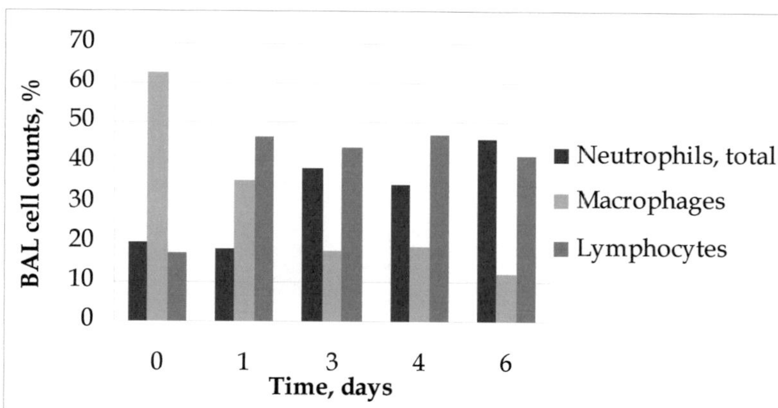

Fig. 1. Bronchoalveolar differential cell counts (macrophages, neutrophils and lymphocytes, %) following NPs challenge (day 0).

The number influx of mature neutrophils from the third day is the robust indicative of acute lung inflammation. Transient fall down of relative macrophage number is the additional evidence of curtain cytotoxicity of iron NPs for macrophages in endocytosis stage. The nanotoxicity for cells of reticulum-endothelial system is implicated with bronchotracheal pathway of delivery of nanoparticulate agent. The observed high absorption of iron NPs by alveolar macrophages is consistent with preliminary results of study in vitro. The cell cultures of human monocytes and macrophages of inbred rat RAW264.7 were reported to uptake superparamagnetic iron oxide NPs (Muller et al., 2007; Stroh et al., 2004) . The particles uptake activates the following cascade of events as the start function of macrophagal lysosomal enzymes, generation of reactive oxidative species (ROS) and signaling system of nuclear transcriptional factor NF-kB that controls expression of a wide variety of pro-inflammatory genes. These features are known to involve in general mechanism of acute inflammation caused not only by nanoparticulate magnetic substance but other external pollutant factors. The ability of micronized particles of various chemical nature and size to induce acute lung inflammation made in some reports supports this notion (Danilov et al., 2003; Siglienti et al., 2006; Fujiia et al., 2002; Schins et al., 2004; Donaldson et al., 2002; Wong et al., 2007).

The cellular profile of BAL compared to corresponding data for whole blood may be interpreted as evidence of general immunoregulatory reaction which develops during few days after NPs delivery to trachea. The data presented in Fig.1 and Table 1 lead to conclusion that inflammation study can be able to explore right from the third day after NPs challenge.

Days after NPs challenge	Neutrophils, mature; %	
	Blood	BAL
0 (control)	25.68	4.37
3	28.30	27.20
4	36.68	16.95
7	23.13	28.00

Table 1. The relative content of granulocytes in blood and BAL at different terms of inflammation.

The inflammation level was estimated by total cell number in BAL and differential content of macrophages, lymphocytes, neutrophils (total and mature). These results are presented in Table 2.

Characteristics of inflammation	Control	Inflammation
Total cells (x106/ml BAL)	0.01±0.00	0.05±0.01
Macrophages,%	50.1±10.3	17.5±7.3
Lymphocytes,%	31.8±5.3	23.8±4.8
Neutrophils, total; %	18.1±6.3	57.4±10.8
Neutrophils, mature; %	4.0±2.3	33.5±9.7

Table 2. Inflammation response of bronchoalveolar system to inhalative challenge by iron NPs according data of cell count in BAL.

The table data after 3-days post exposure intratracheal instillation of iron NPs are appeared to display acute lung inflammation accompanied by the relative 5 times increase of total number of neutrophils and 8 times increase of number of mature lung neutrophils. Accumulation of neutrophils in the BAL is the result of a breakdown of alveolar barrier and impaired lung gas exchange function.

The possible drawbacks of inhalative delivery of redox active agent oliphen had been checked by control of cell composition of BAL, blood and morphological examination of bronchopulmonary ducts. Acute hypoxia changed the cell profile of BAL fluid. Hypoxic challenge provoked lymphocyte influx thus increasing the relative content of lymphoid cells in BAL. No inflammation response or changes in relative quantity of cellular morphological forms in BAL were recorded either for oliphen or for hydroxybutyrate lung delivery both in normoxic and hypoxic conditions. Nontoxic doses determined by separate study of dose-effect dependence in aerosol experiments on BAL data for oliphen were not more than 0.3 mg. Inhalation of oliphen in total doses higher 0.3 mg resulted in toxic accumulative effects and shortage of life-span.

3.1 Proteome analysis of BAL after iron NPs challenge by HPLC and PAAGE

Protein leakage to bronchoalveolar space was assessed by proteome analysis. The results of estimation of total protein content in BALF are presented in Table 3.

Inflammation indicator	Control	Iron NPs administration
Total protein, μg/ml	89.3±3.0	171.3±44.7

Table 3. Influence of inhalative NPs challenge on total protein content in BALF. All values represent mean ± SEM from 5 independent mice in group, for level of significance ($p<0.05$) v.s control (mice without NPs) in accordance with Manna-Whitney and Student tests.

From analysis of total protein follows that iron NPs instilled to trachea give 2-fold increase in BALF total protein content. Altered protein composition of BALF of exposed mice was investigated by HPLC and PAAGE methods. There were two groups of mice (C57BL/6 line, n=10 and outbred mice, n=10), each group contained 5 mice treated with iron NPs and 5 nontreated (control) mice. The analysis of samples of bronchoalveolar lavage supernatants with inflammation caused by NPs at dose 1 mg/mouse in 0.1 ml 0.9 % sodium chloride solution showed the altered protein composition.

The samples numbers 06, 09 from NPs-exposed animals differ from control non exposed samples most prominently for certain as there are new peaks with retention times at ~ 6.8, 7.2, 7.6 min on chromatograms of these samples (Fig. 3, 4). Spectral analysis results showed that all peaks belong to proteins by nature that correlates with data of total protein content in these samples. The chromatogram of a mouse serum sample was made for protein peaks identification.

Comparison of chromatogram of a mouse serum sample with chromatograms of samples numbers 6, 9 leads to conclusion that main peaks retention times of bronchoalveolar lavage sample with inflammation and serum sample were identical. In accordance with samples with inflammation main peaks present serum proteins peaks themselves and don't have any relation to IL-1ra (Fig. 5).

The results obtained for C57BL/6 mice are shown in Fig. 6. The similar results were obtained for outbred mice. Data for outbred mice are not shown. BALF from control and inflammatory mice contain four main spots for proteins with molecular mass 82, 72, 55, 52

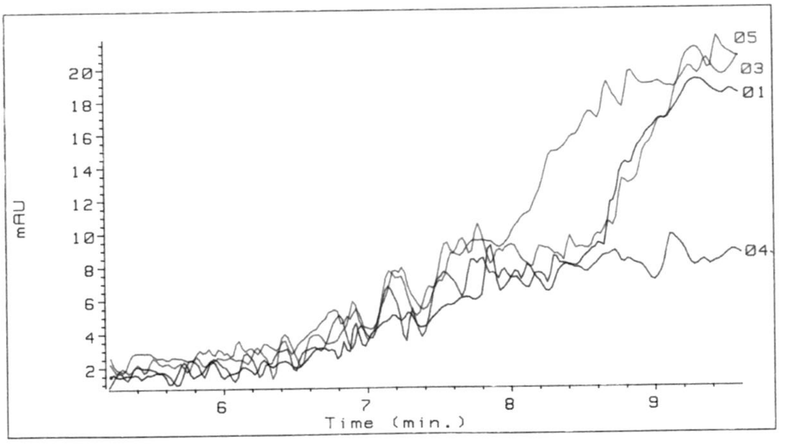

Fig. 3. Chromatograms of bronchoalveolar lavage fluids of C57BL/6 mice nontreated by NPs. 01-05 - BALF samples of nontreated mice (control); wavelength 220 nm.

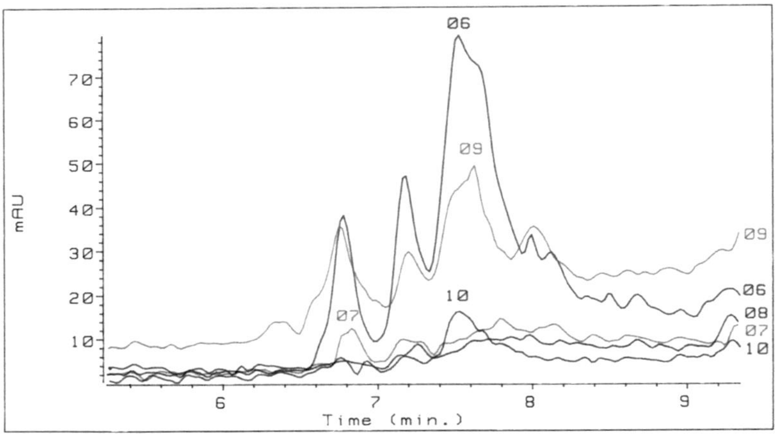

Fig. 4. Chromatograms of bronchoalveolar lavage fluid of C57BL/6 mice treated by NPs. 06-10 - BALF samples of treated mice (with inflammation); wavelength 220 nm.

kDa (Wattiez, R., 2005); furthermore, concentrations of these proteins in inflammatory samples are higher than those in control samples. In addition, one observed the presence of proteins with molecular mass 14, 25, 40 kDa in inflammatory samples. All the above proteins were found in the mouse plasma. The increase in the content of BALF proteins and emergence of low-molecular-mass proteins in BALF (in comparison with control samples) is apparently caused by the increase of capillaries permeability and passive diffusion of plasma proteins.

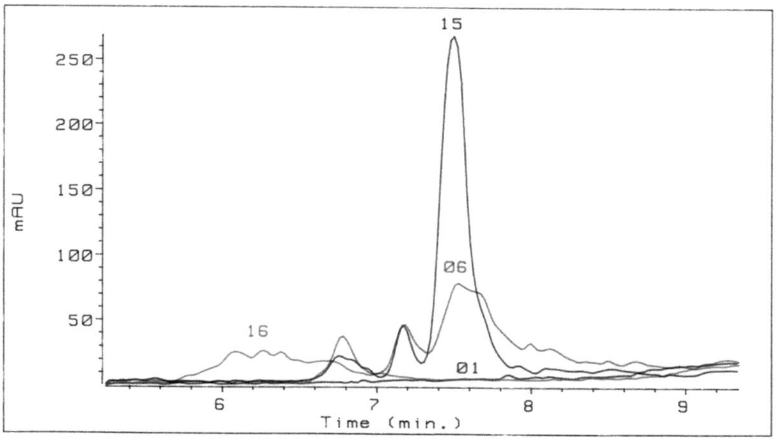

Fig. 5. The comparison of a bronchoalveolar lavage sample with inflammation and a serum sample by HPLC data. 01 - a control BALF sample; 06 - a BALF sample with inflammation; 15 - a control serum sample; 16 - sample of IL-1ra; wavelength 220 nm.

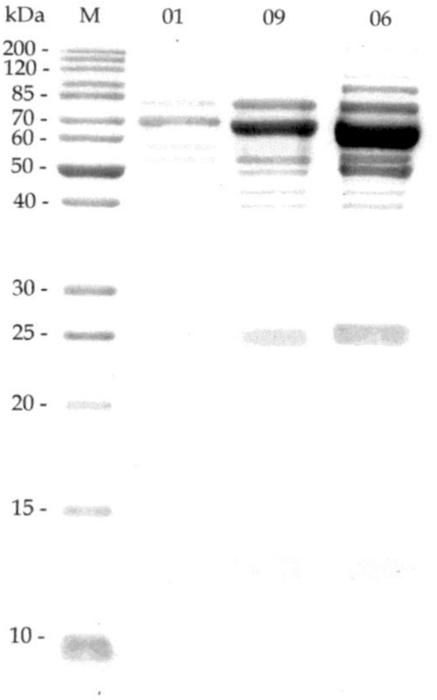

Fig. 6. Gel electrophoresis of BALF proteins obtained from C57BL/6 mice. M – protein molecular weight marker; 01 – BALF sample from nontreated mouse (control); 06 and 09 – BALF samples from mice treated by NPs 1 mg/mouse on the 3rd day after treating.

3.2 NMR analysis

In the ^1H NMR spectra of BAL fluids collected from the control mice there are the signals of lactate (doublet at 1.36 ppm) and acetate (singlet at 1.95 ppm) with the ratio of integral intensities approximately 6:1. In addition, low-intensity signal at 2.4 ppm assigned presumably to succinate and also a broad signal in the range 3.7-3.8 ppm were observed General view of spectra is shown in Fig.7.

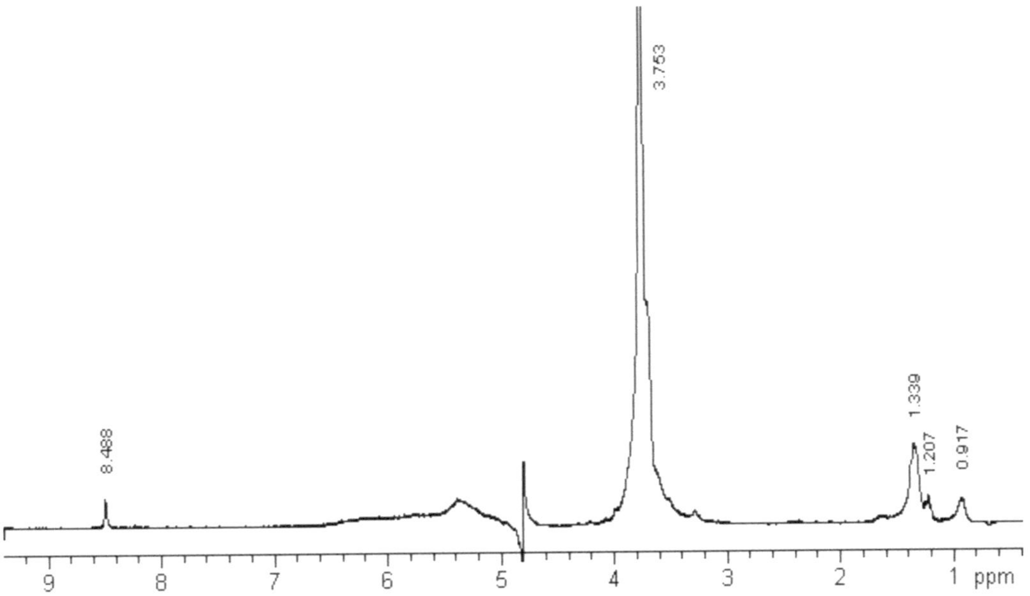

Fig. 7. ^{1}H NMR spectra of BALF of nontreated white outbred mice by NPs.

According to published data (Pottset al., 2001; Azmi et al., 2005) the latter signal was assigned to monosaccharides, particularly glucose, which is characterized by the signals at 3.83, 3.76, and 3.42 ppm. In the spectra of BAL fluids collected from the mice subjected to the action of iron NPs (Fig. 8) there are the signals of lactate and acetate but the signal at 2.4 ppm was not observed.

The main distinction in the spectra of BAL fluids collected from the mice subjected to the action of iron NPs and those collected from the control mice is the presence of a triplet at 1.22 ppm whose intensity is comparable with the intensity of the lactate signal and a multiplet at 3.3 ppm. We assigned these signals to ethanol, which is apparently one of the metabolites appearing under the action of iron NPs on the bronchopulmonary system of a mouse. This assumption was verified by addition of ethanol into the sample of BAL fluid, which resulted in increase in the intensity of the corresponding signals. At the same time, in addition of sodium propionate into the sample of BAL fluid we also observed a triplet and quartet typical for ethyl group but their chemical shifts did not coincide with the chemical shifts of the signals in hand.

The presence of lactate and ethanol in BAL fluids suggests inflammation status of mouse organism, which is accompanied by not only suppression of lung respiration but also insufficient oxygen supply into the cells, i.e. tissue hypoxia. As was noted previously (Nikolaev et al., 2003), an increased level of lactate concentration and the absence of pyruvate signals in the ^{1}H NMR spectra suggest an anoxia of cells. As an example, the ^{1}H NMR spectra of BAL collected from mice subjected to acute hypoxia and acute inflammation caused by bacterial endotoxin (Ischenko, 2007) have signals corresponding to the products of glycolysis similar to those observed in the case of cell hypoxia caused by iron NPs. The emergence of ethanol in BAL of NPs-exposed mice seems to originate from impaired glottic competency and following gastric particle deposition in bronchopulmonary ways. Gastric particulate lung aspiration yields the partial contamination of airways by anaerobic bacteria or fungi. The colonization of the respiratory tract by microorganisms from digestive tract or upper respiratory tract (nose, throat) during inflammation may cause metabolite accumulation of end products of glycolysis in the injured lung. The production of minor quantity of ethanol in the inflamed mice with low level of its oxidation in lung results in BAL ethanol accumulation compared with control mice. A wide range of hypoxia stress tolerant animals use similar adaptive defense strategy switching over to metabolic pathways with synthesis of ketone bodies and low consumption of oxygen. As example, the appearance of ethyl alcohol as a metabolite was also noted for some deep-sea animals, for which metabolism is of anaerobic feature (Leninger, 1982).

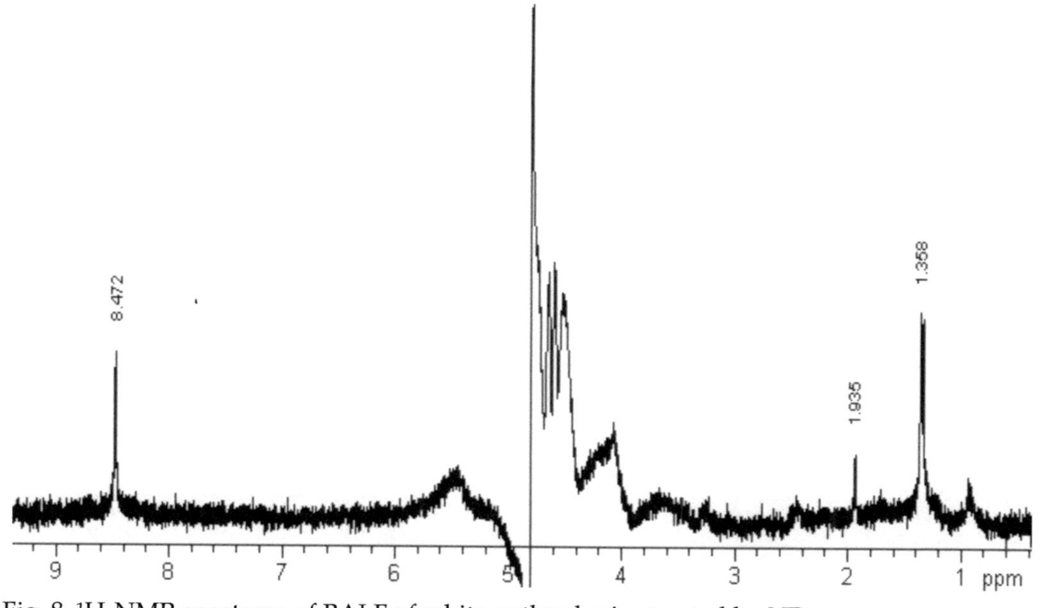

Fig. 8. ^{1}H-NMR spectrum of BALF of white outbred mice treated by NPs.

The data of lactic acidosis were also received by blood spectral analysis of animals in hypoxia stress. NMR spectra of mice blood plasma under acute hypoxia are shown on Figure 9.

The NMR spectra consist of one distorted wide line of resonance of water and groups of lines of saccharides and phospholipids. The resonance of species were assigned in a straitforward manner on the basis of multiplet structure, chemical shift and spectra of individual substances in solution.

The intense quadruplet and doublet lines are the essential characteristics of COH, CH_3 groups of lactate which have the definite position 4.1 and 1.3 ppm in spectra (Aime et al., 2002; Nikolaev et al., 1997). The spectra displayed the appearance of high lactate lines of resonance in hypoxic cases. No traces of pyruvate were observed in spectra. Integral intensity of lines was proportional to concentration of soluble lactate in plasma. Acute hypoxia induced the growth of lactate level up to 6.7 mM (normoxic value 2.3 mM).

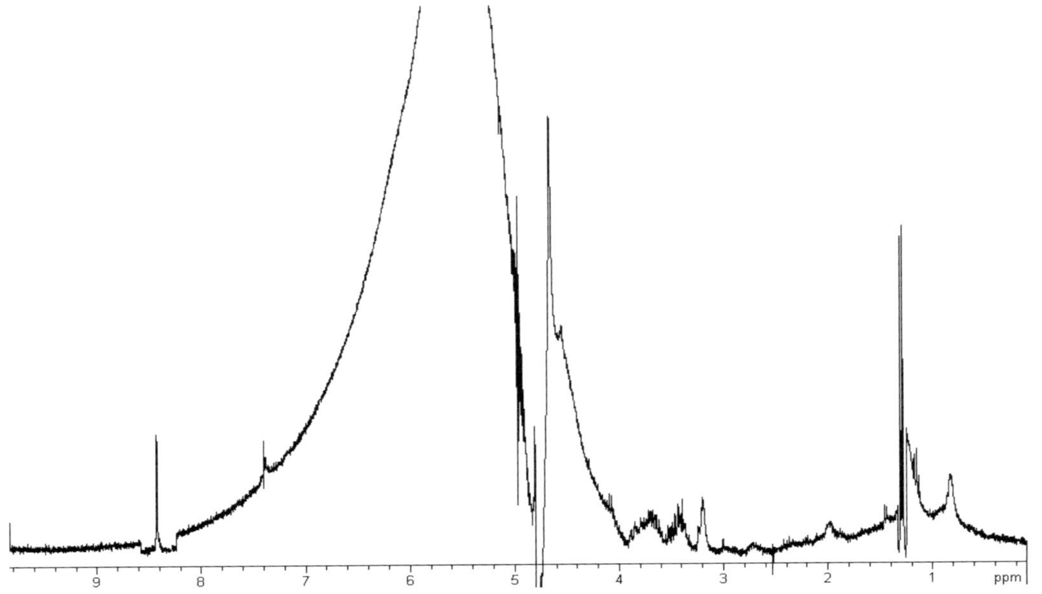

Fig. 9. ^{1}H-NMR spectrum of blood plasma of mouse after acute hypoxia at 4.2% pO_2 during 10 min.

3.3 The post-exposure treatment with IL-1ra

The anti-inflammatory activity of IL-1ra delivered as aerosol was estimated in the experiments with different doses (1, 10, 100 μg) of preparation in the cycle of three inhalations. The indicators of inflammation measured in inbred mice upon IL-1ra administration using aerosol were the percent of mature neutrophils in BAL relative to total BAL cells and the total protein in BAL fluid. The results of these measurements for different doses of IL-1ra dispersion are presented in Table 4. The experimental data show that inhalation of IL-1ra aerosol at the dose of 100 mkg per mouse in the cycle of three inhalations resulted in reduction of inflammation. IL-1ra treatment attenuated the iron NPs-induced protein increase and prevented the increase in pulmonary vascular permeability. The anti-inflammatory effect is achieved due to the 77% homology between recombinant human IL-1ra and the native murine protein. The blockage of IL-1 receptors on phagocytic monocytes, macrophages and pulmonary epithelium cells inhibits the generalization of inflammation throughout respiratory organs. The anti-inflammatory effect of IL-1ra is appeared to be of dose-dependent character.

3.4 The treatment of hypoxia by redox active agent

Anti-hypoxic action of oliphen was studied in a mouse model of acute normobaric hypoxia. Experimental study of hypoxic response of mice to normobaric acute hypoxia exposure revealed the relationship between the rate of mortality, partial pressure O_2, length of exposure and initial time of prechallenge. LD_{50} served as criteria to select partial pressure 4% pO_2 in chamber for 1 hour exposure. Hypoxic response in assigned mice groups was a function of prechallenge time. The dependence of percentage of mortality from prechallenge time had a maximum value at 15 min. The 10 min was chosen as standard time for introduction to hypoxic state for all trials. The response of mice to acute hypoxia has appeared to be dependent on preliminary aerosol treatment by redox agent. Preliminary inhalation of nontoxic doses of oliphen prolonged the survival of mice under hypoxic conditions only at a certain dosage. The time-course of survival rate is shown in Fig.10.

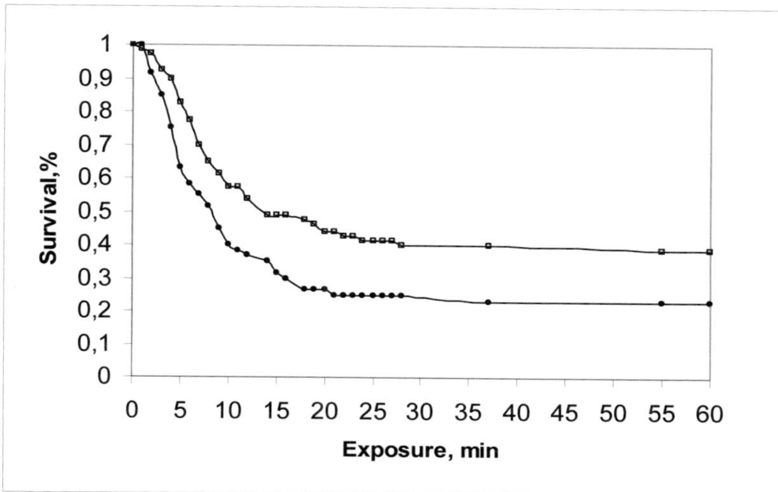

Fig. 10. Survival rate of white outbred mice treated by oliphen (upper line) as compared with control (lower line) in acute normobaric hypoxia for 60 min. Upper line - mice (60 animals) with oliphen (aspiration doses for oliphen 0.05 mg), median of survival 14 min; lower line - nontreated mice (80 animals) without oliphen, median of survival 9 min.

Oliphen prolonged the life-span of mice and increased their survival rate. The median time of life span increased from 9 min to 14 min after aerosol treatment by oliphen. Equivalent aspirated doses of oliphen in droplike and spray-dryed aerosols appeared to exert the same protective effect against acute hypoxia. Significant changes in mytosis of bone marrow cells, growth of permeability of erythrocyte membranes and their electrochemical potential $\Delta\psi$ were consistent with appreciable systemic absorption of oliphen in inhalation route. The investigation of oliphen action on mouse red blood cells by pulmonary delivery indicates that particulated oliphen regulates the permeability of erythrocytes membranes. This action is revealed as the decrease of membrane potential and the increase of the permeability for urea. So, the redox agent oliphen not only participates in the ETC functioning of prokaryotic cells but also regulates the permeability of blood cells as oxygen-transfer units. A full compensation of unsufficient supply of oxygen by physiological adaptation lasted for no more than 2-5 minutes. At late stage the symptoms of tissue hypoxia were observed. Suppressed NAD-dependent aerobic pathway of cellular oxidation switches over more rational succinate-oxidase and alternative pentosephosphate way of glucose consumption (Zarubina, 1999; Hochachka et al., 1996; Boutilier et al., 2000). Oxygen lack in blood leads to intensive high energy phosphate utilization. Severe tissue oxygen insufficiency affects oxidative phosphorylation in mitochondria, switching over the alternative glycolytic ways of respiration. The glycolytic reactions become the dominant source of energy production in cells at oxygen lack. Elevated lactate level observed by NMR method and drop of pH in blood were the important symptoms of cell oxygen starvation (Shchukina, 1986). The common cause of lactic acidosis is tissue hypoxia and as consequence - the onset of glycolytic conversion of glucose to lactate. When there is the sufficient supply of cells by carbon substrate, glycolysis provides dramatic accumulation of lactic acid. Along with the change of pH and lactate level there were a low down shifts of hemoglobin, decrease of hematocrit and the increase of deformability of red blood cells (Table 4). The reticulocyte count in blood after hypoxic challenge diminished too.

	Parameters	Respiration state	
		Normoxia (n=6)	Hypoxia, (n=5)
	Hemoglobin (Hb), %	15.0±1.4	*13.4±0.4
	Mean corpuscular hemoglobin (MCHb), pg/red cell	15.1±1.2	13.9±2.5
	Erythrocyte sedimentation rate (ESR), mm/h	0.7±0.4	0.6±0.2
	Hematocrit (Hct), %	93.8±1.6	*90,0±2,9
	Whole blood pH	7.4±0.1	*6.9±0.2
Red blood cells	Red blood cells (RBC) count, mln/mm^3	9.9±0.6	9.6±1.4
Red blood cells	Red cell deformability (Def.), %	60.3±20.4	*98.0±4.5
Red blood cells	Reticulocytes, %	3.2±1.7	2.6±1.8

Table 4. Hematological parameters of mice in normoxia and hypoxia. The values represent mean±standard error of the mean.*P<0,05 in relation to the control group.

Treatment of hypoxemia by oliphen inhalations was shown to be due to systemic activation of erythropoiesis in red marrow. Short-time exposure to aerosolized oliphen initiated the proliferation of proerythroid cells immediately after challenge (Rodrhguez et al., 2000; Klausen et al., 1996). Adaptive reaction of blood system to drug challenge in normoxia have been recorded in 2 times growth of mitotic index of bone marrow cells for time 48 h. The possible synthesis of erythropoietin (EPO) de novo may be responsible for rise of oxygen capacity of blood initiated by systemic influence of oliphen. The pattern of polyxemia response has some similar features with intermittent hypoxic training, resulting in increase of hematocrit. Indeed these data are in accordance with observations on production de novo EPO for rodents after delay time 15 min and 2-6 h for humans as a result of hypoxic reaction (Klausen et al., 1996). The considerable advantage of oliphen in inhalation route of administration comprises the possibility to induce polycythemic response and to quench the reactive oxygen species of posthypoxic state at the same time (Cuzzocrea et al., 2001).

The antihypoxic action of aerosolozied oliphen in animal model may be of medical use for treatment of hypoxic syndrome caused by infections, since the severity of disorders correlates with oxygen stress burden for tissue respiration. As shown, combined application of redox active oliphen with recombinant interferon alfa-2b in a form of respirable aerosol resulted in acceleration of normalization of main clinical functions (lung ventilation, heart output and arterial pressure) of blood parameters in complicated influenza and pneumonia (Vasilieva et al., 2002). The experimentally confirmed efficacy, the absence of adverse effects and the feasibility to achieve therapeutically effective doses in noninvasive route of administration is a prerequisite for further investigation of oliphen as a perspective antihypoxic agent.

4. Discussion

The study of the effect of magnetic iron NPs on rodent model showed that inhalation administration of iron NPs is accompanied by the general inflammatory response. Direct intratracheal administration of iron NPs results in the total inflammatory response for third day, which is reliably confirmed by the increase in the content of neutrophils, alveolar macrophages, and proteins in the bronchoalveolar lavage fluid. It was established that iron NPs synthesized by the gas-phase procedure and microemulsion precipitation method is a pro-inflammatory agent in administration via the respiratory ways. Administration of iron NPs activates macrophage nonspecific lung protection. In the BAL samples there are a large number of macrophages engulfing dispersed iron. The relative population of macrophages decreases in time after a contact with a dispersed iron sample and consequent ingesting. The latter also confirms the toxicity of iron NPs for the mucous epithelium cells. Aspiration of suspended iron NPs through a trachea is aimed for delivery of iron particles to distal area of the lung. For clinical practice ventilation of lung with aqueous aerosols of iron NPs can serve as a method of delivery of contrasting agent for MRI (Rinck, 2001; Wang et al., 2001; Gubin et al 2005; Bonnemain et al., 1998). However, in inhalation procedures a possibility of damage or, in the other words, enhancement of permeability of capillary-alveolar membrane for blood proteins should be considered. The results of studying the protein composition by the HLPC and gel electrophoresis suggest transfer of a part of plasmatic proteins including antibodies into the alveolar area, which is a part of nonspecific protection of the organism caused by administration of NPs into bronchopulmonary tract. As a whole, the pattern of inflammatory processes with initiation of synthesis of pro-inflammatory cytokins and activation of the NF-kB complex corresponds to the pattern of nonspecific response of lung to inhalation of finely dispersed aerosols (Limbach et al., 2007; Oberdörster et al., 2005; Li et al., 1999; Tran et al., 2000; Danilov et al., 2003; Siglienti et al., 2006; Schins et al., 2004; Donaldson K. & Tran Cl, 2002; Zhu et al., 2008). The toxicity of iron NPs can be caused by the chemical composition of materials, the particular features of their size, increased aspiration dose of the preparation, and physiological peculiarities of their delivery through a respiratory tract. In step-by-step estimation of the role of each factor the following circumstances should be considered. The structure of dispersed samples used in trials presents an iron core covered with oxide shell. These preparations as chemical compounds are low toxic. Intravenous injections of suspended Fe_3O_4 and Fe_2O_3 NPs as contrasting agents for MRI are widely used in clinical practice without noticeable drawbacks (Rinck, 2001; Wang et al., 2001). An excess of iron is assimilated with an organism in the form of ferritin and is removed by kidneys in the form of soluble salts. Along with this, a possibility of contamination of NPs with iron ions, which are able to initiate burst-like oxidation processes in macrophage system, should be taken in account. Generation of active oxygen species and radical products catalyzed by the reactions of the Fenton type facilitates the pathogenesis of inflammatory response on respiratory iron NPs impact (Lay et al., 1998; 1999). Through monitoring of feasible contamination should exclude or noticeably reduce inflammatory activity of iron NPs.

The main reasons of inflammatory activity of particulate metal are governed by the spatial organization of NPs at the lung surface. It is likely that the toxicity of iron NPs observed in respiratory delivery is caused by their self-association. This self-association manifests itself in strong broadening of ^{1}H NMR signals under the action of iron NPs, which is explained by appearance of superparamagnetic state in reaching threshold particle sizes (from our estimation, approximately 30 nm) (Wang et al., 2001; Björnerud et al., 2004). Due to the strong magnetic interactions the iron NPs can form micron-sized thread-like structures, which are eliminated with macrophage cells during phagocytos. Formation of thread-like structures is may be the main factor of inflammation development, which is governed by the degree of association and the amount of aspirate in airways. The metabonomic characteristics of BAL fluid (concentrations of lactate, pyruvate and appearance of ethanol-like signal) studied by the ^{1}H NMR suggest that the inflammatory process results in the lack of oxygen and suppression of cell respiration (Nikolaev et al., 2003). As a rule the severe pulmonary tissue hypoxia follows after delivery of metal NPs. The limited time of the experiment (one week) did not allow to solve the problem of persistence of inflammation process initiated by aspiration of iron NPs into the respiratory tract and probability of formation of chronic lesion focuses.

The inflammatory response can be reduced by decrease in the single dose of iron NPs, multiple aspirations with reduced dose of NPs in diluent, modification of the surface of iron NPs with biologically adapted polymers (dextran, alginate, chitosan, etc), desintegration of associates of iron NPs formed in synthesis by mechanical action and addition of surfactants. The adverse effects of iron NPs-inhalation can be also prevented by prudent therapeutic procedures. One of the major drawbacks of iron NPs-inhalation is acute inflammatory response accompanied by hyperproduction of a pro-inflammatory cytokine mediator IL-1. As it has been shown IL-1 receptor antagonist (IL-1ra) suppresses the adverse effects of IL-1 by blocking the related cell receptors making it a promising way for treatment and prevention of a number of pathological inflammatory states (Simbirtsev, 2011). Once in this study IL-1ra was successfully used to reduce iron NPs-induced pulmonary neutrophils infiltration and other inflammation-associated processes in the mouse, the aerosol delivery of IL-1ra may be considered as medicine for treatment NPs-induced lung inflammation. To avoid inflammation effects of iron NPs for MRI assay IL-1ra may be delivered jointly with NPs in the form of aerosol or instillation. It is hopeful application of this substance for iron NPs-enhanced MR imaging of lung and magnetic field-guided drug delivery to the lung.

The therapeutic efficiency of airway administration of IL-1 receptor antagonist is reduced by tissue hypoxia caused by defects in working of ETC. The main sign of tissue hypoxia at late stage is the energetic discharge in the respiratory chain of cells. Severe hypoxic exhaustion of high energy substrates ATP, ADP, phosphocreatine resulted in membrane polarization and uncontrolled influx of calcium, being the possible reason of cell death and the following dangerous inevitable consequences for whole body (Zarubina, 1999; Boutilier et al., 2000). Acute hypoxia exerts a damaging effect on the cells, their membranes and function. According to the study of mouse normobaric hypoxia model, these defects of cell respiration can be effectively removed by delivery of redox active agent of quinone origin into bronchopulmonary tract. Thanks to a concomitant matching between the redox potential of oliphen (0.7 V) and redox state of substrates participating in electron transfer at the NAD/NADH locus of mitochondrial ETC, oliphen performs the role of electron equivalent carrier capable to shunt the injured sites of membrane respiratory complex (Tolstoy & Medvedev, 2000). Administration of oliphen into blood circulation results in restoration of the electron transport function of mitochondrial respiration chain in NADH/NAD site subjected oxidative damage at oxygen deficit (Vinogradov et al., 1973; Smirnov et al., 1992; Popov & Igumova, 1999). The treatment by oliphen recovered the lactate level to the normoxic value and stimulated the rise of hematological characteristics: hemoglobin and reticulocyte count. In conditions of hypoxia the oliphen showed antihypoxic activity at low pO_2 without reoxygenation. The membrane potential of mice red blood cells, measured by degree of absorption of tetraphenylphosphonium cation, decreased from 2.0 to 1.4 mkM ($n=28$; $P<0.05$). The presence of exogenous oliphen in interstitial liquids maintained the adequate transmembrane potential for transfer of protons across membrane. These data are consistent with our early reported results on study of oliphen participation in anaerobic cell respiration in vitro (Yakovleva et al., 2002).

The physiological cell response to single aerosol oliphen challenge retained activity for 3-4 houres. The main mass of inhaled soluble oliphen was removed from circulation by the liver and kidney. The particulate oliphen had been withdrawn also by bronchocillary clearance system. Airway delivery of dry particulate oliphen stimulated nonspecific defence reaction. As electron microscopy data showed macrophages engulfed particulate oliphen. Phagocytosis is known to be the source of pro-inflammatory impact which easily spread over the whole body (Cuzzocrea et al., 2001). The marked absence of irritation at the site of deposited particles was associated with powerful antioxidant activity (Driscoll et al., 2002). Oliphen inhibits the pathogenetic products of oxidative burst of macrophages by scavenging active forms O_2, H_2O_2 and other reactive products into nontoxic species. Antioxidant action of oliphen prevents development of inflammation in epithelial tissue (Cuzzocrea et al., 2001). The influence of oliphen on tolerance to hypoxia was more pronounced as compared with the action analogous nootropic and actoprotective drugs such as sodium hydroxybutyrate and sodium succinate (Smirnov et al., 1992; Zarubina, 1999). Inhalation of sodium hydroxybutyrate in a form of liquid aerosol essentially reduced the life-span (resulted in death of all tested mice at 10 min of exposure) at aspiration dose of 2.0 mg, which stimulate central nervous system. High sedative aspiration doses of sodium hydroxybutyrate (9 mg) improved resistance to acute hypoxia, but its effect was less than that elicited by oliphen inhalations. Inhalation of dry powder of sodium succinate induced additional resistance of mice to acute hypoxia at high aspiration doses 0.06 mg about in two times. Human insufflations of sodium succinate had some drawbacks as cough and irritation of respiration ways, but adverse effects were not observed in mouse model. Aerosol treatment with all of these drugs 2-2.5 times increased the initial time of life. This initial time is considered to be associated with adaptation of organism via increased ventilation of lung, cardiac output, mobilizing erythrocytes depot and allocation of pathways of oxygen consumption in more rational schemes.

So, on base of our experimental study the metal NPs exposure generate inflammation, oxidative stress and hypoxia of respiratory system. The respiratory disease induced by NPs challenge may be effectively treated by airway delivery of IL-1 receptor antagonist and of redox active preparation of oliphen. The applied IL-1ra improve the anti-inflammatory resistance of airway epithelium to iron NPs exposure. The adverse consequences of acute tissue hypoxia associated with pulmonary inflammation may be successfully overcome by inhalative application of the redox active hydroquinone oligomer in form of drug "Oliphen". The positive effect of antihypoxic treatment by oliphen is related with restoration of normal functioning of alveolar and epithelium mitochondria ETC. The combined application of recombinant antagonist IL-1 and redox active agent "Oliphen" may be the new possible strategy of therapeutic defensive measures against toxic hazards of production and application of metal NPs.

5. Conclusion

A tracheobronchial administration of iron NPs is shown to induce an acute inflammatory response in mouse model, which is similar to inflammatory pattern of inhaled endotoxin with respect to cellular and biochemical characteristics. The inflammatory response to inhalation contact with iron NPs is characterized by hypoxic secretion of partial oxidation products (ethanol and lactate) into bronchoalveolar lavage fluid. Receptor antagonist IL-1 aerosol administration may prevent the development of the acute inflammation induced by inhalation of iron NPs. Inhalation of redox agent in the form of fine respirable oliphen positively corrected the hypoxic response. Short-time exposure to aerosolized oliphen enhanced nonspecific defence against acute hypoxia through activation of mitochondrial respiration and stimulation of erythropoesis. Inhalation aerosol treatment by micronized oliphen could be realized in effective antihypoxic doses without adverse effects of inflammation and irritation of respirative tract. So, the pulmonary contact with iron NPs causes acute pulmonary inflammation, but IL-1β receptor antagonist and redox active agent oliphen may be used for it's suppression. A tracheobronchial administration of iron NPs combined with anti-inflammatory medicine IL-1 receptor antagonist is feasible for delivery into lungs as a contrasting agent for MRI.

6. Acknowledgments

The work was supported in part by research grant № 40.002.10.0 of Federal Medical-Biological Agency of Russian Federation. We would like to thank Varyushina E.A., Alexandrov G.V. and Senkevich I.E. for their assistance in histology assay; Tolochko O.V., Vasilieva E. for providing of iron NPs; Vorobeichikov E.V. for fruitful discussions of experimental design; Kotova T.V., Chernyaeva E.V. for aerosol experiments; Tolparov Yu. N. for technical assistance in NMR experiments

7. References

Aime, S.; Botta, M.; Mainero, V.; Terreno, E. (2002). Separation of Intra- and extracellular Lactate NMR Signals Using a lanthanide Shift Reagent. *Magnetic Resonance in Medicine*, No.47, pp. 10-13

Azmi, J.; Connelly, J.; Holmes, E. (2005). Characterization of the biochemical effects of 1-nitronaphthalene in rats using global metabolic profiling by NMR spectroscopy and pattern recognition. *Biomark*, Vol.10, No.6, pp. 401-416

Björnerud, A.; Johansson, L. (2004). The utility of superparamagnetic contrast agents in MRI: theoretical consideration and applications in the cardiovascular system. *NMR Biomed*, No.17, pp. 465-477

Bonnemain, B. (1998). Superparamagnetic agents in magnetic resonance imaging: physiochemical characteristics and clinical applications. A review. *Drug Target*, No.6, pp. 167-74

Boutilier, R.G.; St-Pierre, J. (2000). Surviving hypoxia without really dying. *Comp.Biochem.Physiol*, part A No.126, pp. 481-490

Buttgereit, F.; Burmester, G-R.; Brand, M-D. (2000). Bioenergetics of immune functions: fundamental and therapeutic aspects. *Immunology Today*, Vol.21, No.4, 192-199

Capek, I. (2004), Preparation of metal nanoparticles in water-in-oil (w/o) microemulsions. *Adv. Colloid Interface Sci*, No.110, pp. 49–74

Choi, C.J.; Tolochko, O.V.; Kim, B.K. (2002). Preparation of iron nanoparticles by chemichal vapour condensation. *Mater. Lett*, No.56, pp. 289-294

Cuzzocrea, S.; Riley, D.P.; Caputi, A.P.; Salemini, D. (2001). Antioxidant therapy: A New Pharmacological Approach in Shock. Inflammation and Ischemia/Reperfusion Injury. *Pharm.rev.*, Vol.53; No.1, pp. 135-159

Danilov, L.N.; Lebedeva, E.S.; Dvorakovskaya, I.V.; Simbirtsev, A.S.; Ilcovich, M.M. (2003). Effect of Interleukin 1 receptor antogonist on the development of oxidative stress in lungs. *Cytokines and Inflammation* (in Russian). No.4, pp. 14-20

Donaldson, K.; Tran, Cl. (2002). Inflammation caused by particles and fibers. *Inhal. Toxicol*, No.14, pp. 5-27

Driscoll, K.E.; Carter, J.M.; Borm, P.J.A. (2002). Antioxidant defense mechanisms and the toxicity of fibrous and nonfibrous particles. *Inhal. Toxicol*. No.14; pp. 101-118

Dziurla, R.; Gaber, T.; Fangradt, M.; Hahne, M.; Tripmacher, R.; Kolar, P.; Spies, C.M.; Burmester, G.R.; Buttgereit, F. (2010). Effects of hypoxia and/or lack of glucose on cellular energy metabolism and cytokine production in stimulated human CD4+ T lymphocytes. *Immunology Letters*. No.131, pp. 97-105

Fujiia, H.; Yoshika, K.; Berliner, L. J. (2002). In vivo fate of superparamagnetic iron oxides during sepsis. *Magnetic Resonance Imaging*. No.20, pp. 271–276

Gubin, S.P.; Koksharov, Yu.A.; Chomutov, G.B. (2005). Magnetic nanoparticles, manufacturing, structure and characterization (in Russian). *Adv. Chem*. No.74, pp. 539-574

Gupta, A.K.; Gupta, M. (2005). Cytotoxicity suppression and cellular uptake enhancement of surface modified magnetic nanoparticles. *Biomaterials*. No.26, pp. 1565-1573

Haage, P.; Karaagac, S.; Spuntrup, E.; Truong, H.T.; Schmidt, T.; Gunther, R.W. (2005). Feasibility of pulmonary ventilation visualization with aerosolized Magnetic Resonance contrast media. *Invest.Radio*. No.40, pp. 85-88

Hochachka, P.W.; Buck, L.T.; Doll, C.J.; Land, S.C. (1996). Unifying Theory of Hypoxia Tolerance: Molecular/Metabolic Defence and Rescue Mechanisms for Surviving Oxygen Lack. *Proc.Natl.Acad.Sci.USA*. Vol.93, No.18, pp. 9493 - 9498

Ischenko, A.M.; Nikolaev, B.P.; Kotova, T.V.; Vorobeychikov, E.V.; Konusova, V. G.; Yakovleva L.Yu. (2007). IL-1 Receptor Antagonist as an Aerosol in Inflammation. *J.Aerosol Med*. No.4, pp. 445-459

Kim, D.; Vasilieva, E.S.; Nasibulin, A.G.; Lee, D.W.; Tolochko, O.V.; Kim, B.K. (2007). Aerosol synthesis and growth mechanism of magnetic iron nanoparticles. *Mater. Sci. Forum.* No.9-12, pp. 534-536

Klausen, T.; Christensen, H.; Hansen, J.M.; Nielsen, O.J.; Andersen, N.F.; Olsen N.V. (1996). Human erythropoietin response to hypocapnic hypoxia, normocapnic hypoxia, and hypocapnic normoxia. *Eur.J.Appl.Physiol.* No.74, pp. 475 - 480

Laemmli, U.K. (1970). Cleavage of structural proteins during the assembly of the head of bacteriophage. *Nature.* Vol.4, No.227, pp. 680–685

Lay, J.C.; Bennett, W.D.; Ghio, A.J.; Bromberg, P.A.; Costa, D.l.; Kim, C.S.; Koren H.S.; Devlin, R.B.D. (1999). Cellular and biochemical response of the human lung after intrapulmonary instillation of ferric oxide particles. *Am. J. Respir. Cell Mol. Biol.* Vol.20, No.4, pp. 631-642

Lay, J.C.; Bennett, W.D.; Kim, C.S.; Devlin, R.B.D.; Bromberg, P.A. (1998). Retention and intrecellular distribution of instilled iron oxide particles in human alveolar macrophages. *Am. J. Respir. Cell Mol. Biol.* No.18, pp. 687-695

Lehninger, A. (1982). *Principles of Biochemistry.* Worth, New York

Li, X.Y.; Brown, D.; Smith, S.; MacNee, W.; Donaldson, K. (1999). Short-term inflammatory responses following intratracheal instillation of fine and ultrafine carbon black in rats. *Inhal. Toxicol.* No.11, pp. 709-731

Limbach, L.; Eterwick, P.; Usmanser, P.I.; Grass, R.N.; Riebruinink, A. and Stark, W.J. (2007). Exposure of engineered nanoparticles to human lung epithelial cells: Influence of chemical composition and catalytic activity on oxidative stress. *Environ. Sci. Technol.* No.41, pp. 4158-4163

Montagne, F.; Mondain-Monval, O.; Pichot, C.; Mozzanega, H.; Ela' Ëssari, A. (2002). Preparation and characterization of narrow sized (O/W) magnetic emulsion. *J. Magn. Mater.* No.250, pp. 302–312

Muller, K.; Skepper, J.N.; Posfai, M.; Trivedi, R.; Howarth, S.; Corot, C.; Lancelot, E.; Thompson, P.W.; Brown, A.P.; Gillard, J.H. (2007). Effect of ultrasmall supermagnetic iron oxide nanoparticles Ferumoxtran-100 on human monocyte-macrophages in vitro. *Biomat.* No.28, pp. 1629-1642

Nicholls, D.G.; Ward, M.W. (2000). Mitochondrial membrane potential and cell death: mortality and millivolts. *Trends Neurosci.* No.23, pp. 166–174

Nikolaev, B.P.; Schlakov, A.M.; Chalenko, V.V. (1997). Blood plasma and serum of patients with hard traumatic pathology. *Aids, cancer and related problems.* Vol.1, No.1, p.241

Nikolaev, B.P.; Yakovleva, L.Yu.; Kotova, T.V.; Manyakhin, E.E.; Vorobeichikov, E.V.; Kontorina, N.V.; Chernyaeva, E.V.; Tolparov, Yu.N. (2003). The evidences of antihypoxic action of medicine "Oliphen" in aerosol inhalative route administration studied in normobaric hypoxia mouse model. *Hypoxia Med. J.* No.1-2, pp. 9-14

Nikolaev, B.P. (2009). The pulmonary toxicity induced by airway exposure of magnetic iron nanoparticles in mice, In: Programme and Abstracts. 4th Intrenational conference on Nanotechnology - Occupational and Environmental Health, ISBN 978-951-802-927-7, Helsinki, august 2009

Oberdörster, G.; Finkelstein, J.N.; Johnston, C.; Gelein, R.; Cox, C.; Baggs, R.; Elder A.C.P. (2000). Acute pulmonary effects of ultrafine particles in rats and mice, In: *HEI research report.* August 2000, www.healtheffects.org/pubs-research.htm (2000).

Oberdörster, G.; Maynard, A.; Donaldson, K.; Castranova, V.; Fitzpatrick, J.; Ausman, K.; Carter, J.; Karn, B.; Kreyling, W.; Lai, D.; Olin, S.; Monteiro-Riviere, N.; Warheit, D.; Yang, H.: (2005). Review. Principles for characterizing the potential human health effects from exposure to nanomaterials: elements of a screening strategy. *Part. Fibre Toxicol.* doi:10.1186/1743-8977-2-8

Mashkovsky, M.D. (2002). *The Drugs. Reference book* (14 edition), «New Wave», Moscow (In Russian)

Popov, V.G.; Igumova, E.M. (1999). Sodium salt ((poly-(2,5-dihydroxyphenilene))-4-thiosulphuric acid of linear structure as regulator of cell metabolism and production method thereof. *US Patent.* No.6, 117 970

Potts, B.C.M.; Deese, A.J.; Stevens, G.J.; Reily, M.D.; Robertson, D.J.; Theiss, J. (2001). NMR of biofluids and pattern recognition: assessing the impact of NMR parameters on the principal component analysis of urine from rat and mouse. *J. Pharm. Biochem. Analysis.* No.26, pp. 463-476

Rinck, P.A.: (2001). *Magnetic Resonance in Medicine.* Vienna, Blackwell Wissenschafts-Verlag, Berlin

Rodrhguez, F.A.; Ventura, J.l.; Casas, M.; Casas, H.; Padgis, T.; Rama, R.; Ricart, A.; Palacios, L.; Viscor, G. (2000). Erythropoietin acute reaction and haemotological adaptations to short intermittent hypobaric hypoxia. *Eur.J.Appl.Physiol.* Vol.82; No.3; pp. 170-177

Schins, R.P.F.; Lightbody, J.H.; Borm, P.J.A.; Shi, T.; Donaldson, K.; Stone, K. (2004). Inflammatory effects of coarse and fine particulate matter in relationto chemical and biological constituents. *Toxicol. Appl. Pharm.* No.195, pp. 1–11

Shchukina, Mia (1986). Concentration of lactic acid in the blood and erythropoiesis during exposure to hypoxia. *Fiziol Zh SSSR Im I M Sechenova.* Vol.72, No.56, pp. 668-672 (In Russian)

Siglienti, I.; Bendszus, M.; Kleinschnitz, C.; Stoll, G. (2006). Cytokine profile of iron-laden macrophages. Implications for cellular magnetic resonance imaging. *J. Neuroimmunol.* No.173, pp. 166 – 173

Simbirtsev, A.S. (2011). *Interleukin -1.* Foliant, St.Petersburg

Skulachev, V.P. (1996). Role of uncoupled and non-coupled oxidations in maintenance of safely low levels of oxygen and its one-electron reductants. *Q. Rev. Biophys.* No.29, pp. 169–202

Smirnov, A.V.; Aksenov, I.V.; Zaitseva, K.K. (1992). The correction of hypoxic and ischemic states with the use of antihypoxants (a review of the literature). *Voen.Med.Zh.* No.10, pp. 36-40 (In Russian)

Stroh, A.; Zimmer, C.; Gutzeit, C.; Jakstadt, M.; Marschinke, F.; Jung, T.; Pilgrim, H.; Grune, T. (2004). Iron oxide particles for molecular magnetic resonance imaging cause transient oxidative stress in rat macrophages. *Free Radic. Biol. Med.* Vol.36, No. 8, pp. 976-984

Tolstoy, A.D.; Medvedev, Yu.V. (2000). Hypoxia and free radicals in pathogenesis of organism. *Terra-calendar and Promotion.* Moscow (In Russian)

Tran, C.L.; Buchanan, R.T.; Cullen, R.T.; Searl, A.; Jones, A.D.; Donaldson, K. (2000). Inhalation of poorly soluble particles II influence of particle surface area on inflammation and clearance. *Inhal. Toxicol.* No.12, pp. 113-1126

Vasilieva, I.A.; Kiselev, O.I.; Chepic, E.B.; Sventitsky, E.N.; Nikolaev, B.P.; Manyakhin, E.E. (2002). The experience of aerosol application of the combined preparation of interferon with "Olifen" in complex therapy of influenza complicated by pneumonia. *Med.Acad.J.* Vol.2, No.3, pp. 73-78. (In Russian)

Vasilieva, E.S.; Tolochko, O.V.; Yudin, V.E.; Kim, D.; Lee, D.W. (2006). Production and application of metal-based nanoparticles. *European Nano System 2006 (ENS'06) conference: Collection of papers. TIMA editions, Paris, France, 2006*

Vidal-Vidal, J.; Rivas, J.; Lopez-Quintela, M.A. (2006). Synthesis of monodisperse maghemite nanoparticles by the microemulsion method. Colloids and Surfaces A: *Physicochem. Eng.* Asp. No.288, pp. 44–51

Vinogradov, V.M.; Aleksandrova, A.E.; Pastushenkov, L.V. (1973). Biochemical prerequisites to development of remedies increasing the organism resistance to hypoxia. *Urgent Problems of Neuropathology and Neurosurgery.* Antonov I.P.; ed. Nauka I Teknika.Issue 6. Minsk, pp. 33-49 (In Russian)

Wang, Y.X, Hussain, S.M, Krestin, G.P.: (2001). Superparamagnetic iron oxide contrast agents: physicochemical characteristics and applications in MR imaging. *Eur Radiol.* No.11, pp. 2319–31

Wattiez, R.; Falmagne, P. (2005). Review. Proteomics of bronchoalveolar lavage fluid. *J. Chromatogr.* No.815, pp. 169–178

Wong, W.S.; Zhu, H.; Liao, W. (2007). Cysteinyl leukotriene receptor antagonist MK-571 alters bronchoalveolar lavage fluid proteome in a mouse asthma model. *Eur. J. Pharmacol.* No.575, pp. 134–141

Yakovleva, L.Yu.; Toropov, D.K.; Chernyaeva, E.V.; Vorobeychikov, E.V.; Nikolaev, B.P. (2002). The investigation of antihypoxant oliphen action on E.coli and red blood cells by ionometric and spin-probe methods. *53rd Annual Meeting of the International Society of Electrochemistry "Electrochemistry in Molecular and Microscopic Dimensions".* Dusseldorf, Germany, 15-20 September 2002

Zarubina, I.V. (1999). Biochemical aspects of hypoxic cell injury. *Hypoxia Medical J.* V.7, No.1-2, pp. 2-9

Zhu, M-T.; Feng, W-Y.; Wang, B.; Tian-Cheng, W.; Gu, Y-Q.; Wang, M.; Wanga, Y.; Ouyang, H.; Zhao, Y-L.; Chai, Z-F. (2008). Comparative study of pulmonary responses to nano- and submicron-sized ferric oxide in rats. *Toxicology.* No.247, pp. 102–111

Part 5

Nanoelectrochemistry

11

Novel Synthetic Route for Tungsten Oxide Nanobundles with Controllable Morphologies

Yun-Tsung Hsieh[1], Li-Wei Chang[1], Chen-Chuan Chang[1], Bor-Jou Wei[2], and Han C. Shih[1,3*]
[1]*Department of Materials Science and Engineering, National Tsing Hua University, Hsinchu, Taiwan,*
[2]*Department of Materials Science and Engineering, National Chung Hsing University, Taichung, Taiwan,*
[3]*Department of Chemical and Materials Engineering, Chinese Culture University, Taipei, Taiwan, R.O.C.*

1. Introduction

The formation and characterization of one-dimensional nanomaterials have attracted considerable attention because of their unique physical and chemical properties [1–6]. For example, they can be applied in organic light-emitting diodes (OLEDs), solid lubricants, DNA analysis, and electrochromic devices [7–10]. Because of their semiconducting properties and high surface areas, transition metal oxide nanostructures have been used in a variety of research projects that developed device-oriented nanostructures. For example, tungsten oxide nanostructures such as nanowires, nanorods, nanotubes, nanobelts, and nanofibers have well-known electrical, mechanical, gasochromic, and photoelectrochromic properties that make them very useful for various applications, including field-emission devices, electrochromic devices, light-emitting diodes, and gas and chemical sensors [11–15]. Recently, nanostructured tungsten oxide materials have been of great scientific and applied interest.

Bulk properties such as piezoelectricity, chemical sensing, and photoconductivity of these materials are enhanced in their quasi one-dimensional (Q1D) form. Interestingly, Q1D structures can be used as template for the growth of other nanostructures resulting in novel hierarchical nanoheterostructures with enhanced functionality. Nanosized tungsten oxide particles have been found useful in fabricating gas sensors for the detection of nitrogen oxides, ammonia, and hydrogen sulfide. Current research, however, has focused on the use of polycrystalline tungsten oxide systems for these applications, and thus important sensor requirements such as high sensitivity and reproducibility, which can be obtained only by using size-controlled pure nanomaterials, have not been accomplished.

Cao et al. [16] synthesized WO_3 nanowire by thermal evaporation method, and reported gas sensing tests revealed that the sensor based on the WO_3 nanowire array had the capability of detecting NO_2 concentrations as low as 50 ppb, demonstrating a promising application in the field of low concentration gas detection. Kim [17] has investigated the dependence of gas-sensing characteristics on thermal treatment conditions in a tungsten oxide nanorod system that demonstrated the facile detection of various analysts at ambient temperature. Thermal treatments under the O_2-containing active environments were found to result in bad reproducibility in sensor response compared with the inert N_2 conditions. As a result, the recommendable thermal treatment conditions for $WO_{2.72}$ sensors were an annealing temperature of 300–500 ◦C under inert N_2 or Ar ambience. The annealing temperature within the range could be utilized as a parameter to regulate a relative population ratio between W^{5+} and W^{6+} states without any noticeable change in morphology or crystalline structure. Sen et al [18] reported that nanowire hierarchical hetero-structures of SnO_2:$WO_{2.72}$ have been prepared by thermal evaporation technique. The density of $WO_{2.72}$ nanowires was found to depend on partial pressure of oxygen and source temperature. Single wires of heterostructures have been aligned between two electrodes to make gas sensors that show high sensitivity and selective response to chlorine gas at room temperature. Improvement in selectivity is attributed to transfer of electrons from $WO_{2.72}$ to SnO_2 on formation of heterojunctions. The study shows the potential of semiconductor oxide hetero-junctions for application to sensors and other electronic devices.

Field emission (FE) (also known as electron field emission) is an emission of electrons induced by external electromagnetic fields. Field emission can happen from solid and liquid surfaces, or individual atoms into vacuum or open air, or result in promotion of electrons from the valence to conduction band of semiconductors. Field emission involves the extraction of electrons from a solid by tunneling through the surface potential barrier. The emitted current depends directly on the local electric field at the emitting surface, E, and on its work function, ϕ, as shown below. In fact, a simple model (the Fowler-Nordheim model, can be written as $J = (E^2\beta^2 / \phi) \exp(-B\ \phi^{1.5} / E\beta)$, where J is the current density, E is the applied electric field, ϕ is the work function (eV), and β is the field enhancement factor) shows that the dependence of the emitted current on the local electric field and the work function is exponential-like. As a consequence, a small variation of the shape or surrounding of the emitter (geometric field enhancement) and/or the chemical state of the surface has a strong impact on the emitted current [19,20].

Tungsten oxide nanowires are one-dimensional nanostructures, with diameters of 10–100 nm and a length of about 1 µm. Since the nanowires have high aspect ratios and are easily fabricated, they have attracted considerable attention as promising materials for field emitters of the field-emission displays (FEDs) [21-23]. Furubayashi et al. [24] have demonstrated that the tungsten oxide nanowires obtained from sputtered tungsten films have good field emission properties. The number density of nanowires was discovered to be an important parameter in determining the field emission properties. A suitably low density resulted in good field emission properties owing to the concentration of the electric field. Concentrating an electric field at the tip of tungsten oxide nanowires increases their emission current. Moreover, they synthesized tungsten oxide nanowires on 1x1 μm^2 islands with various pitches. The lengths of tungsten oxide nanowires were 600–1000 nm and the number of nanowires per island was about 100. It found that the sample with the island pitch of 5µm exhibited the highest field emission current among the samples with pitches of 2, 5, 10, 20 and 30µm.

Among the tungsten oxides, tungsten oxide (WO_3), which is a versatile semiconductor material with a wide bandgap ranging from 2.5–3.6 eV, has strong potential for many interesting applications [25–28]. Methods of synthesizing WO_3 with various morphologies and phases using either physical or chemical routes have been the subject of considerable research. Xu et al. recently synthesized tungsten oxide nanowires using a flame with an air jet impinging on an opposing jet of nitrogen-diluted methane at atmospheric pressure [29]. Klinke et al. found that tungsten oxide nanowires could be formed in a gas mixture of argon, hydrogen, and methane [30]. Wilson et al. reported on a synthesis method that uses tungsten probes 1 m in diameter inserted into an opposed flow of methane oxy-flame with ethane [31].

However, these techniques for making tungsten oxide nanomaterials on substrates require external catalysts or reactants, as well as complicated sample pretreatment and preparation conditions. Furthermore, the importance of substrate temperature in the synthesis of WO_3 nanomaterials has not been considered in previous work. For practical purposes, it is also desirable to develop a simple method for fabricating WO_3 nanobundles of different morphologies, in which appropriate morphology control can be achieved by adjusting the growth parameters. Such a method would yield a thorough understanding of the relationship between the morphologies and properties of the obtained nanobundles. Fabrication of tungsten oxide nanobundles of different morphologies under controlled growth conditions has always been a challenge. As a result, efficient, large-scale fabrication of morphology-controllable tungsten oxide nanobundles is a difficult task.

The thermal chemical vapor deposition (CVD) process offers significant advantages: simple experimental equipment, high homogeneity, very short processing time, cost effectiveness, high efficiency, easy synthesis, and controllability. High quality can be easily obtained with good reproducibility by controlling the CVD parameters [32–34]. Therefore, in the present work, we explore a simple, more economical thermal CVD method for large-scale fabrication of WO_3 nanobundles with controllable morphologies on silicon (100) substrates with no additional catalysts. By heating tungsten powder to 1100°C in vacuum (6.13 Pa) in a two-step process, WO_3 nanobundles with controllable morphologies were produced in high yield; nanowires, nanobars, and nanobulk were produced in furnace temperature ranges of 250–350°C, 450–550°C, and 650–750°C, respectively. Furthermore, in a series of experiments, we successfully achieved room-temperature blue emission from the as-synthesized WO_3 nanobundles, which can be attributed to the band–band indirect transitions of the WO_3 nanobundles.

2. Experimental

In our experiment, we synthesized tungsten oxide nanobundles using thermal CVD. WO_3 nanobundles were synthesized in a conventional horizontal tube furnace made from quartz. Tungsten powder (0.05 g, Alfa AESAR; particle size, 12 μm; purity, 99.99%) acted as the source material; it was deposited on a ceramic boat and placed in the constant-temperature zone of the furnace. A silicon (100) wafer, which acted as the substrate, was subjected to ultrasonic cleaning in ethanol for 30 min and then placed in different temperature zones ranging from 100°C to 800°C, which is about 5 cm downstream from the source. After the tube was pumped to the required vacuum of 0.67–0.8 Pa , the temperature of the furnace was raised from room temperature to 800°C at a ramping rate of 30°C/min. The flow rates of a mixture of argon and oxygen gases were maintained at 10 sccm and 1 sccm, respectively, and were controlled by a flow meter. The pressure was maintained at 6.13 Pa, and then the temperature of the furnace was increased from 800°C to 1100°C. After this temperature was maintained for 1.2 h, the furnace was cooled naturally to room temperature. Samples formed at substrate temperatures of 250–350°C, 450–550°C, and 650–750°C were removed for characterization.

The deposited nanobundles were then characterized and analyzed. A scanning electron microscope (SEM; JEOL JSM-6500F) was used for morphological analysis. X-ray analysis was performed using a Shimadzu Lab XRD-6000 diffractometer equipped with a graphite monochromator. The Cu Kα radiation had a wavelength of λ = 0.154056 nm; it was operated at 43 kV and 30 mA. Transmission electron microscopy (TEM) and energy-dispersive x-ray spectroscopy (EDS) were conducted using a JEOL 2010 transmission electron microscope operated at 200 kV. Raman spectroscopy was performed using a micro-Raman setup (LabRAM; Dilor) equipped with a He–Ne laser emitting radiation at 632.8 nm.

The optical properties of the as-prepared nanobundles were measured by cathodoluminescence (CL) spectrometry on a JEOL-JSM-7001F field-emission scanning electron microscope (FESEM) at room temperature. A 15-keV electron beam was used to excite the sample. The CL light was dispersed by a 1200-nm grating spectrometer and detected by a liquid-nitrogen-cooled charge-coupled device.

3. Results

Figure 1 shows SEM images of numerous high-density tungsten oxide nanobundles formed at different substrate temperatures. The morphology of the WO_3 nanobundles on the silicon substrate clearly reveals the formation of nanowires, nanorods, and nanobulk material. The SEM image in Fig. 1a clearly demonstrates that the nanowires obtained on the substrate at 250–350°C have uniform one-dimensional morphologies and are produced at high density and large scale. The nanowires had diameters of 20–30 nm and lengths of up to several nanometers. The as-synthesized nanowires have a high aspect ratio similar to that reported previously [35]. Figure 1b shows nanorods prepared on the substrate at 450–550°C. The average lengths of these nanorods were 400–500 nm, and their diameters were 100–150 nm, and they have polygonal cross sections. It is apparent that the nanorods are formed by an increase in the diameter of the nanowires. Figure 1c shows nanobulk material that covered the entire surface of the substrate at 650–750°C. The material consists of polyhedrons with an average diameter of up to 300 nm and lengths of up to hundreds of nanometers.

Fig. 1. SEM images of tungsten oxide nanobundles formed at different substrate temperatures: (a) 250–350°C, (b) 450–550°C, (c) 650–750°C.

The phases of the as-prepared WO_3 nanobundles were identified by XRD. Figure 2 shows XRD spectra of WO_3 nanobundles fabricated at three different substrate temperatures. All the spectral peaks were indexed well to monoclinic WO_3 in accordance with the Joint Committee on Powder Diffraction Standards (JCPDS) card No. 43-1035 (lattice constants: a = 0.7297 nm; b = 0.7539 nm; c = 0.7688 nm; β = 90.91°). No diffraction peaks corresponding to tungsten oxides other than WO_3 could be detected in the spectra. Strong, sharp diffraction peaks also indicate good crystallinity in the as-synthesized product. The (002) diffraction peak shows the strongest diffraction, indicating that [002] is the major growth direction of the nanobundles. Moreover, the intensity of the (002) peak was found to increase with increasing substrate temperature. This may reflect the increasing volume of the nanobundles with increasing substrate temperature. However, the distribution and alignment of the nanobundles whose spectra appear in Fig. 2 may vary with substrate temperature. This change in distribution and alignment may also change the relative intensities of the (002), (200), (120), and (140) peaks in the XRD spectra. Further, the intensities of the (022), (202), and (114) diffraction peaks also increased with increasing substrate temperature. This behavior indicates that the [002] growth direction also influence all three peaks with increasing substrate temperature.

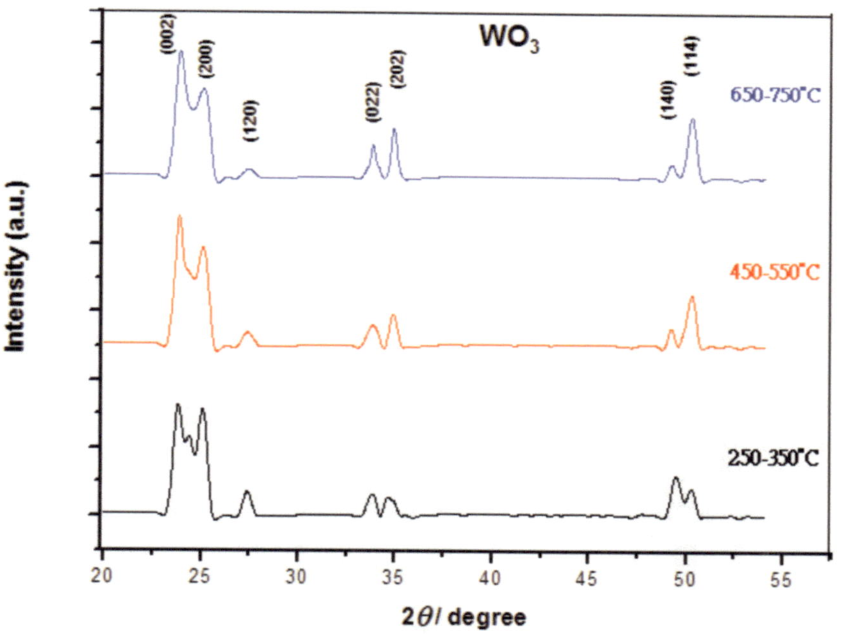

Fig. 2. XRD spectra of the nanobundles fabricated at substrate temperatures of 250–350°C, 450–550°C, and 650–750°C.

We further studied the morphologies of the as-prepared nanobundles by TEM. Figure 3a shows a typical low-magnification TEM image of a straight nanowire with a uniform diameter (25 nm). The crystal structure and growth direction of individual nanowires are further investigated by high-resolution TEM (HRTEM), as shown in Fig. 3b. The inset in this figure shows the selected-area diffraction (SAD) pattern along the [100] zone axis. It reveals that the nanowire has a single-crystal structure. The lattice spacings are 0.385 and 0.379 nm along the two orthogonal directions, which correspond to the (002) and (020) planes of monoclinic WO_3, respectively. The SAD pattern shown in the inset of Fig. 3b also confirms that the nanowires exhibit the monoclinic WO_3 phase (JCPDS 43-1035). Figure 3c shows a typical low-magnification TEM image of the obtained nanorods (width: 120 nm). The HRTEM image shown in Fig. 3d shows the crystal structure and growth direction of the individual nanorod in Fig. 3c. The measured lattice spacings along the two orthonormal directions are 0.385 and 0.379 nm, which correspond to the (002) and (020) planes of monoclinic WO_3, respectively. The measured lattice spacing and SAD pattern show that the nanobundles are single crystalline and that the major growth direction is [002].

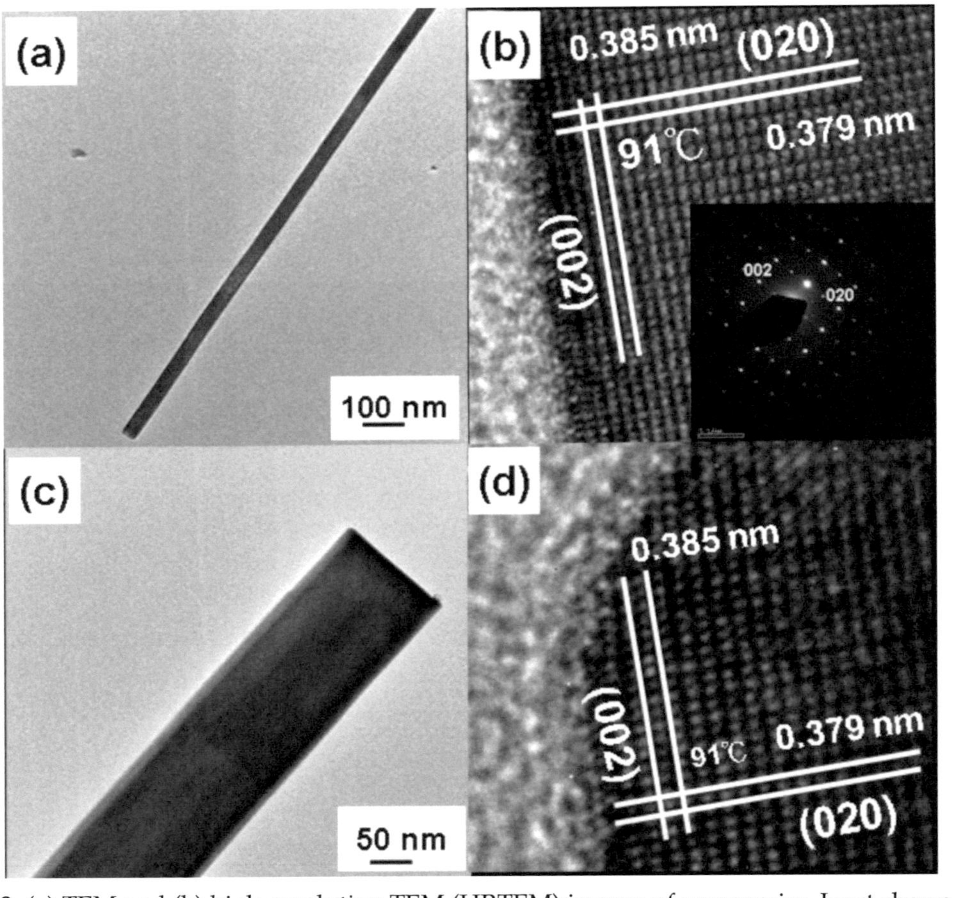

Fig. 3. (a) TEM and (b) high-resolution TEM (HRTEM) images of a nanowire. Inset shows the corresponding selected-area diffraction (SAD) pattern. (c) TEM and (b) HRTEM images of a nanorod.

Figure 4a shows the corresponding elemental line-scan mapping of the nanowires, revealing that they contain only O and W atoms. Typical EDS data recorded for a single nanowire and a single nanorod (Fig. 4b and 4c) confirm that both these nanobundles are made of WO3 and that the C and Cu signals can be attributed to the Cu grids used for our TEM measurements. Therefore, the WO3 nanobundles fabricated in this study are confirmed to have high purity.

An analysis of individual nanobundles shows that the calculated atomic ratio of W to O is approximately 1:3, which is consistent with the XRD and TEM results obtained for WO3.

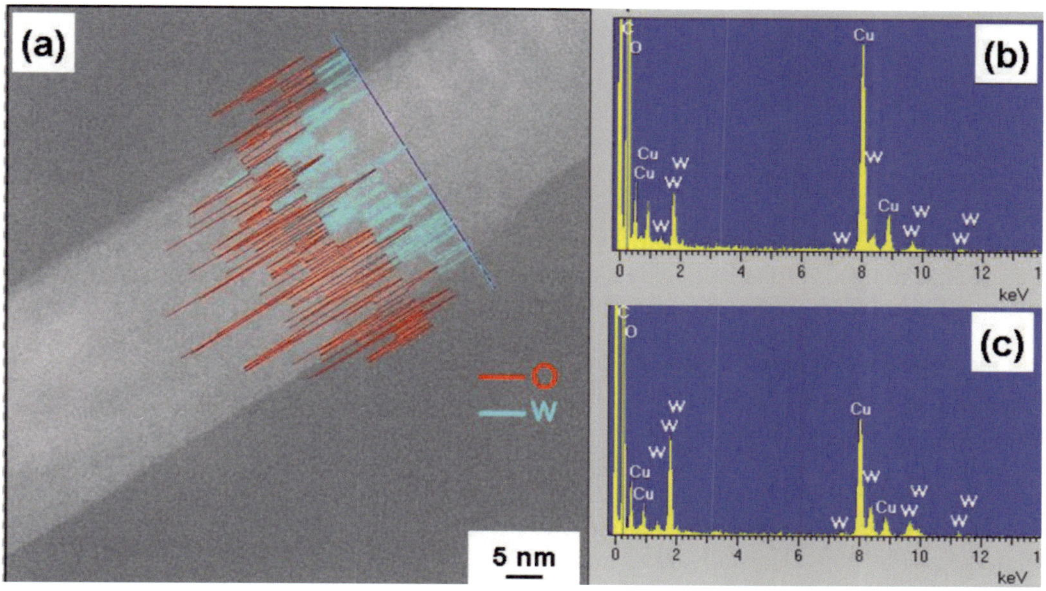

Fig. 4. (a) Energy-dispersive X-ray spectroscopy (EDS) elemental line profile of nanowire shown in Fig. 3a. (b) and (c) EDS spectra for nanobundles fabricated at substrate temperatures of 250-350°C and 450-550°C, respectively. Nanobundles contain only W and O atoms. Peaks due to C and Cu signals can be attributed to copper TEM grids.

To examine the necessary conditions for the growth of WO_3 nanobundles in more detail, we prepared samples using the same method but forming samples on the silicon substrate at different temperatures. Figure 5a, 5b, and 5c shows SEM images of samples synthesized using the same methods and conditions as those for growing the samples shown in Fig. 1; however, the substrate temperatures were 150-250°C, 350-450°C, and 550-650°C, respectively. In Fig. 5a, where the substrate temperature was 150-250°C, almost no nanowires are visible, but many unknown WO_x nanoparticles appear on the substrate. This is because nucleation is difficult with such a low surface energy. In Fig. 5b, where the substrate temperature was 350-450°C, although tungsten oxide nanomaterials were fabricated with high yield, the morphology is disordered. At the substrate temperature of 550-650°C used to fabricate the samples shown in Fig. 5c, the surface energy increased with the substrate temperature and enhanced nucleation. However, the growth environment was still not optimal for producing the required nanobundles but instead yielded unknown tungsten oxide nanomaterials. Thus, these findings unambiguously demonstrate the significant effects of the temperature of the silicon substrate on the WO_3 nanomaterials with controllable morphology.

Finally, the optical properties of the synthesized tungsten oxide nanobundles were characterized by room temperature CL spectroscopy. Figure 6 shows typical CL spectra of the as-prepared WO_3 nanobundles. The strongest CL emission peaks are centered at 351 nm, 350 nm, and 349 nm for substrate temperatures of 250-350°C, 450-550°C, and 650-750°C,

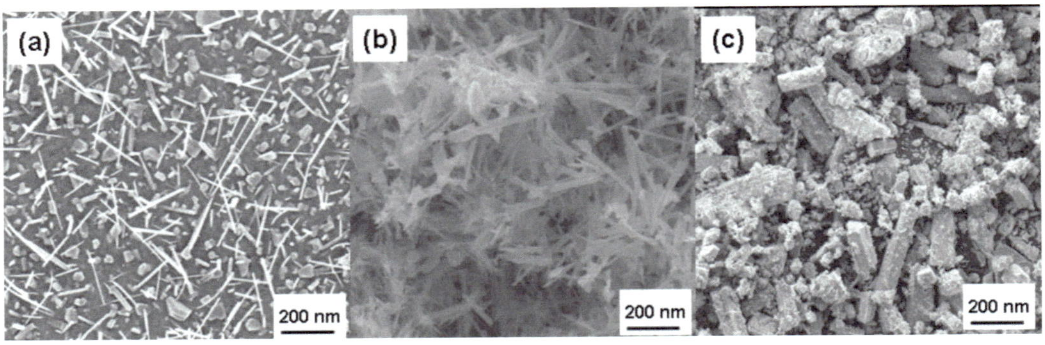

Fig. 5. SEM images of tungsten oxide nanobundles formed at different substrate temperatures: (a) 150–250°C, (b) 350–450°C, (c) 550–650°C.

respectively. Tungsten trioxide is an indirect band gap semiconductor, and its crystals or thin films do not produce such strong luminescence [36]. Single peak at the wavelength has been detected by Lee et al. [37] and Feng et al. [36] They suggested that the peak is due to the intrinsic band-band transition emission induced by quantum confinement effects in nanomaterials with an ultrafine diameter (<5 nm) of individual nanomaterials within each bundle. It is interesting that the nanobundles exhibited a blueshift caused by the increase in diameter with increasing substrate temperature. On the other hand, other origins such as foreign matter or native defects in the structure cannot be ruled out.

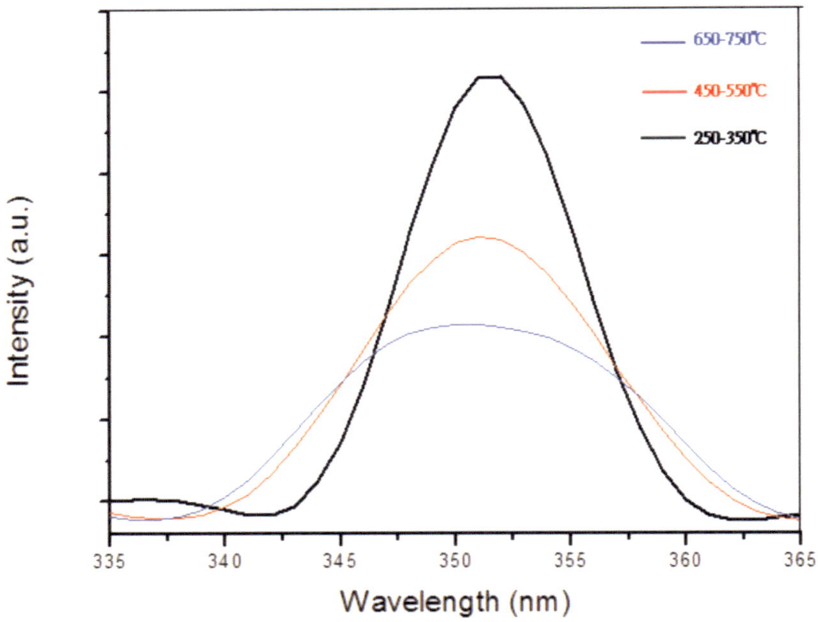

Fig. 6. Cathodoluminescence (CL) spectra of the as-prepared WO_3 nanobundles formed at different substrate temperatures.

4. Discussion

The vapor–solid mechanism is responsible for the growth of WO_3 nanobundles in this experiment, since no catalysts were used [38]. Tungsten powder begins to sublimate from the quartz boat when the temperature is increased to 800°C, and this process is highly enhanced at a temperature of 1100°C. The sublimated tungsten vapor reacts if sufficient oxygen is present. Tungsten trioxide vapor subsequently flows to the lower-temperature zone where the silicon substrate is located and becomes supersaturated, with nucleation of small clusters and subsequent growth of nanobundles. A previous report revealed that when the vacuum is not high enough, the mean free path of the vapor in the furnace is small; this affects the degree of supersaturation over the substrate and thereby obstructs the nucleation process [39]. On the other hand, when the vacuum becomes significantly higher, such that the mean free path of the tungsten trioxide vapor is highly enhanced, its partial pressure, and hence supersaturation, increases in the substrate zone in which the nucleation of WO_3 on the silicon substrate has been enhanced. In this growth process, if the furnace has a vacuum of intermediate strength, nucleation on the silicon substrate is rare. Increasing the temperature and selecting a suitable gas flux and substrate temperature promote nucleation by modifying the adsorption and diffusion characteristics of the surface. This increases the efficiency of nucleation, thereby promoting nanobundle growth. It is extremely interesting to compare our results with those of studies in which WO_3 nanostructures are fabricated on a silicon substrate using different types of catalyst [40, 41].

For reasons of practicality and efficiency, we tried to produce tungsten oxide nanobundles while controlling fewer experimental parameters. We found that different substrate temperatures can produce different nanomaterials, including nanowires, nanorods, and nanobulk, under a suitable growth environment. However, the mechanism responsible for the formation of different tungsten oxide nanobundles is not yet fully understood. Ye et al. [42] reported that the growth rate of the crystals was generally determined by gas-phase supersaturation of the species of the growth material, and that the shape of the final crystal was determined by the surface energy of the planes of the growing surface. These conclusions can also explain the growth process in our study. To our knowledge, the observed types of tungsten oxide nanobundle, i.e., nanowires, nanorods, and nanobulk, may be directly related to the principal factor in our experiment, substrate temperature. During growth, the surface energy increases with increasing substrate temperature, facilitating easy and efficient nucleation. Therefore, the growth of various nanobundles was influenced by changing the substrate temperature.

5. Conclusions

We fabricated WO_3 nanobundles by thermal chemical vapor deposition via a two-step heating process with no catalyst. By simply varying the substrate temperatures, several types of uniform, single-crystalline WO_3 nanobundles can be formed, such as nanowires, nanorods, and nanobulk. This method is simple, effective, catalyst-free, and easily repeatable; further, it affords the desired nanobundles in very large yields. Thus, this method is highly suitable for the fabrication of WO_3 nanobundles. The fabricated nanobundles can be used in gas and chemical sensors, field-emission devices, electrochromic devices, and light-emitting diodes.

In a future study, we intend to apply this more efficient and controllable fabrication method to the synthesis of metal-semiconductor composite nanoarchitectures with promising structural, morphological, and electrochemical properties.

6. Acknowledgments

This research was supported by the National Science Council through Grant No. 96-2221-E-034-006-MY2 and No.98-2221-E-034-007-MY2.

7. References

[1] P. E. Hovsepian, A. P. Ehiasarian, Y. P. Purandare, R. Braun and I. M. Ross, Plasma Process Polym., 6 (2009) S118

[2] L. Gao, R. L. Woo, B. Liang, M. Pozuelo, S. Prikhodko, M. Jackson, N. Goel, M. K. Hudait, D. L. Huffaker, M. S. Goorsky, S. Kodambaka and R. F. Hicks, Nano Lett., 9 (2009) 2223

[3] M. Stuber, H. Leiste, S. Ulrich, H. Holleck and D. Schild, Surf. Coat. Tech., 150 (2002) 218

[4] K. Fadenberger, I. E. Gunduz, C. Tsotsos, M. Kokonou, S. Gravani, S. Brandstetter, A. Bergamaschi, B. Schmitt, P. H. Mayrhofer, C. C. Doumanidis and C. Rebholz, Appl. Phys. Lett., 97 (2010) 144101

[5] F. Pinakidou, M. Katsikini, P. Patsalas, G. Abadias and E. C. Paloura, J. Nano Res., 6 (2009) 43

[6] M. McNallan, D. Ersoy, R. Zhu, A. Lee, C. White, S. Welz, Y. Gogotsi, A. Erdemir and A. Kovalchenko, Tsinghua Sci. Technol., 10 (2005) 699

[7] C. J. Chiang, S. Bull, C. Winscom and A. Monkman, Org. Electron., 11 (2010) 450

[8] D. Medaboina, V. Gade, S. K. R. Patil and S. V. Khare, Phys. Rev. B, 76 (2007) 205327

[9] X. Zhang, B. Luster, A. Church, C. Muratore, A. A. Voevodin, P. Kohli, S. Aouadi and S. Talapatra, Appl. Mat. Interfaces, 1 (2009) 735

[10] B. M. Venkatesan, B. Dorvel, S. Yemenicioglu, N. Watkins, I. Petrov and R. Bashir, Adv. Mater., 21 (2009) 2771

[11] X. Li, G. Zhang, F. Cheng, B. Guo and J. Chen, J. Electrochem. Soc., 153 (2006) H133

[12] Y. D. Huh, J. H. Shim, Y. Kim and Y. R. Do, J. Electrochem. Soc., 150 (2003) H57

[13] K. Huang, Q. Pan, F. Yang, S. Ni and D. He, Physica E, 39 (2007) 219

[14] S. Sen, P. Kanitkar, A. Sharma, K. P. Muthe, A. Rath, S. K. Deshpande, M. Kaur, R. C. Aiyer, S. K. Gupta and J. V. Yakhmi, Sensor. Actuat. B-Chem., 147 (2010) 453

[15] R. Seelaboyina, J. Huang, J. Park, D. H. Kang and W. B. Choi, Nanotechnology, 17 (2006), 4840

[16] B. Cao, J. Chen, X. Tang, and W. Zhou, "Growth of monoclinic WO_3 nanowire array for highly sensitive NO_2 detection", *J. Mater. Chem.*, 19, 2323 (2009).

[17] Y. S. Kim, *Sensor.* "Thermal treatment effects on the material and gas-sensing properties of room-temperature tungsten oxide nanorod sensors", *Actuat. B-Chem.*, 137, 297 (2009).

[18] S. Sen, P. Kanitkar, A. Sharma, K. P. Muthe, A. Rath, S. K. Deshpande, M. Kaur, R., "Growth of SnO_2/WO2.72 nanowire hierarchical heterostructure and their application as chemical sensor", *Sensor. Actuat. B-Chem.*, 147, 453 (2010).

[19] K. Huang, Q. Pan, F. Yang, S. Ni, and D. He, "Synthesis and field-emission properties of the tungsten oxide nanowire arrays", *Physica. E*, 39, 219 (2007).

[20] Y. M. Zhao, Y. H. Li, I. Ahmad, D. G. Mccartney, and Y. Q. Zhu, "Two-dimensional tungsten oxide nanowire networks", *Appl. Phys. Lett.*, 89, 133116 (2006).

[21] K. Huang, Q. Pan, F. Yang, S. Ni, and D. He, "The catalyst-free synthesis of large-area tungsten oxide nanowire arrays on ITO substrate and field emission properties", *Mater. Res. Bull.*, 43, 919 (2008).

[22] S. Jeon, H. Kim, and K. Yong, "Deposition of tungsten oxynitride nanowires through simple evaporation and subsequent annealing", *J. Vac. Sci. Technol. B*, 27, 671 (2009).

[23] Y. H. Lee, C. H. Choi, Y. T. Jang, E. K. Kim, and B. K. Ju, "Tungsten nanowires and their field electron emission properties", *Appl. Phys. Lett.*, 81, 745 (2002).

[24] M. Furubayashi, K. Nagato, H. Moritani, T. Hamaguchi, and M. Nakao, "Field emission properties of discretely synthesized tungsten oxide nanowires", *Microelectron. Eng.*, 87, 1594 (2010).

[25] M. Boulova and G. Lucazeau, J. Solid State Chem., 167 (2002) 425

[26] J. H. Ha, P. Muralidharan and D. K. Kim, J. Alloy. Compd., 475 (2009) 446

[27] Y. Wu, Z. Xi, G. Zhang, J. Yu and D. Guo, J. Cryst. Growth, 292 (2006) 143

[28] L. G. Teoh, J. Shieh, W. H. Lai, I. M. Hung and M. H. Hon, J. Alloy. Compd., 396 (2005) 251

[29] F. Xu, S. D. Tse, J. F. Al-Sharab and B. H. Kear, Appl. Phys. Lett., 88 (2006) 243115

[30] C. Klinke, J. B. Hannon, L. Gignac, K. Reuter and P. Avouris, *J. Phys. Chem.*, 109 (2005) 17787

[31] M. M. Wilson, S. A. Saveliev, W. C. Jimenez and G. Salkar, Carbon, 48 (2010) 4510

[32] P. C. Chang, Z. Fan, D. Wang, W. Y. Tseng, W. A. Chiou, J. Hong and J. G. Lu, Chem. Mater., 24 (2004) 5133

[33] Q. G. Fu, H. J. Li, X. H. Shi, K. Z. Li, J. Wei and Z. B. Hu, Mater. Chem. Phys., 100 (2006) 108

[34] J.H. Yen, I.C. Leu, M.T. Wu, C.C. Lin and M.H. Hon, Diam. Relat. Mater, 14 (2005) 841

[35] X. L. Li, J. F. Liu and Y. D. Li, Inorg. Chem., 42 (2003) 921.

[36] M. Feng, A. L. Pan, H. R. Zhang, Z. A. Li, F. Liu, H. W. Liu, D. X. Shi, B. S. Zou and H. J. Gao, Appl. Phys. Lett., 86 (2005) 141901

[37] K. Lee, W. S. Seo, J. T. Park, J. Am. Chem. Soc., 125 (2003) 3408

[38] H. Wang, X. Quan, Y. Zhang and S. Chen, Nanotechnology, 19 (2008) 065704

[39] Z. R. Dai, Z. W. Pan and Z. L. Wang, Adv. Funct. Mater., 13 (2003) 9

[40] K. Hong, M. Xie, R. Hu and H. Wu, Nanotechnology, 19 (2008) 085604

[41] J. Zhou, L. Gong, S. Z. Deng, J. C. She and N. S. Xu, J. Appl. Phys., 87 (2005) 223108
[42] C. Ye, X. Fang, Y. Hao, X. Teng and L. Zhang, J. Phys. Chem. B, 109 (2005) 19758.

12

Electrochemical Methods in Nanomaterials Preparation

B. Kalska-Szostko
University of Bialytsok, Institute of Chemistry, Poland

1. Introduction

Nanotechnology has become the most fashionable science since the end of the last century. Since then, a lot of effort has been made to achieve numbers of multifunctional materials with simple synthetic procedure. At the same time, investigations of easy processing for subsequent applications have become more and more popular as well. Chemical and electrochemical preparation methods are one of the possible and powerful options for the fabrication of a new class of nanomaterials. Many popular methods used for nanostructures fabrication need to involve very expensive devices like: UHV chambers equipped with MBE or sputtering, etc. It is useful to develop methods which are less expensive and lead to similar quality final products. From the application point of view, methods which allow to perform mass production in a relatively easy way were sought. Such a method is electrochemistry, which allows us to deposit a large variety of materials in many different forms from various solutions. Chemical methods also provide an opportunity to obtain nanomaterials in big quantities. Recently, hybrid nanomaterials and nanocomposites have been studied extensively because of new opportunities of the fabrication of a novel class of materials which use nanostructures as building blocks. Such subsequent hierarchical ordering of constituent nanoelements often enhance their needed magnetic, electrical, optical, structural and mechanical properties and extinct unneeded one [1]. In many cases, the tubular or elongated structures have a significant advantage over the round ones due to a possible selective interaction with the environment with or without linkage chemistry.

The nanostructures presented in this chapter were obtained by electrochemical methods [2] in matrixes possessing proper pores. Fabricated nanowires and nanotubes were deposited in anodic porous alumina template (AAO) with a pore diameter ranging from 40 to 230 nm [3].

Electrochemical deposition is a very attractive method because the process is simple and effective. There is a huge variety of possibly reduced ions, and it has no limitation as far as sample shape and size are concerned. Deposition of the layer can be done both in constant, also called direct (DC), and accelerating current (AC) modes depending on the needs, possibilities or applications and wires characteristics. In the first case, constant current is applied to electrodes whereas in the second one the potential of working electrode is controlled. Electrochemical deposition of nanowires is a technique which combines either bottom-up or top-down approaches. This is due to the fact that the wires have grown atom by atom manner and it can be obtained in the matrixes which were subjected to anodization process during which the nanopores are obtained in bulk material.

Before the templates can be used for the deposition of, e.g., nanowires or nanotubes, it should obey few requirements:

- it should be chemically inactive in a particular process;
- deposited material should wet the pores wall;
- deposition process should not be chaotic and start from the bottom or from the wall of the pore and go upwards or to the centre, respectively;

2. Preparation of AAO matrix

The most effective method to fabricate nanowires in large quantities is their deposition in porous matrixes. One of possible templates is nanoporous anodic alumina oxide (AAO). The AAO membranes can be obtained with various pores size and distance between them. The process condition such as: temperature of anodization process, composition of the used solution and current parameters, determine matrix parameters (pore diameter and distance between each other). The quality of alumina templates meaning the ordering of nanopores can be improved by a number of repeated anodization processes. The more repetitions of anodization processes are used, the better self organization of the pores at the surface is obtained. An example of such dependence is depicted in Fig. 1, where pore diameter dependence from current potential or temperature of the processes is shown.

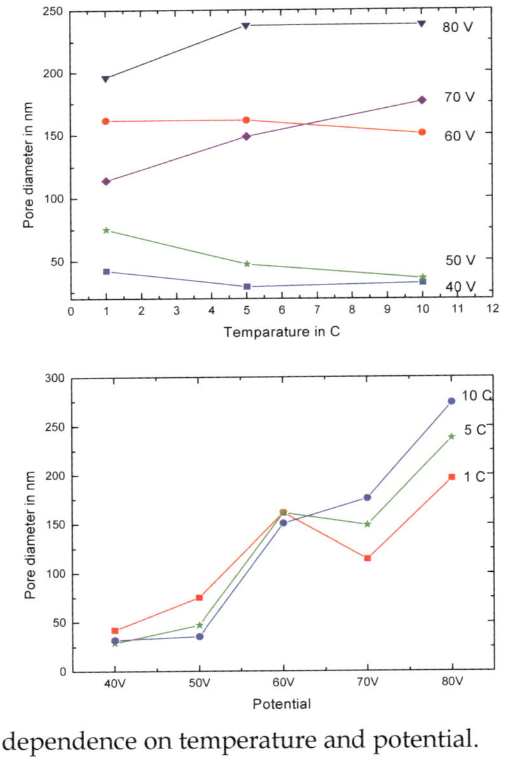

Fig. 1. Pores diameter dependence on temperature and potential.

Such parameters as pore diameter and average distance between them as well as length strongly depend on electrolyte composition besides temperature and current conditions. Most popular solutions used as electrolytes are phosphoric acid, oxalic acid and sulfuric aide. According to the equation presented below, there is dependence between anodization potential U and distance between pores D_c (1) [4]:

$$D_c = d + 2\alpha U \tag{1}$$

Where: d - pores diameter, α - proportional coefficient (its value vary between 2,5 - 2,8nm/V)

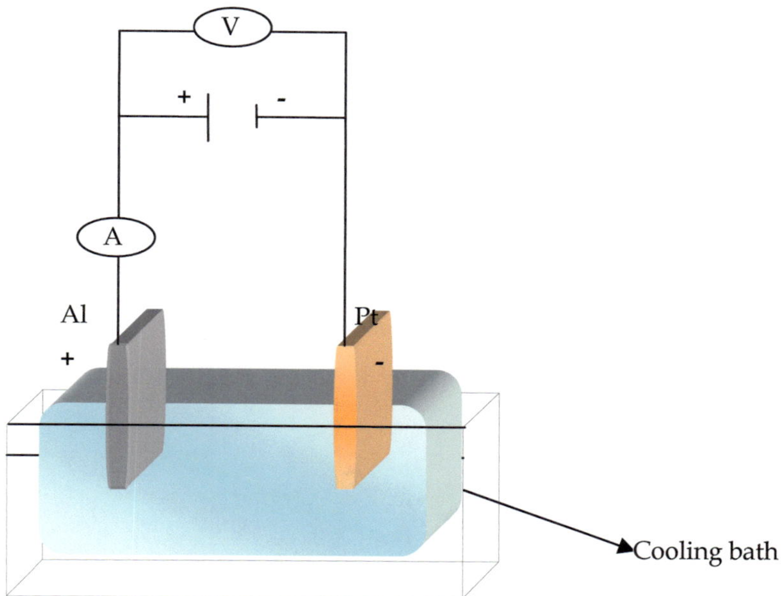

Fig. 2. Simple sketch of electrochemical double electrode cell.

In Fig. 2, a simple sketch of electrochemical double electrode cell is depicted. Here the positive electrode anode is aluminum sheet and negative electrode - Pt plate. The whole cell with the working solution is placed in a cooling bath.

To obtain good quality anodic alumina templates, few steps process should be performed.

Before the proper anodization process is done, initial degreasing, heating and electropolishing are performed. Dissolution of organic grease, oils and dust originating from technological processes from the plate surface is necessary to perform proper anodization process. It can be done, for example, by bathing Al plates for 10 min at room temperature in acetone.

Aluminum is a metal which covers immediately in contact with atmospheric oxygen or water by passive oxide layer according to the reactions (2-3) [5]:

$$2Al + \frac{3}{2}O_2 \rightarrow \alpha Al_2O_3 \tag{2}$$

$$2Al + 3H_2O \rightarrow \alpha Al_2O_3 + 3H_2 \quad (3)$$

High negative value of Gibbs free energy ($\Delta G°$ = -1582kJ/mol and $\Delta G°$ = -871kJ/mol, respectively) proves high efficiency of spontaneous oxidation process [6].

This layer can be removed by dissolution in mixture of HNO_3:HCl:H_2O in molar ratio 10:20:70 [7], at room temperature for minimum 5 min time according to the following reaction (4):

$$Al_2O_3 + 6HNO_3 \rightarrow 2Al(NO_3)_3 + 3H_2O \quad (4)$$

Such process is called surface activation and can be also performed in alkaline solution at elevated temperature around 50-60°C according to the equation (5):

$$Al_2O_3 + 2NaOH + 3H_2O \rightarrow 2Na[Al(OH)_4] \quad (5)$$

All these initial steps are required to obtain good regularity pores in matrix and/or better attachment of deposited material.

Electropolishing process takes place when electrode surface is covered by a sticky layer which has resistance much higher than electrolyte. In such instance, part of the Al surface which is higher is covered by a much thinner layer and the current density will be higher in this places. For parts where valley are present at the surface, the high resistant layer will be thicker and much less current will penetrate through (Fig. 3). In such a way, the hills will be dissolved and in effect the surface will become less rough.

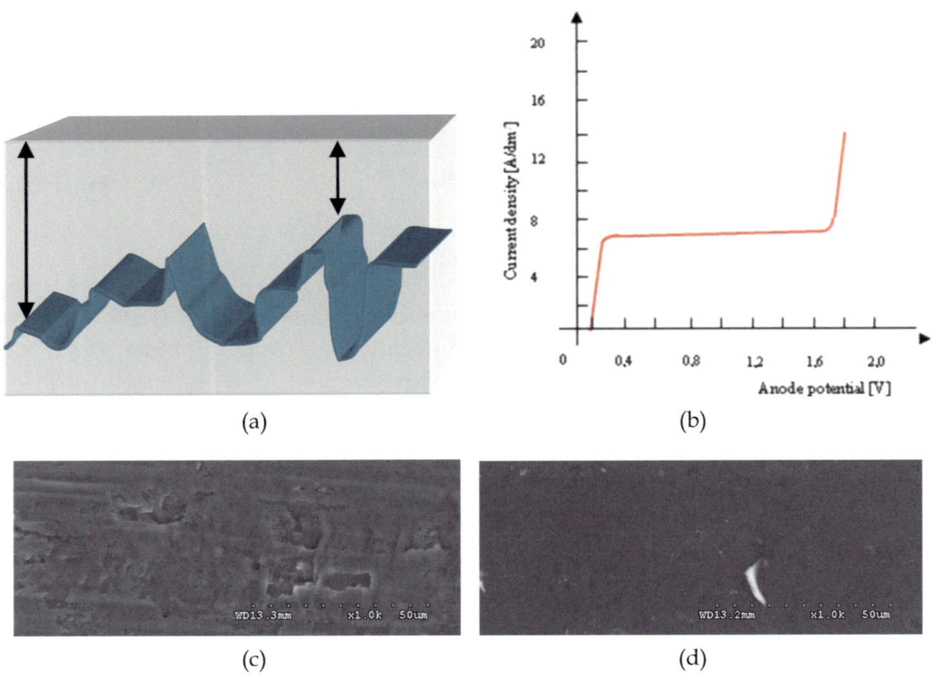

Fig. 3. a) principle of polishing process; b) polarization principles c) SEM image of non-polished and d) polished Al surface after 10 min process at room temperature in mixture of $HClO_4$:C_2H_5OH in molar ratio 75:25 [7b].

As the next step, heating at 200° - 500 °C for minimum 3 hours in inert atmosphere is needed to diminish surface roughness, stress, microcracks or other type defects caused by the production process of aluminum sheet. At elevated temperature the metal recovers and recrystallization process takes place [7a].

To find out what current parameters should be used for the polishing process, one should run polarization curves and mark out plateau where current density will change very little and the polishing process should be stable.

As it can be seen above, the resulting surface after the polishing process is characterized by much less surface defects and roughness, is more shiny, shows better reflection and better tribological properties. Removing technological origin microstructures guarantees better self-organization of pores at the Al surface.

Anodization process is anodic oxidation of the electrode material. In case of anodic alumina oxide, the initially continuous surface of Al plate becomes structured and has a unique ordered porous structure. The order of pores and their quality can be tuned by the process condition and current parameters. The effects of the conditions changes on the matrix structural parameters are depicted in Fig. 4.

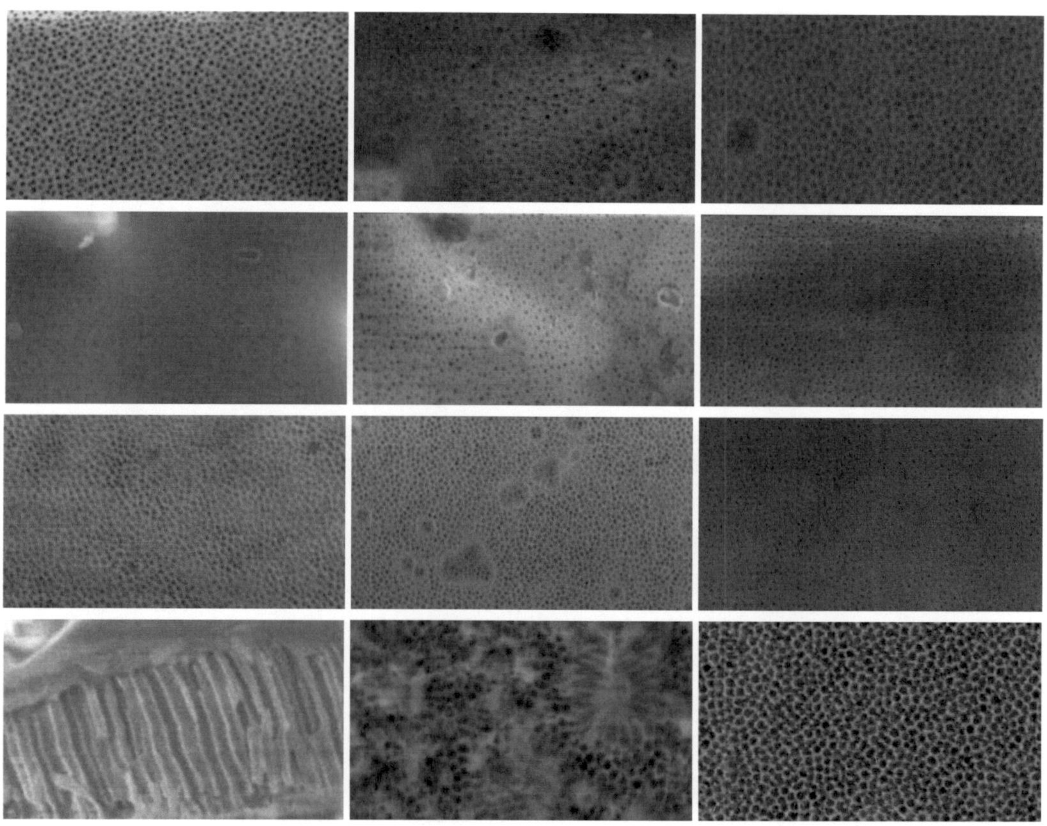

Fig. 4. Examples of porous alumina oxide surface topography and pores cross section.

The images of the pores presented above prove that in large range pores diameter and distance between them can be tuned by process and current parameters. The obtained example cross section image of that presents that the template pores are rather even and well defined.

During anodization process two parallel reactions are taking place:

- the first one at the interface metal/oxide, where Al^{3+} ions migrate to oxide layer (6) [8]:

$$Al \rightarrow Al^{3+} + 3e^{-} \tag{6}$$

- the second one at the interface oxide layer/electrolyte, from where oxide penetrates the metal layer (7):

$$\frac{3}{2}H_2O \rightarrow 3H^{+} + \frac{3}{2}O^{2-} \tag{7}$$

Formed ions Al^{3+} and O^{2-} migrate to the interface metal/oxide and form Al_2O_3.

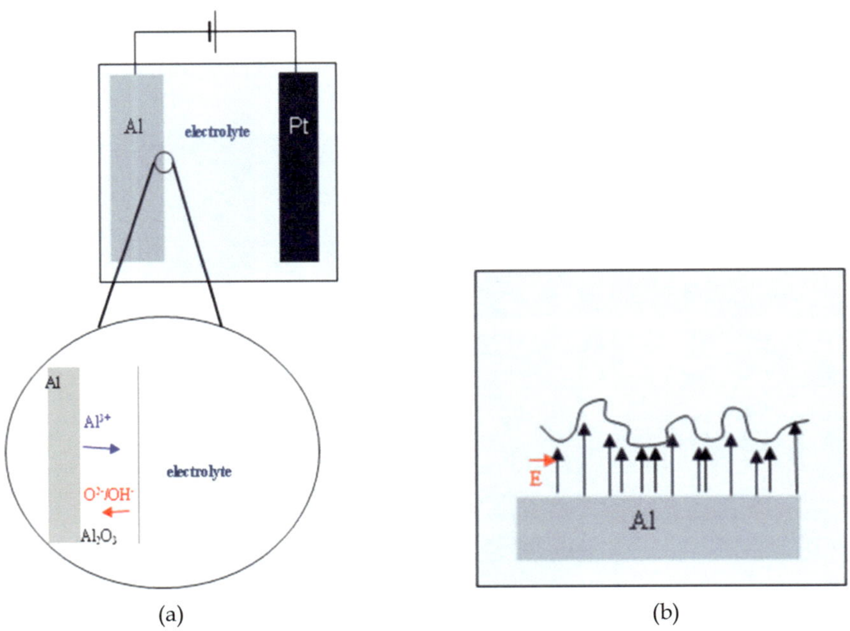

Fig. 5. Schematic sketch of: processes proceed at the electrode surface (a), alumina oxide quality (b).

Parallel to the reaction mentioned above, dissolution of the oxide takes place according to the scheme (8):

$$\frac{1}{2}Al_2O_3 + 3H^{+} \rightarrow Al^{3+} + \frac{3}{2}H_2O \tag{8}$$

This shows that both processes creation and dissolution of the oxide run at the same time, and to obtain needed porous oxide layer, the first one should dominate over the second one, respectively. The best ratio between the both processes is 7:3 [9].

The temperature of anodization electrolyte should be lower than room temperature to avoid too fast alumina dissolution process. Low temperature also prevents local overheating of the Al plate, which can cause cracking and pill of the alumina layer. Such unwanted effect is presented in Fig. 6 a.

Anodization potential influences not only pores size but also their ordering. However, applying to a large value of potential can effect in surface degradation, which is shown in Fig. 6 b.

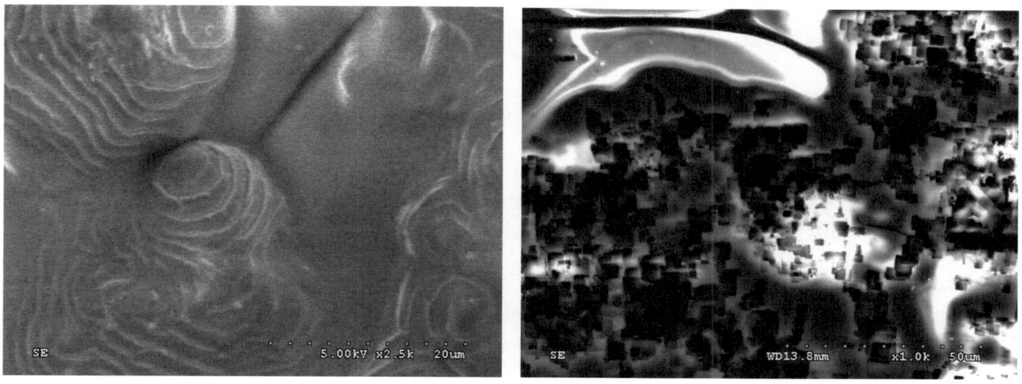

(a) (b)

Fig. 6. a) effect of too high temperature of solution; b) effect of too high anodization potential.

At first stage, Al surface is covered by flat barrier oxide layer which is conducting very little ($10^{10} \sim 10^{12}\Omega$cm) [10]. At the next stage, this layer becomes more and more rough and corrugated, which results in local increasing of the current density, which therefore causes the increase of temperature and dissolving of the oxide layer also locally. In such a way, the pores are created. Prolongation of anodization process influences the diameter of obtained pores, a distance between them and their ordering.

The matrix prepared in such a way is not yet ready to use because at the surface the barrier oxide layer is present. This layer contains impurity from the solution and should be dissolved in one of the following solutions: mixture of 6% H_3PO_4 and 1,8% $H_2C_2O_4$, 5% solution of H_3PO_4, mixture of 1M H_2SO_4 and 2M H_3PO_4, solution of $HgCl_2$ (saturated) [11]. To illustrate the effect of this process, please look at the last row of Fig. 4.

Finally, a perfect ordered porous alumina layer is obtained (see Fig.7).

3. Deposition of the materials

Two kinds of deposition processes can be considered with controlled potential or current. Current controlling demands two electrode system with voltmeter and ampere-meter devices.

Controlling of the potential adjusted to a stable value is performed in three electrode set-ups, where, besides working and reference, a counter electrode is needed. This electrode should be chosen in such a way that it has a significant surface so that the current obtained

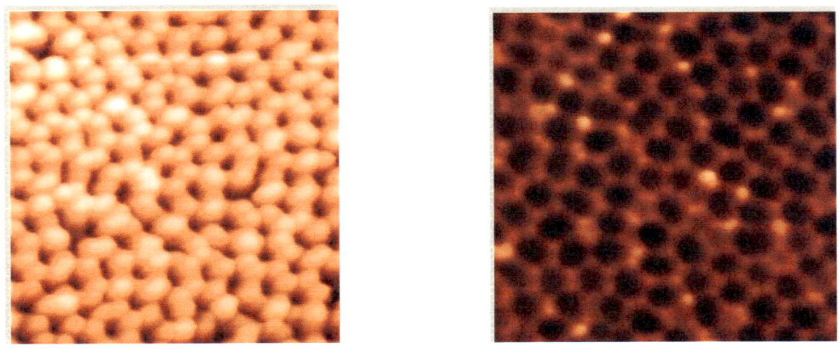

(a) (b)

Fig. 7. AFM images of final matrixes: a) oxalic acid, b) phosphoric acid.

in the electrolysis process is not changing its potential. This electrode is applied to the working electrode's constant current, which stabilizes constant potential in the reference electrode. A potential value of the electrode is controlled by potentiostat.

Accelerating current mode demands a two electrode device, where working and reference electrodes can be made from the same material.

Constant current deposition can be carried out without adjusting other current parameters before the deposition process. In case of deposition with the controlled potential, its value should be calculated from the equations described below to assure large enough process efficiency. It is also possible to run voltoamperometric or chronoamperometric curves and on its ground a characteristic value of the deposition process is established. Such specific potential has to be calculated for every type of electrolyte and electrodes.

Electrodeposition is a method by which ions from the solution are deposited at the surface of the cathode (working electrode). This process can run parallel to electrolysis, and deposited material can form a continuous layer, wires or tubes when prestructured matrixes are used. The amount of the deposited material, thickness of the layer or wires length depend on the deposition time matrix structure.

Electrolysis is a number of processes taking place in electrolyte and at electrode interfaces while electrical current is applied from external sources. In electrolysis the change from electrical energy to chemical potential takes place. The most important reactions proceed at the interfaces electrode metal/solution. When the electrode is dissolved and in oxide form is present in the solution, this electrode we call anode. The opposite electrode - cathode, is one where ions are reduced and metallic surface is formed. During the electrolysis process, large gradients of ions concentrations are observed. The reactions taking place at the electrodes are described in the following form [12]:

$$\begin{aligned} C(-) &: Ox_A + z_A e \rightarrow \mathrm{Re}\,d_A \\ A(+) &: \mathrm{Re}\,d_B \rightarrow Ox_B + z_B e \end{aligned} \tag{9}$$

In a two electrode system, both electrodes are connected to a proper potential source. Current starts to increase when the value of decomposition potential U_r is exceeded (Fig. 8). After that, the value of potential electrolysis can be used in a practical application.

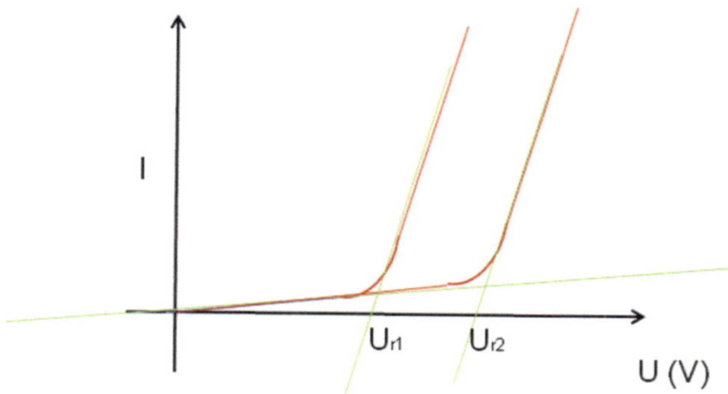

Fig. 8. Drawing of potential changes during electrodeposition process.

Decomposition potential can be described in the following way (12):

$$U_r = SEM + IR \tag{10}$$

Where: *SEM* – electromotive force of cell, *I* - current, *R* - resistance

Theoretically, this potential should increase by the value (E_A – Ec), where E_A – anode potential, Ec – cathode potential, which is needed to obtain current *I* and resistivity.

In reality, decomposition potential between two electrodes is higher compared to the theoretical one by the value of overpotential η [12]:

$$U = (E_A - E_C) + IR + \eta \tag{11}$$

Overpotential is present due to deviation of electrodes potential from equilibrium value during current flow. The reasons why such phenomenon is present are the following: to slow ion diffusion from solution to electrodes, to slow velocity of redox reactions, delay in crystallization processes and electrode surface, too big difference of ions concentration in electrolyte. The value of overpotential depends on: a type of electrode, electrode surface quality, current density, and temperature of electrolyte.

Electrolysis process is a few stage process: (i) ion transportation in electrolyte from one electrode to the opposite one; (ii) chemical reaction with electrical charge transfer at the interface electrode/solution; (iii) other type of surface reactions – adsorption, desorption, crystallization.

Faraday law defines dependence between current flow through solution and amount of material reduced at the electrodes. The first law – mass of material deposited at electrodes is proportional to the amount of current flow through electrolyte, which is described by the equation [13]:

$$m = k \cdot Q = k \cdot I \cdot t \tag{12}$$

where: *m* – mass of material reduced at electrode;
k – electrochemical equivalent
Q – charge
I – current
t – time

The second Faraday law - mass of different substances deposited at electrodes during flow of equal charge is proportional to its electrochemical equivalent according to the equation [14]:

$$\frac{m}{Q} = \frac{1}{F} \cdot \frac{M}{n} \text{ or } F = \frac{Q}{m} \cdot \frac{M}{n} \quad (13)$$

where: F -Faraday constant (96 500 C)
M - molar mass
n - number of electrons in the process

Before each deposition process it is preferable to run voltamperometric or chronovoltamperometric curves to optimize current parameters of the process. Voltamperometric dependence is current function of potential which changes in a way shown in Fig. 9 (a). Intensity of peak and at the same time current value can be calculated according to Randles - Sevčik equation [15]. In cyclic curve two peaks can be observed: cathodic reduction and anodic oxidation of the ions Fig. 9 (b). To obtain deposition with reasonable effect, deposition should be conducted in a range of cathodic reduction peak.

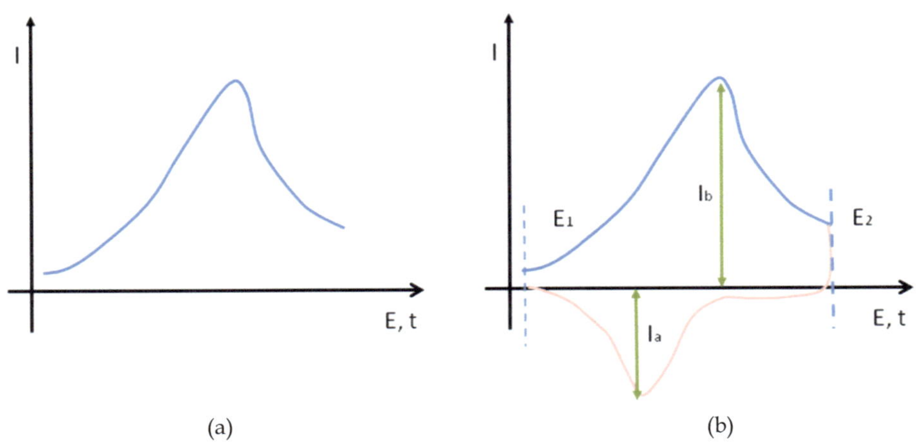

Fig. 9. (a) Typical voltamperomertic curve I=f(E), (b) typical cyclic voltoamperometric curve.

In addition to these two methods, chronoamperometric dependence $I=f(t)$ should be done, where the value of current is monitored as a function of potential. Such measurement gives us additional qualitative and quantitative information about electrolyte composition, speed of electrode processes and chemical reaction taking place at the electrodes and in electrolyte. These curves inform about electrode deposition process [15]. Chronoamperometric curves should be registered at potential exceeding overpotential value. By collecting these curves for few potential values best deposition condition can be determined.

Current parameters for each particular deposition process should be estimated for each electrochemical cell and electrolyte.

Electrochemically deposited material has in most cases a crystalline structure but the quality of the obtained layers can be different due to the various speed of crystal growth and crystal seeds creation. Modification of few parameters such as: current density, type and concentration of electrolyte, pH of electrolyte, temperature, presence of surfactants, matrix properties, etc., has significant influence on the quality of deposited material. For each material a range of parameters giving best quality surface can be assigned.

Current density is a ratio between total current value and electrode surface. This is probably the most important parameter defining quality of deposited material. For low current density, its value coarse surface can be obtained because the speed of nucleation centers creation is much lower compared to crystal growth of the existing seeds. For high current density, quite often a decrease of ions concentration close to the deposited surface is observed, or hydrogen evolution takes place and the obtained layers become non-continuous, spongy and porous. To avoid non-equal current density in deposition process, a parallel arrangement of electrodes in the centre of vessels should be assured. In case of nanowires, deposition proper current density should be well matched to protect hydroxide and cracks or discontinuous inclusion of the wire creation. This avoids too fast degradation of the wires during fabrication processes and assures good electrical conductivity [12].

pH variation also influences surface morphology heavily. Low pH causes hydrogen evolution, which penetrates the layer, and deposited material becomes harder, and it introduces a lot of stress to the layer. pH also influences magnetic properties of deposited material. Because of that, electrodeposition process should be conducted at a stable value of pH, any variation can be compensated by the addition of alkaline or acidic media [12].

Electrolyte concentration - increase of electrolyte concentration, effects in continuously grown and well attached layers. Growth speed in-plane dominates growth out-of-plane, which avoids high roughness [12]. Higher ion concentration allows to work with higher current density without the risk of hydrogen evolution. A very similar effect is observed in case of solution stirring because it eliminates the concentration of gradients in the solution.

Temperature - its increase, influences two opposite phenomena. First of all, it helps in diffusion process, which prevents rough and spongy layer growth even for high current density. On the other hand, it increases the speed of crystal growth and causes the presence of bigger crystals and hydrogen evolution. At lower temperatures, the first scenario dominates and better quality layers are obtained compared to higher temperatures. Moderate heating of electrolyte influences positively on the quality of deposited layers but too large temperature elevation gives rather bad quality surfaces [12].

The presence of surfactants in electrolyte decreases surface tension at the electrode surface and influences the release of hydrogen from the electrode surface, which is produced in hydrogen evolution process [12]. The addition of surface active substances and their adsorption at crystal surfaces stops the growth of existing agglomerates and affects nucleation of new ones. Such modification of the process helps to fabricate layers build up from small crystalline material. Very alike phenomenon is observed in case of complex ions.

Extremely important parameter which governs good quality of obtained films is used matrix, which often influences and modifies a crystal structure at least in the first layers of deposited material.

In aqueous solution of metal M^0 salts M^{z+} ions are present. Placing two electrodes connected with current source into it originates the following process [13]:

$$M_{aq}^{z+} + ze^{-} = M^{0} \tag{14}$$

The mechanism of layer deposition is quite complex and is illustrated in Fig. 10. Metal ions are covered by hydration shell, which helps in easier diffusion, convection and migration processes. These phenomena cause transport of hydrated ions in cathode direction. Close to cathode party dehydration process takes place and released ion can be placed in a crystal structure of a growing layer.

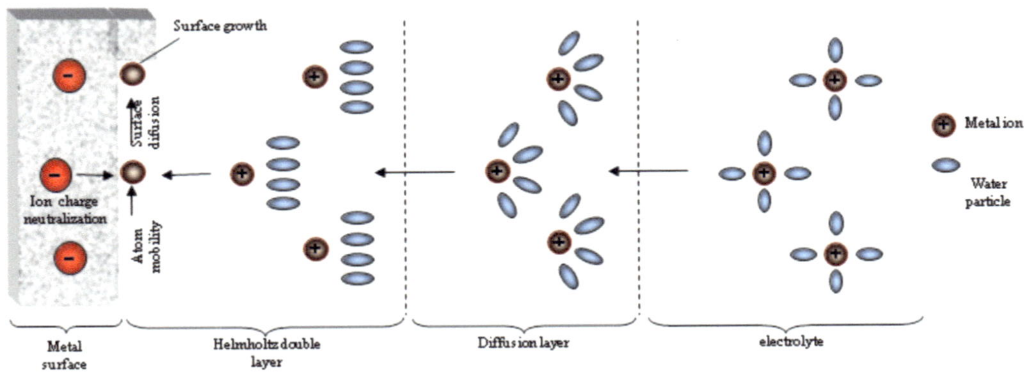

Fig. 10. Schematic drawing of the process taking place close to electrode.

In case of metal ions, deposition at electrode surface (other metal) incorporation of its into layer is not an immediate process after landing at the surface because of energetic differentiation between the sites (Fig.11). Normally, atoms migrate at the electrode surface until free defects sites with lowest energy meet.

Elecrodeposition process can be divided into two main steps: (i) seeds nucleation (when M_{aq}^{z+} ion approaches the surface, loose solvation shell and its charge is reduced by charge transfer. In such a way, neutral ad-atom M^0 is formed (Fig.11). Ad-atom can migrate at the surface as long as preferred site will be reached. Most energetically favorite site is hole, next node, dislocation and least flat surface.

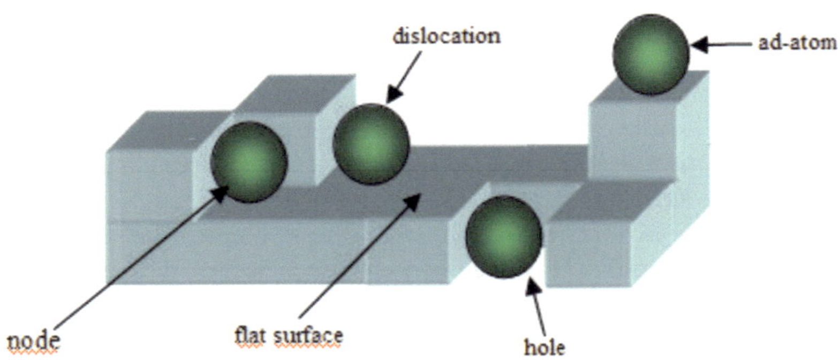

Fig. 11. Drawing of the atoms location in deposition process

These principles define electrodeposition of the material in prestructured or porous matrixes (e.g. anodic alumina oxide), where at first atoms are deposited at the bottom of the pores.

Next, the collection of ad-atoms in agglomerates nucleates crystallization seeds of a new phase, which grows upwards. Such a scenario is very profitable because the whole pore can be used as a template for wire growth and no stuffing of the pore takes place.

Using the above phenomena, Cu, Ni, Co and Fe were deposited to the porous AAO matrix characterized by a different pores diameter [7]. As it was already mentioned, the deposition conditions influence strongly the quality and length of the obtained nanowires. Such structures as nanowires and nanotubes can be obtained via constant current deposition (DC) method when well distributed nanowires standing perpendicularly to the surface plane are relatively easily obtained. The acceleration current method (AC) in most cases (for long enough time of deposition) ends up with long wires which are randomly oriented at the surface due to strong magnetic interaction between them.

AC platting is often performed in a two electrode cell where working electrode is AAO plate and reference electrode Pt plate or wire. The example of the electrolyte composition used in the process was: 120 g/l $NiSO_4$ + 40 g/l H_3BO_3, electrolyte: 50 g/l CoSO4·7H2O + 25 g/l H3BO3 + 20 g/l glycerin, 0.9 mol/l $FeSO_4$ $7H_2O$, 0.15 mol/l $FeCl_2$ $4H_2O$ and 0.43 mol/l NH_4Cl for Ni, Co, Fe AC deposition at current condition 20V, 16V and 10V, respectively. DC deposition required a three electrode cell and besides, two already mentioned counter electrodes as Ag wire is present [2].

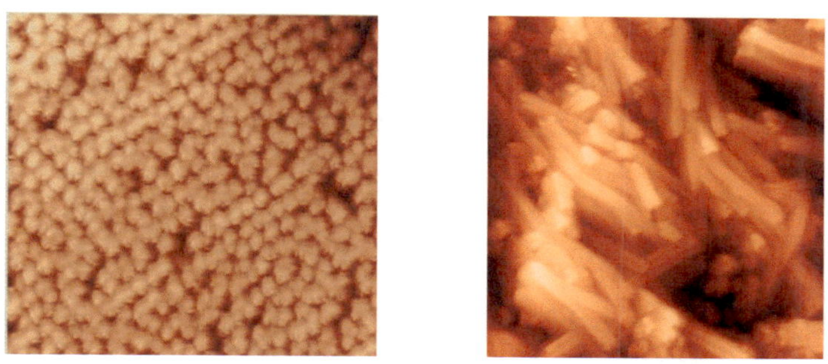

(a) (b)

Fig. 12. Different quality of obtained wires by (a) DC and (b) AC modes

A choice of deposition mode depends on the needs and effectiveness of deposition processes. In DC mode, whole time while current is on deposition process takes place. AC plating combines deposition and dissolution processes which are running after each other. Firstly, the material is deposited and in the opposite current phase this layer is dissolved. Effective platting is only in this case when deposition dominates dissolution process.

The principles mentioned above should be modified and few additional phenomena must be considered. In general, kinetic parameters of deposition process are quite complex and depend on many parameters, which can be seen in the following equations (15-20). There are still many unknown points in deposition process which demand big effort of scientists and consume much time to be resolved. In many cases the proper and most accurate values describing the observed processes are obtained by different models. In case of porous membranes, one possible model is developed by Osterle called Space-Charge model [16]. There are modifications of this and some points are improved. In general, there are few equations which can be considered for the explanation of the problem and few equations must be fulfilled to understand the deposition phenomena [17]. The Nernst-Planck equation - which describes molar flux density (J) present in the tube:

$$J = -D\nabla C - KZCQ\nabla\phi + Cu \tag{15}$$

Where: D – diffusion coefficient (diffusivity), C – ion concentration, K - mobility Z – valence of the ion, Q – , Φ – electrical potential, u – fluid velocity

The Einstein-relation which combines dependence between mobility and diffusivity of ions [17]:

$$K = \frac{D}{RT} \tag{16}$$

The Navier-Stokes equation – describing the fluid flow in a cylindrical tube [17]:

$$-\nabla p - F\sum_{i=1}^{2} ZC\nabla\phi + u\nabla^2 u = 0 \tag{17}$$

Where: p – hydrostatic pressure,

F - Faraday constant

The Poisson-Boltzmann equation – which takes into account a relation between charge density and electrical potential in cylindrical coordinates [17]:

$$\frac{\varepsilon_0}{r}\frac{\partial}{\partial r}\left(\varepsilon r\frac{\partial\varphi}{\partial r}\right) = \sum_{i=1} ZFc^b \exp\left[-\frac{ZF\varphi(r)}{RT} - \frac{AF}{eRT}\left(\frac{1}{\varepsilon} - \frac{1}{\varepsilon_p}\right)\right] \tag{18}$$

Where: e – electric charge, r – hydrated radius, ε – permittivity of free space, A – hydration constant, ε – dielectric constant.

Surface charge density – fraction of the total number of surface sites per unit area that are unbounded/charged [17]:

$$\sigma_s = e\left(\left[AH_2^{2+}\right]_s - \left[A^-\right]_s\right) \tag{19}$$

Where: -A^- - fixed charges, s –surface quantity, H^+ - mobile ions,

Finite size ion effects for strongly hydrated electrolytes which consider the case of dissociated ions and full penetration in case of binary electrolytes [17]:

$$\frac{n}{n^b}\left(\frac{N - n^b}{N - n}\right) = \exp\left[-\frac{z\,F\Psi(r)}{RT} - \frac{A\,F}{eRT}\left(\frac{1}{\varepsilon} - \frac{1}{\varepsilon_b}\right)\right] \tag{20}$$

Where: n – number of density of ions near the charged wall, n^b - number of density of ions in the bulk, N – total number of vacant sites (N=1/v here v is the volume of the hydrated ion)

The movement of the ions in cylindrical tubes which are characterized by small radius is impeded in proximity of the pore wall. This fact decreases ionic diffusivity, which can be described as D_i :

$$D_i = R_i D_i \tag{21}$$

Where: R_i - ion specific hydrodynamic retardation factor.

Pore fluid conductivity of the capillary tube (c_p) can be calculated by dividing the fluid conductivity coefficient K by cross-section area (a^2) of the pore [17].

$$C_p = \frac{K}{\pi a^2} \tag{22}$$

In principle, pore conductivity and the electrical potential gradient should increase with the increase of ion concentration.

Apart from that, simplified assumptions (small Debye length, small surface charge density, constant surface charge and surface potential) can be used to describe electrokinetic phenomena to derive phenomenological coefficients. Simplified equations can also be based only on terms of forces and fluxes.

The characteristic parameters of electrical double layer such as electrical potential and profile of ion concentration are often modified by dielectric saturation effects, which, on the other hand, depend on the strength of electric field, and this derives from the interaction between fixed charges at, e.g., pore wall and the mobile ions present in electrolyte [18].

4. Variation of deposited materials

Even if a theoretical approach to electrodeposition can be done in many ways, quite often experimental results are difficult to understand because quite many parameters can vary at the same time and the variation of one of them in other order leads to completely opposite results. This is a huge problem for scientists and makes nanomaterials become very ardent but also extremely difficult topic to study.

Not only single element wires are interesting but even more promising are these which are fabricated from more than one element. Such elements can be distributed in wire body in few manners: i) alloy-like composition, ii) layered structures, iii) core-shell structures (see Fig. 13).

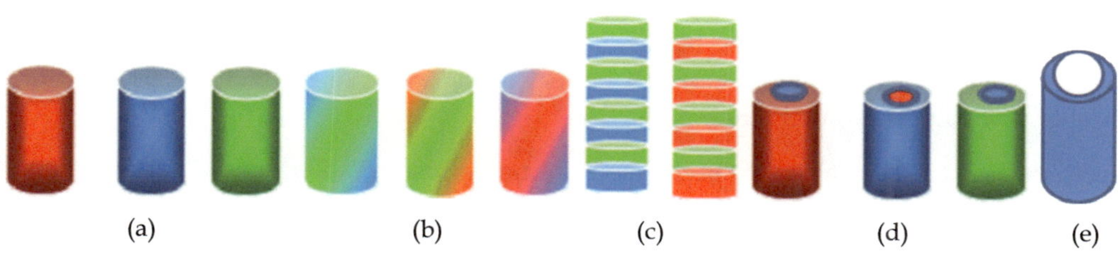

Fig. 13. Examples of possible structure modification of electrodeposited nanowires (a) single phase, (b) alloy like, (c) layered, (d) core-shell, (e) tubes.

Depending on the application, one of these are more interesting compared to other one. The application of noble metal wires such as Ag, Au, Pt and Cu are interesting in bioapplication but they are almost of no use in electronic devices or sensor application. On the other hand, Co and Ni oxides are often interesting to use in electronic and sensing industry and rather no application in biotechnology can be found. For example, if one considers modification of the wire surface by a compound which is active only with metal oxides, the surface of the wire should be covered intently by such. At the same time, affinity to thiols and amine groups will show wires which posses layered structure of Ag and Fe. Such structures require modification of electrodeposition process to obey both material deposition conditions (Fig. 14).

Among elongated structures nanotubes are placed and they also have quite a large variety of possible applications. Variation of the inert structure of nanomaterials allows to estimate the best composition which obeys the best productivity of surface functionalization processes.

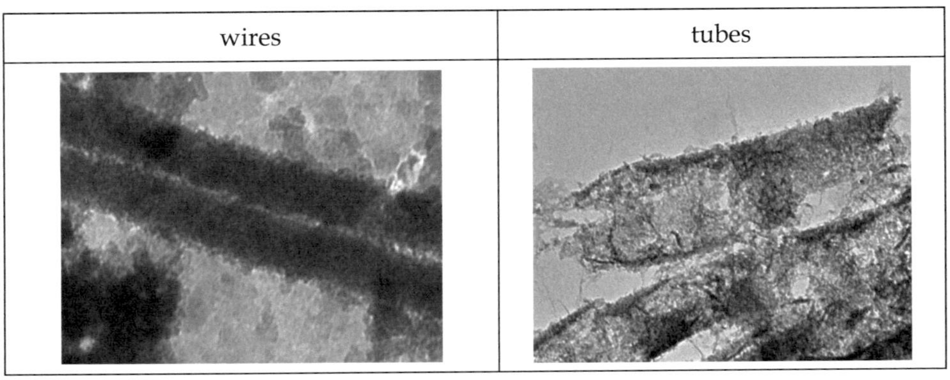

Fig. 14. TEM images of electrodeposited nanowires and nanotubes

Effectiveness of deposition process strongly depends on elements which are deposited. This problem can be faced when deposition of elements such as Co, Ni and Fe in AAO matrix is compared. On the one hand , the diffusion of the ions is almost the same because we use the same type of the matrix and the radius of the ions is very alike. This suggests that the value will rather be the same for the considered elements and the same matrix. However, the difference in a growing rate of the two compared cases is observed. This uneven process can be due to dissimilar crystal growth modes in the pores. The evolution of H_2 can also be not the same and it can sustain each element differently. When the surface is smoother, there are fewer nucleation centers and the atoms are worse adsorbed on the surface. On the other hand, the quality of the wires is much better compared to one which possesses a large number of the nucleation centers [19, 20].

Electrodeposition can also be performed in non-aqueous solution, which is valuable especially for very expensive or rare elements. Such deposition process can be done in, for example, acetone electrolyte by AC platting method, where one electrode (working) is matrix at which a layer should be deposited, and the second electrode from which the element is released (dissolved). [21]

Wires obtained in matrixes can be released by dissolving the alumina layer by such solutions as: 0,52 M H_3PO_4 or 1 M NaOH. Dissolution of matrix can be performed up to the stage when only a part of or until the whole template is removed from the plate surface. It depends on the particular application of fabricated nanowires [22].

5. Composite preparation

An interesting type of composites were these which had elongated nanostructures as nanowires or nanotubes in the structure. A special type of these are the nanostructures which are made of pure metal or metal oxide. Many of them have magnetic and at the same time conducting properties or selectively only one of them. This causes that the elements or whole composite materials can be easily manipulated by external magnetic field, which extends their application possibility. A special kind of nanocomposites are bionanocompisites, where the structure is constructed in such a way that bimolecule is already attached to the composite or can be caught by free active bonding. All non organic elements can be connected with biomolecules directly or via linkage chemistry (see Fig. 15.).

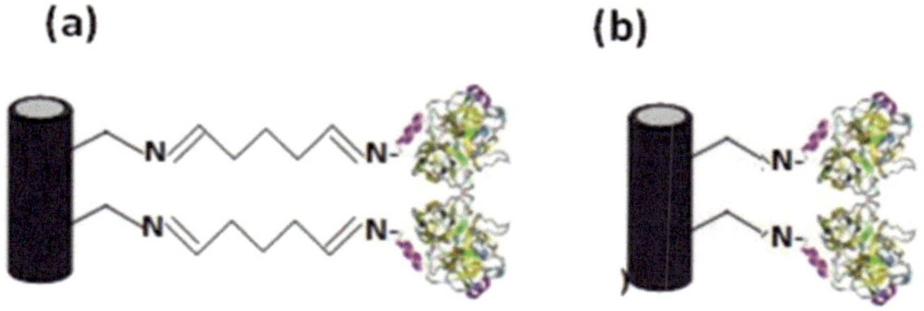

Fig. 15. Examples of biomolecule attachment to composite (a) via linkage chemistry, (b) directly.

Variation of the nanowires composition causes that properties of a particular composite can be modified in such a way that enhance needed one and weakened useless. Modification of the wires composition also influences affinity to certain linkage chemistry, which determines a particular application.

6. Possible application of described materials

Nanomaterials with high aspect ratio are particularly interesting due to the possibility of combining easily two kinds of totally different species. It allows to modify them by linkage chemistry, which is sensitive to various molecules. The array of nanowires can be used as chip for catching specific chemical materials. In external magnetic field manipulation of separate wires to construct long chains is possible to obtain.

Deposition of core-shell nanowires is particularly interesting when prevention from oxidation process is expected. This is particularly important when any modification of magnetic core is critical and application due to oxidation process is limited. Controlling of the oxide layer opens a new field in fabrication and application of electrochemically obtained nanowires.

Elongated and differentiated particles combine possible immobilization at differently active parts of particle various active particles. This opens a quite broad field of application in biology and medicine.

The possibilities in constructing new nanomaterials is huge. There is an enormous number of structures which are described in the literature. However, there is also a gigantic number of unknown structures and a lot of hard work must be performed to define and describe their properties.

7. Acknowledgement

I would like to thank all my students and co-workers for contribution to this work. This work was partly supported by Polish National Grant No. N N204 246435.

8. References

[1] P. G. Romero, C. Sanchez, Functional Hybrid Materials, (wiley, Weinheim, 2003) p 86; Y. Wu, J. Xiang, C. Yang, W. Lu, M.C. Liber, Nature 430 (2004) 61; B. H. Kim, J.H. Jung, S.H Hong, J. Joo, A.L Epstein, K. Mizoguchi, J.W. Kim, H.J. Choi, Macromolecules 35 (2002) 1419; W.U. Huynh, J.J.Dittmer, A.P. Alivisatos, Science 295 (2002) 2425;

[2] B. Kalska-Szostko, E. Brancewicz, W. Olszewski, K. Szymański, P. Mazalski, J, Sveklo, Acta Phys. Pol. A 115 (2009) 542

[3] B. Kalska-Szostko, E. Brancewicz, W. Olszewski, K. Szymański, P. Mazalski, J, Sveklo, Solid State. Phenom. 151 (2009) 190; B. Kalska-Szostko, E. Brancewicz, E. Orzechowska, P. Mazalski, T. Wojciechowski, Mater. Sci. For. 674 (2011) 231

[4] G.D. Sułka „ Wysokouporządkowane nanostruktury na anodyzowanym aluminium" Instytut Chemii Fizycznej, Uniwersytet Jagielloński

[5] P. Tomassi „Ochrona aluminium przed korozją" Instytut Mechaniki Precyzyjnej, W-wa 2006

[6] J. Choi „Fabrication of monodomain porous alumina using nanoimprint lithography and its applications" Mathematisch-Naturwissenchaftlich-Technischen Fakultät der Martin-Luther-Universität Halle-Wittenberg 2004

[7] a) B. Kalska, E. Orzechowska Mater. Chem. Phys (2011), Curr. Apel. Phys. (2011); b) H. Adelkhani, S. Nasoodi, A. H. Jafari "A study of the Morphology and Optical Properties of Electropolished Aluminum in the Vis-IR region" Int. J. Electrochem. Sci., 4 (2009) 238 - 246.

[8] A.P.Li, F. Müller, A. Birner, K.Nielch, U.Gösele "Hexagonal pore arrays with a 50-420 nm interpore distance formed by self-organization in anodic alumina", J. Appl. Phys., vol.84, p 6023, 1998

[9] F. Li, L. Zhang, R. M. Metzger "On the growth of highly ordered pores in anodized aluminum oxide", Chem. Mater. 1998, 10, 2470-2480, A.P.Li, F. Müller, A. Birner, K.Nielch, U.Gösele "Hexagonal pore arrays with a 50-420 nm interpore distance formed by self-organization in anodic alumina" J. Appl. Phys., vol.84,(1998) p 6023,

[10] B. Vanderlinden, H.Terryn, J. Vereecken "Investigation of anodic aluminum-oxide layers by electrochemical impedance spectroscopy", J. Apll. Electrochem., vol. 20, p. 798, 1990

[11] Ch.-G.Wu, H.L.lin, N.-L.Shan ''Magnetic nanowires via template electrodeposition'', J.Solid State Electrochem. 2006, 10,198-202; S.inoue, S.-Z.Chu, K. Wada, D.Li ''New roots to formation of nanostructures on glass surface through anodic oxidation of sputtered aluminum'' Science and Technology of Advanced Materials 4, 2003, 269-276; Y.Sui, B.Z.Cui, L.Martinez, et all.''Pore structure, barrier layer topography and matrix alumina structure of porous anodic alumina film'', Thin Solid Films 406 :1-2, 2002, 64-69; S.Lazarouk, S.Stanovski, et all.''Porous alumina as low-ε insulator for multilevel mettalization'', Microelectronic Engineering 50 (2000) 321-327

[12] W. Szczepaniak "Metody instrumentalne w analizie chemicznej" Wydawnictwo naukowe PWN W-wa 2005, „Poradnik galwanotechnika". Praca zbiorowa. Wyd. II poprawione, WN-Tech. W-wa 1985

[13] A. Cygański: "Metody elektroanalityczne" Wydawnictwo Naukowo-Techniczne Warszawa 1991

[14] J. Garaj: Fizyczne i fizykochemiczne metody analizay; Wydawnictwo Naukowo-Techniczne Warszawa 1981

[15] Herman N.A., Kalestyoski A., Widomski L. "Fizyka dla kandydatów na wyższe uczelnie I studentów", wyd. PWN, Warszawa 1995, P. M. S. Monk, R. J. Mortimer and D. R. Rosseinsky, Electrochromism and Electrochromic Devices, Cambridge University Press 2007

[16] R.J Gross and J.F Osterle, Membrane transport characteristics of ultrafine capillaries. J. Chem. Phys., 49 (1968), pp. 228–234

[17] S. Basu, M.M. Sharma, J. Memb.Scien. 124 (1997) 77, F. Wang, S. Arai, M. Endo, Electr. Comm. 7 (2005) 674, A. Vicenzo, P.L. Cavalloti, Electr. Acta 49 (2004) 4079, M. Motoyama, Y. Fukunaka, T. Sakka, Y.H. Ogata, Electr. Acta 53 (2007) 205

[18] M. A. V. Devanathan and B. V. Tilak, The Structure of the Electrical Double Layer at the metal-solution interface 1965

[19] C.L. Chiena, L. Sun, M. Tanase, L.A. Bauer, A. Hultgren, et al. *Electrodeposited magnetic nanowires: arrays, field-induced assembly, and surface functionalization.* Journal of Magnetism and Magnetic Materials 249 (2002) 146–155

[20] M.S. Dresselhausa, Y.M. Lin, O. Rabin, M.R. Black, G. Dresselhaus. *Nanowires.* January 2, 2003

[21] *B. Kalska-Szostko, E. Orzechowska, W. Olszewski.* Electrochemical deposition of Fe nanowires from nonaquaeous solution submitted for publication

[22] A. Jagminas, R. Juškėnas, I. Gailiūtė, G. Statkutė, R. Tomašiūnas "Electrochemical synthesis and optical characterization of copper selenide nanowires arrays within the alumina pores" Journal of Crystal Growth 294(2006) 343-348

13

Novel Electroless Metal Deposition - Oxidation on Mn – Mn_xO_y for Water Remediation

José de Jesús Pérez Bueno[1,*] and Maria Luisa Mendoza López[2]
[1]*Centro de Investigación y Desarrollo Tecnológico en Electroquímica, S.C., Parque Tecnológico Querétaro-Sanfandila,*
[2]*Instituto Tecnológico de Querétaro, Av. Tecnológico s/n Esq. M. Escobedo Col. Centro México*

1. Introduction

Water pollution not only increases in amount every year but it increases in higher toxic contaminants. Humanity has acquired, with harsh and tragic experience, a consciousness of care for the environment. New laws and norms throughout the world prove this. The quest for new processes, as an alternative to those already in use, has been testing many different physical and chemical procedures.

A conventional procedure for the removal of metal ions is chemical precipitation of hydroxides by adjusting the pH (Zouboulis et al., 2002). During the last decade, zero-valent iron has been used for the removal of metals such as zinc and mercury (Dries et al., 2005; Oh et al., 2007; Noubactep, 2010). This procedure is part of the electroless deposition.

The electroless metal deposition has been used in many industries especially due to polymers metallization. Electroless is comprised into three different procedures; autocatalytic, cementation and galvanic deposition.

The cementation process offers the possibility of removing contaminants from water considering its standard electrode potential according to the electromotive series without an external current. Ions of a relatively nobler element are reduced using electrons in exchange for a less nobler metal that gets oxidized. Usually, this process occurs over the entire surface, which concludes by blocking the deposition. This through that limits the applications of the process.

The cementation process has been used for the removal of gold, e.g., using magnesium. The deposition rate of gold is directly proportional to its initial concentration and time (Kuntyi et al., 2007).

Electroless metal deposition has been used for the synthesis of self-assembling silicon nanowires. In this case, it was assumed that the simultaneous growth of metal dendrites, accompanied with the etching of the silicon wafer, was important for the formation of 1D nanostructures (Peng, 2003).

The quantity of oxide formed on the cementating surface depends on pH, the quantity of the cementation agent and the stir variation. The reactivity of Fe^0 depends on the porosity of the oxides formed, the water salinity and the nature of the contaminant. A generation of thicker oxides was identified as being beneficial for the removal of contaminants and their potential reaction products can progressively be trapped in the matrix and remain very stable (Noubactep, 2010).

The removal of metallic ions in aqueous solution by zero-valent iron has been studied in research papers during the last decade (Noubactep, 2009). This route is fundamentally a cementation electroless process. It has been identified as having a synergic effect between adsorption, co-precipitation and the reduction of the contaminant on Fe^0/H_2O. Those studies associate metal removal through a combination of adsorption and co-precipitation onto iron corrosion products and reduction by using Fe^0, Fe^{II} or H_2/H.

This work considers the remediation of water by using electroless going through the formation of oxides immediately after the reduction of the metallic specie on the active surface. A high porosity of the deposited oxide may result, but not necessarily, fundamental in the process of forming the deposit.

2. Methodology

This work used standard ions solutions for water remediation evaluation because of the influence of manganese-manganese oxide. Silver nitrate, aluminum nitrate and lead nitrate (JT Baker) were used for the 1000ppm, 500ppm, and 100ppm solutions in deionizer water.

The crystallinity of manganese-manganese oxide substrates (Alfa Aesar, 99%) and that of deposits were characterized by X-ray diffraction (D8 Advance, Bruker AXS). The deposits weights were obtained by weight differences between the deposit-substrate and the substrate.

The pieces of manganese-manganese oxide used as substrates, with weight and area previously measured, were immersed in 50mL of a standard ion solution. The substrates had two different surfaces, one of them flat and the other irregular; for this study, the depositions were always on the flat surfaces.

When each test concluded, the solution was removed carefully with a pipette from the flask were the test were conducted, leaving the deposit-substrate samples untouched. The deposit-substrate samples were dried at room temperatures. The weight of the deposits was determined by weight difference between the deposit-substrate sample and the original substrate.

The characterizations of morphology and microanalysis of deposits and substrate were performed in a scanning electron microscopy (SEM, JEOL model JSM-5400LV) coupled to a microanalysis (EDS, Kevex-TermoNORAN).

The absorbance and reflectance studies of deposits were determined by Ocean Optics -USB 2000 spectrometer with a coupled integrating sphere in the range of 190 and 850nm.

The profiles of the deposits were analyzed by Veeco profiler, model Dektak 8, for obtaining the surfaces roughness.

3. Results and discussion

Manganese dioxide adsorbs ions of metals in a solution. Also, it undertakes the role of a cathode, which accepts electrons, and which consequently acts as an oxidizing agent. On the other hand, the surfaces of the manganese chips were usually constituted by alternating the areas of oxide and metal. The predominance of any of them depends on the preparation of the surface. The proposed sequence in the process, considers this nature of metal - oxide surface as anode - cathode, respectively. Micro-electrochemical cell requires favourable ΔG, standard electrode potential, and electric conductivity in order to initiate and continue with the deposition.

Each element deposition possesses particular characteristics but then also share similarities. The deposition structures vary among each element from soft dendritic branches of silver to compact layers of aluminium or columnar grains of lead. The similarities can lead to these conditions required for deposition, such as: a) electrical conductivity between active deposition sites and the substrate, b) a higher rate of deposition than the rate of dissolution, c) a diffusion of ions on the active sites of deposition.

In the case of lead, there was no visible bubble formations associated with water electrolysis. In the case of silver, there were bubble formations and their liberation caused the lost of electrical contact between metal substrate and the active deposition sites, which particularly favoured the soft structure. In the case of aluminium, with a very low standard electrode potential, there were prolific formations of bubbles but the layers deposition prevents any significant effect.

3.1 Lead ions transformed into oxides by Electroless deposition

The process of electroless deposition of lead ions, from aqueous solution on the system manganese metal - manganese oxide, is described in this section.

Electrolytically refined manganese chips, covered with oxide (Figure 1a) or polished (Figure 1b), were either used to illustrate deposition and for comparison. Pieces of manganese metal covered with a layer of manganese oxide, formed naturally under normal environmental conditions (amorphous MnO_2), and pieces free of rust were immersed separately in aqueous solutions with 100 ppm of lead during 24 h and one week. The oxide layer of the piece of Mn was removed by using a SiC No. 400 grit. The solutions were kept at room temperature (25 °C) without agitation. At the end of those periods, both samples were removed from the lead solutions, which were then filtered and stored for later analysis. The substrates with deposits were analyzed by SEM.

The following analysis corresponds to deposits from solutions containing around 100ppm of lead, which depends on the surface preparation (metallic manganese or manganese oxide) or the deposition period (middle and long term; 24 h and one week).

The Figs. 2a-b show the deposit covertures after 24 h of immersion in a 100ppm lead solution of a flat Mn chip covered with oxide (Fig. 2a) and a polished Mn (Fig. 2b). These tests were conducted at room temperature (25 °C) without stirring. It was evident, during the tests, that the polished surface had a delay of around half an hour, allowing for the formation of manganese oxide clusters, which posteriorly contribute in starting deposition. Finally, there was a thicker deposit on the manganese oxide surface. The waviness in Fig 2b corresponded to original formations of the electrolytic Mn chip.

(a)

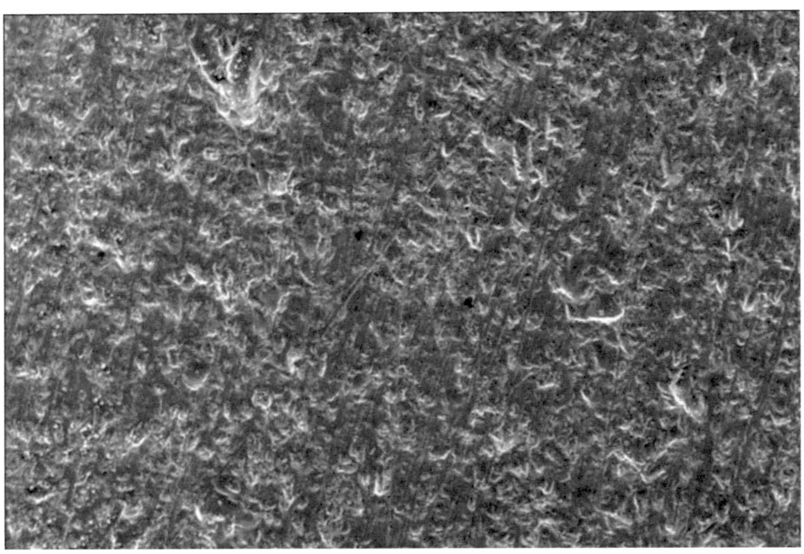

(b)

Fig. 1. **a.** SEM image using 79X of an oxide layer formed naturally on the surface of a Mn chip; **b.** SEM image using 79X on the surface of a polished Mn chip.

(a)

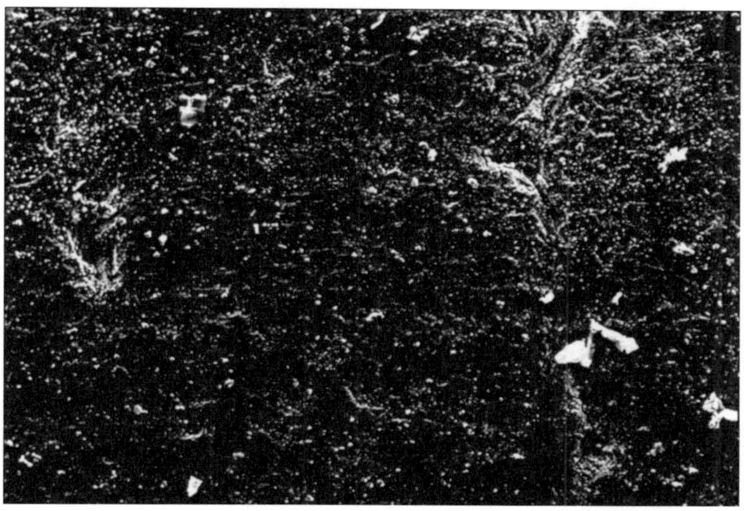

(b)

Fig. 2. **a**. SEM image, using 80X, from a flat Mn chip covered with oxide immersed in a 100ppm lead solution during 24 h.; **b.** SEM image, using 80X, from a polished Mn chip immersed in a 100ppm lead solution during 24 h.

The Mn pieces showed uniform covertures on the surface with the deposits (Fig. 3a), on which there was profuse growth. Fig. 3b shows a magnification, which resulted notorious for having frequent growths of hexagonal platelets. Along with a basis of columnar growth, there were regular hexagons with some dots on their sides. Moreover, Fig. 3b has the peculiarity of hexagonal growths on the edge of other hexagons.

(a)

(b)

Fig. 3. **a.** SEM image, using 79X, from a flat Mn chip covered with oxide immersed in a 100ppm lead solution during one week; **b.** SEM image, using 1269X, from a flat Mn chip covered with oxide immersed in a 100ppm lead solution during one week.

When analyzing the surface near the edge of the substrate, shown in Fig. 4, there were columnar formations made of numerous granules that resembled dendrites. Those growths covered the whole surface with heights that could be visible to the naked eye. The growing rate diminishes with time. This is not only due to a decrease in lead ions concentration but mainly because electrical resistivity was increasing due to lead oxide and the increasing number of grain frontiers. This characteristic of grain pillaring was frequently observed in lead oxide deposits.

Fig. 4. SEM image, using 5077X, of dendrites stacked near an edge of the polished Mn chip.

In the non polished Mn piece, a uniform deposit was observed, on which, hexagonal lamellar formations were also present and had very symmetric profiles (Fig. 5). In the image, rod-like formations appear to be located at the center and at the left inferior corner, which were made of irregular packed granules. The rod-like formations were similar to those shown in Fig. 4 and had multiple granules which were formed concurrently through nucleation and growth.

The effect of tunneling electrons, used for imaging in the SEM, was repeatedly observed in several of the thin hexagonal platelets at Fig 5. The image shows their thickness and size were not closely associated. The electrons tunneling through hexagonal formations having translucent effect allow seeing an inner surface.

EDX analyses were conducted in both samples, polished (Fig. 6a) or covered by manganese oxide (Fig. 6b). The identified elements were: lead, manganese, oxygen and carbon. The percentages of these elements in the deposit are presented, only in an illustrative form, in Table 1.

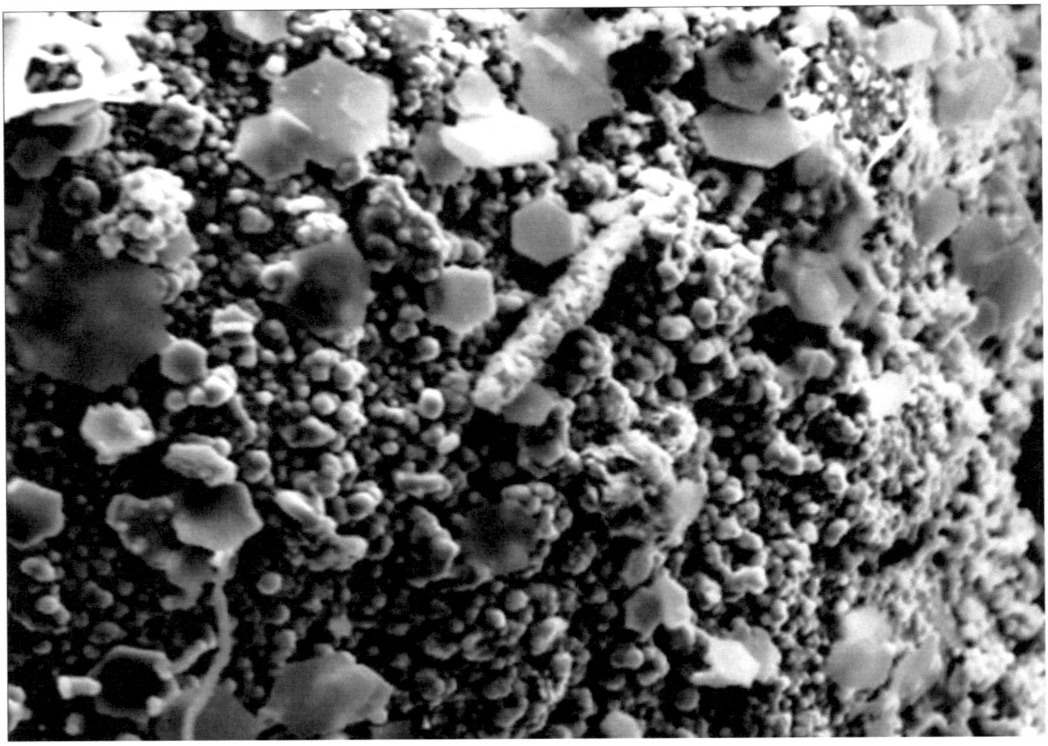

Fig. 5. SEM image at 669X from polished Mn chip immersed in the Pb solution. There were thin hexagonal formations, some of which allow transmission of the electrons from the microscope beam making them look translucent.

The EDX analysis was semi-quantitative, consequently, it is considered here illustrative of chemical composition of the deposit. The exposure to air may considerably change those compositions and change the surface appearance as manganese oxides tend to form. Considering the information from both samples, there was a close relationship between lead and oxygen content. The condition of the analyzed surfaces, there were lead oxides with manganese that eventually tended to be oxidized in the air. In other conducted experiments, lead manganese oxides tend to form ($Pb_xMn_yO_z$).

Element	Sample a (%)	Sample b (%)	Sample c (%)	Sample d (%)
Lead	52.44	39.35	37.95	1.41
Manganese	15.64	42.54	36.34	64.64
Oxygen	24.04	14.62	21.59	1.41
Carbon	7.88	3.5	4.12	4.95

Table 1. Semi-quantitative EDX percentages of elements on the surfaces: a) initially covered by manganese oxide and b) polished Mn; both during 24 h. c) Initially covered by manganese oxide and d) polished Mn; both during one week.

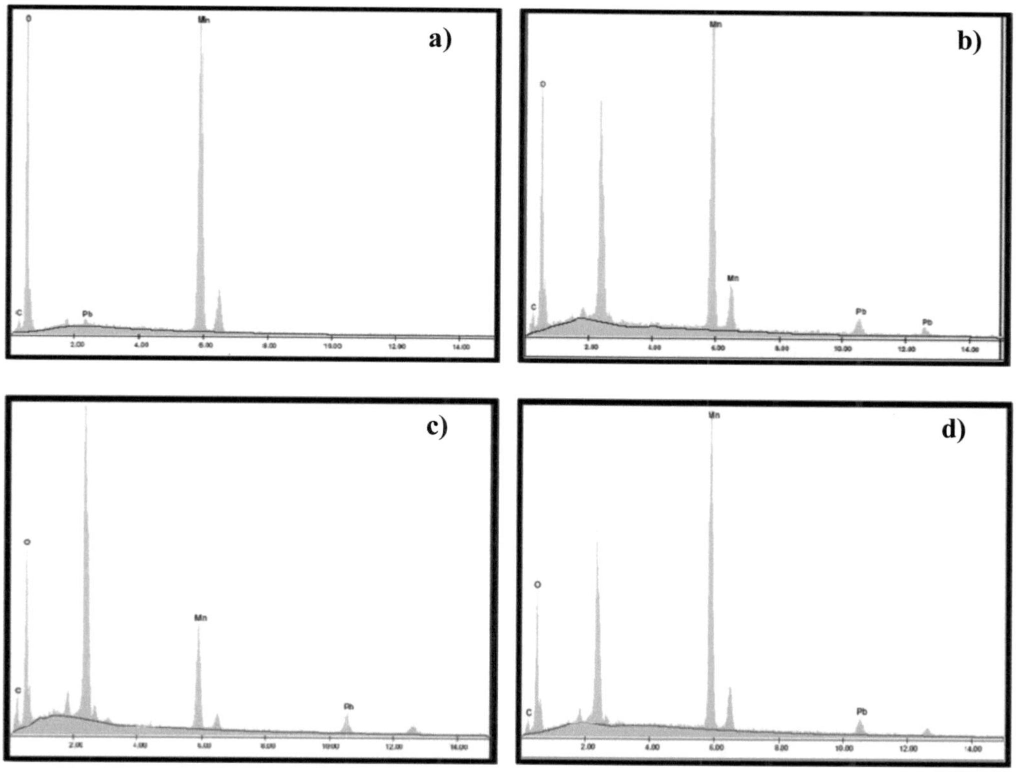

Fig. 6. EDX analysis conducted on surface: a) initially covered by manganese oxide and b) polished Mn; both during 24 h. c) Initially covered by manganese oxide and d) polished Mn; both during one week.

Table 2 presents the lead removal of samples showed after a week in contact with the solution at room temperature (25 ° C) and without agitation. ICP analysis was conducted on the solutions after the tests. There were greater depositions on Mn pieces covered by oxide than in polished Mn pieces during both periods. Also, the total lead removal after a week reached ninety percent.

The deposition was influenced by both, time and surface preparation. Nonetheless, there were other principal variables in the process: Pb concentration, the area of deposition, and the temperature. The lead deposit structure was then strong enough to apply a flow or stirring to the solution during tests. Also, this may increase the deposition rate causing a time reduction or a rise in the lead removed percentage.

X-ray diffraction analyses were conducted in order to identify the crystal composition of the deposits. The diffractograms of these samples identify a crystalline content of Pb_5O_8 (Fig. 7).

In order to observe the lead concentration influence on the structure of the deposits a 500 ppm of lead solution was used for deposits. In this series of tests, the deposition process was conducted three times, without changing the solution, but only substituting the Mn plate. Fig. 8 shows this sequence. Also, only one half of the Mn pieces were polished and subsequently immersed in the lead solution. The immersion time for samples A and B was 24h, but sample C remained for 64h, since the formation of a deposit was not evident initially.

Time	Sample	Pb initial content (ppm)	Pb content after test (ppm)	Pb removal (%)	Area (cm^2)	Pb removal (ppm_{Pb}/cm_2)
24 h	Surface covered by manganese oxide	125	115.44	7.65	0.55	92.01
	Surface of Polished Mn	125	118.45	5.24	0.58	69.73
1 week	Surface covered by manganese oxide	125	10.82	91.344	0.31	368.96
	Surface of Polished Mn	125	19.15	84.68	0.40	267.84

Table 2. Removal of Pb by Mn pieces after 24 h and one week in contact with the solution, analyzed by ICP.

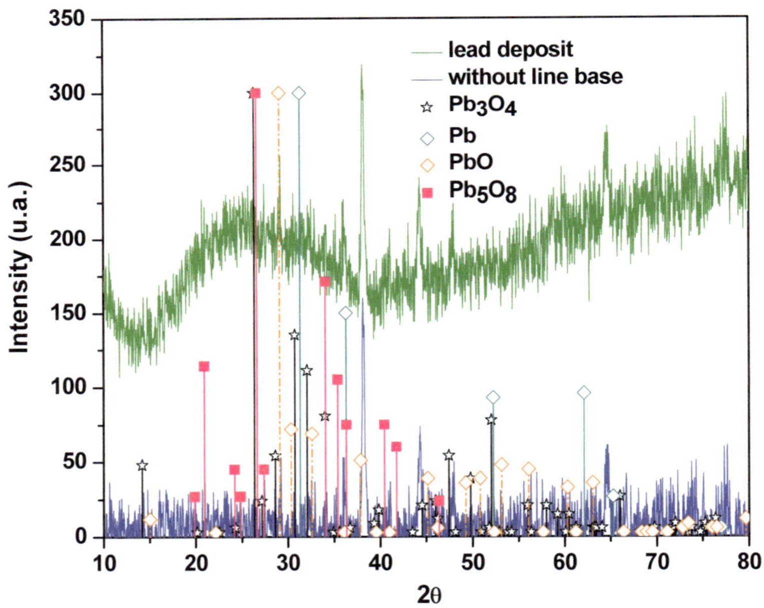

Fig. 7. X-ray diffraction of a lead deposit.

Fig. 8A shows the image corresponding to the first immersed plate into the 500 ppm lead solution A), after drying in air. The deposit was very visible and it was apparently formed by accumulations of dendrites. Under similar conditions, a second Mn plate immersed into the solution showed a still visible deposit, with thin columnar-dendritic growths and thickness (Fig. 8B). Finally, sample C showed that the deposit was present as a homogeneous

(a)

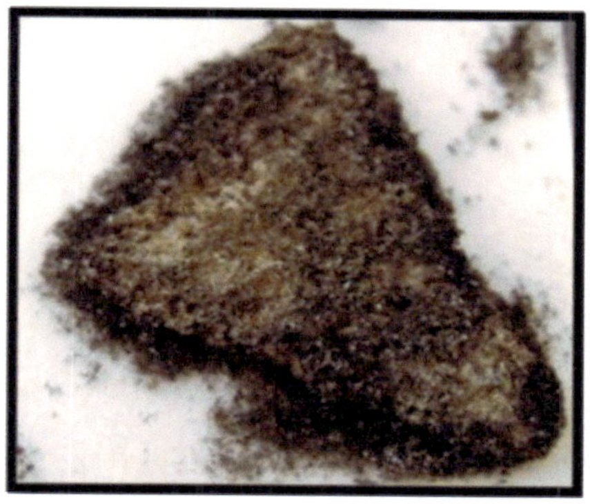

(b)

(a)

Fig. 8. a) Picture of sample A, b) picture of sample B, c) picture of sample C.

layer without a prominent accumulation on its surface (Fig. 8C). Only in the third sample there was evidence of difference between the half polished area and the oxide covered surface, which appeared as a dark area on the right hand side.

These pieces were analyzed by X-ray diffraction, which identified a predominance of Pb5O8 along with small signals indicating the presence of Pb and PbO (Fig. 9).

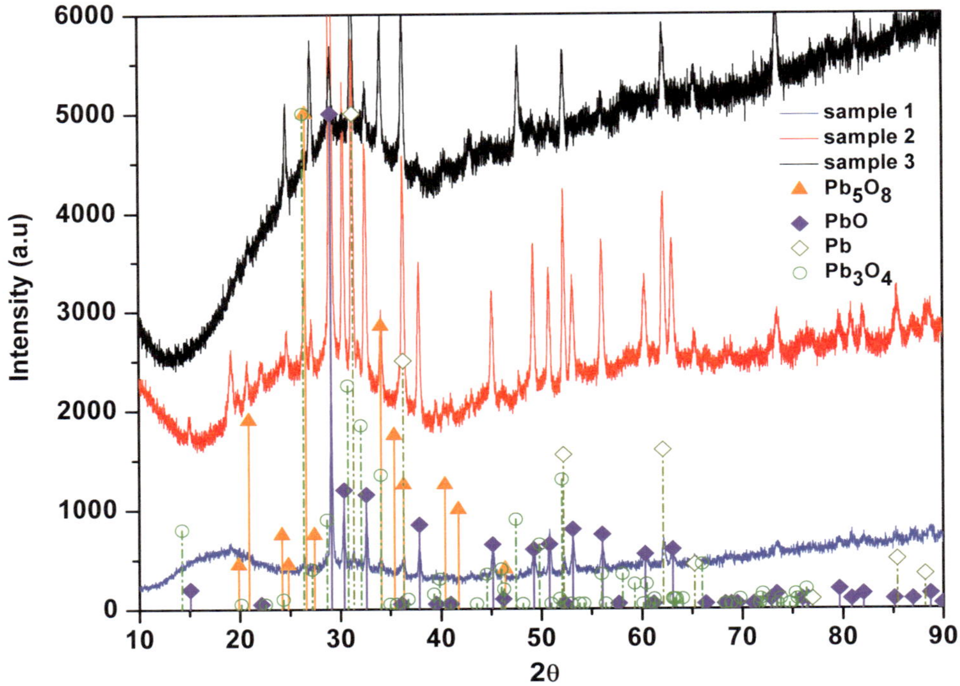

Fig. 9. Difractogram of Mn samples with deposition from a 500 ppm lead solution. A) First plate, B) second plate, C) third plate.

The kinetics of deposition of lead on manganese depends on the Pb ion concentration in solution. High concentration promotes a fast deposition and growing in a dendritic structure. The intermediate deposition rate combines semi-dendritic columns with grain nucleations. Slow deposition rates propitiate nucleations with spherical shaped grains. Nevertheless, some cases were observed with different deposition configurations.

Usually, lead deposition in the edges of a manganese substrate tended to generate the combined form of deposition. Fig 10 presents such projections from an edge of a manganese piece. It possessed a semi-dendritic column formed by grain nucleations. In this infrequent case, it was coronated by forms in a leaves-like shape.

Figure 11 shows the surface of a manganese substrate cover with a lead-based deposit after its immersion was prolonged during a month in a lead containing aqueous solution. Figure 12 shows the element EDS mapping indicating the chemical composition on surface of the lead-based deposit. The manganese covering predominantly the surface (top right). The mapping of Pb (bottom left) and O (bottom right) were more sensible to features on the surface topography. The top coverage presented an intricate needle formation. This kind of deposit covered the exposed surface of the substrate. In this case, the prolonged aging of the deposit permitted those formations as a middle point between those factors that allow the deposit and those that limited its extent.

Fig. 10. Image from a lead deposition in an edge of a manganese plate with a semi-dendritic column constituted by grain nucleations and leaves-like forms on top.

There were some factors limiting the extent of growth, such as the reduction in Pb ions concentration in solution and electric contact with the substrate. Since there were such limitations, the initial lead concentration can be reduced by increasing the area, increasing the time of contact or by successive stages of deposition.

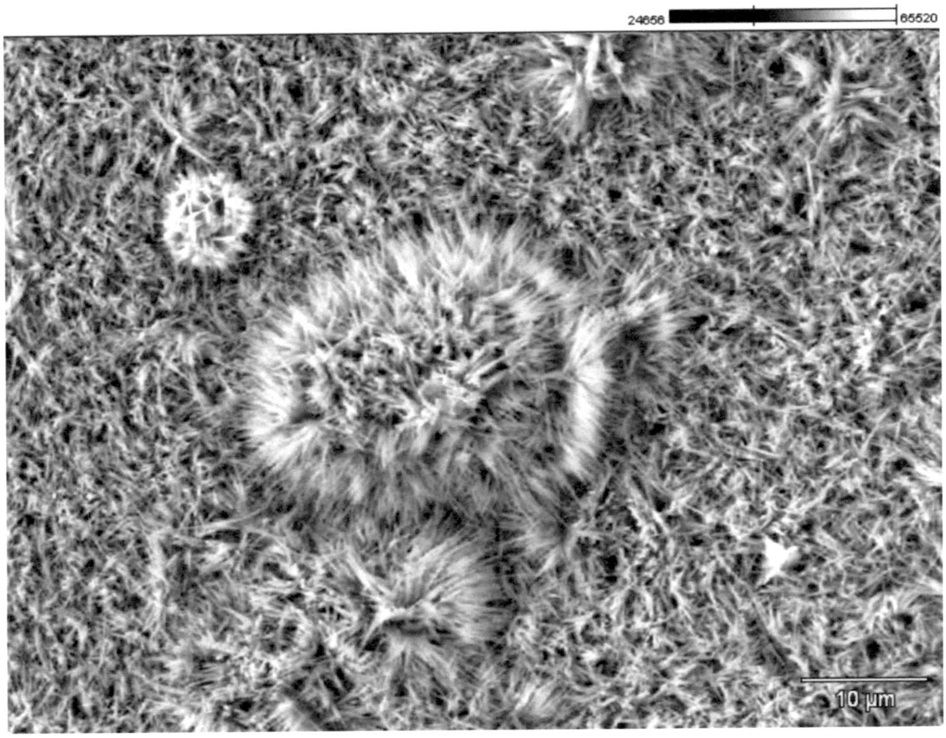

Fig. 11. SEM image of a manganese substrate covered with an intricate needle formation of a lead-based deposit after its immersion was prolonged during a month in a lead containing aqueous solution.

3.2 Electroless deposition from an aqueous solution containing Silver ions

The first step was the preparation of an aqueous solution with a silver content of 100 ppm. One half of a manganese plate was polished (grit 400), on its superior flat area, exposing a portion of the bare metal. The other half of the plaque was left with its manganese oxide which was naturally formed in air. The fundamental assumption was that there were micro-domains on the surface acting as anodes and cathodes. The purpose of this test was to discern the role that each area may play as predominantly anodic or cathodic. Both areas had oxide and metallic regions, but initializing the test in each area clearly predominated one of them.

The polished area began to be covered by a dark homogeneous layer immediately after immersion. The dendritic deposition became perceptible to the naked eye after four minutes of contact with the solution of Ag. The deposit spread to the entire surface and edges of the Mn piece after twenty minutes. The deposits initial colour was dark gray, but after seven hours, the deposit colour changed to a silvery metallic lustre. Posterior observations through a microscopy made it evident that those dendritic deposits began immediately after immersion of the Mn plate. That deposit was studied by X-ray diffraction. However, the corresponding difractogram only revealed the presence of metallic silver and it did not show any crystalline form of oxides.

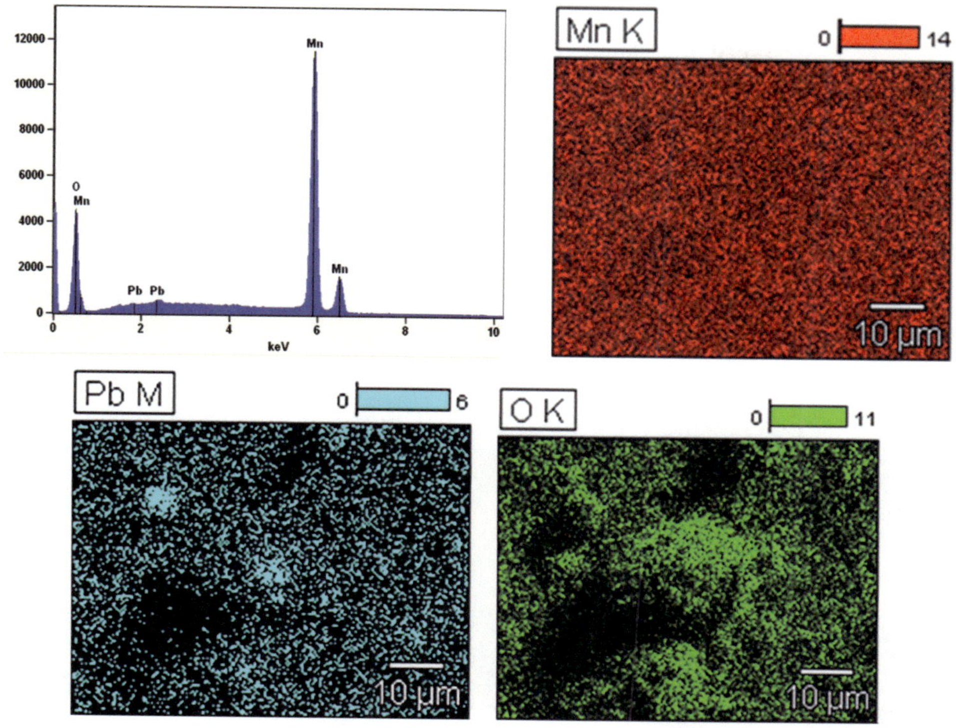

Fig. 12. EDS mapping indicating the chemical composition on surface of the lead-based deposit shown in Fig. 11. The manganese covering predominantly the surface (top right), which also contains Pb (bottom left) and O (bottom right).

The second step was the preparation of an aqueous solution with a silver content of 500 ppm. As in the previous case, one half of a manganese plate was polished and the other one left with manganese oxide.

On this occasion, as soon as the piece came into contact with the solution, the metallic part was covered with a black homogeneous deposit. Meanwhile, in the other half of the piece that still had its oxide layer a broad white spot appeared. The Mn piece was covered with a deposit of gray moss-like colour in less than a minute, which turned dark after three minutes. Five minutes after initializing the test, it was obvious that at the edges of the piece the appearance of deposit was silvery and in the form of dendrites. Throughout the whole surface, the piece was acquiring this colour.

In the case of water containing silver there were some features that differed to lead and other metals. The deposit's rate of formation is by far much faster, with a very long extent of growth, more dendritic and, hence, openly structured with small volumetric density. On the other hand, the only identified crystalline formation was metallic silver. There are two main causes for this behaviour, the standard electrode potential for silver and its electrical conductivity. The first one is related to a favourable ΔG for deposition. The second one allows the electrons transmission from the active reduction sites to the substrate. Also, under normal environmental conditions due to their electronegativity, silver patina is formed by silver sulfide rather than silver oxides.

The process of silver deposit growth showed three stages. In the first, the deposit began with a dark appearance that later changed to clearer as dark gray. In the second, it continued generating an algae or moss-like structure consisting of dendritic formations. Subsequently, the silver crystals grew into a lustrous white metallic colour in the form of dendrites with long needle-shaped branches. Those formations were projected from the sponge-like formation with extensions of around seven millimeters.

Finally, the previous stages were covered by a dendritic growth with a dark and closely packed cauliflower-like formation. This formation was extended beyond the Mn piece

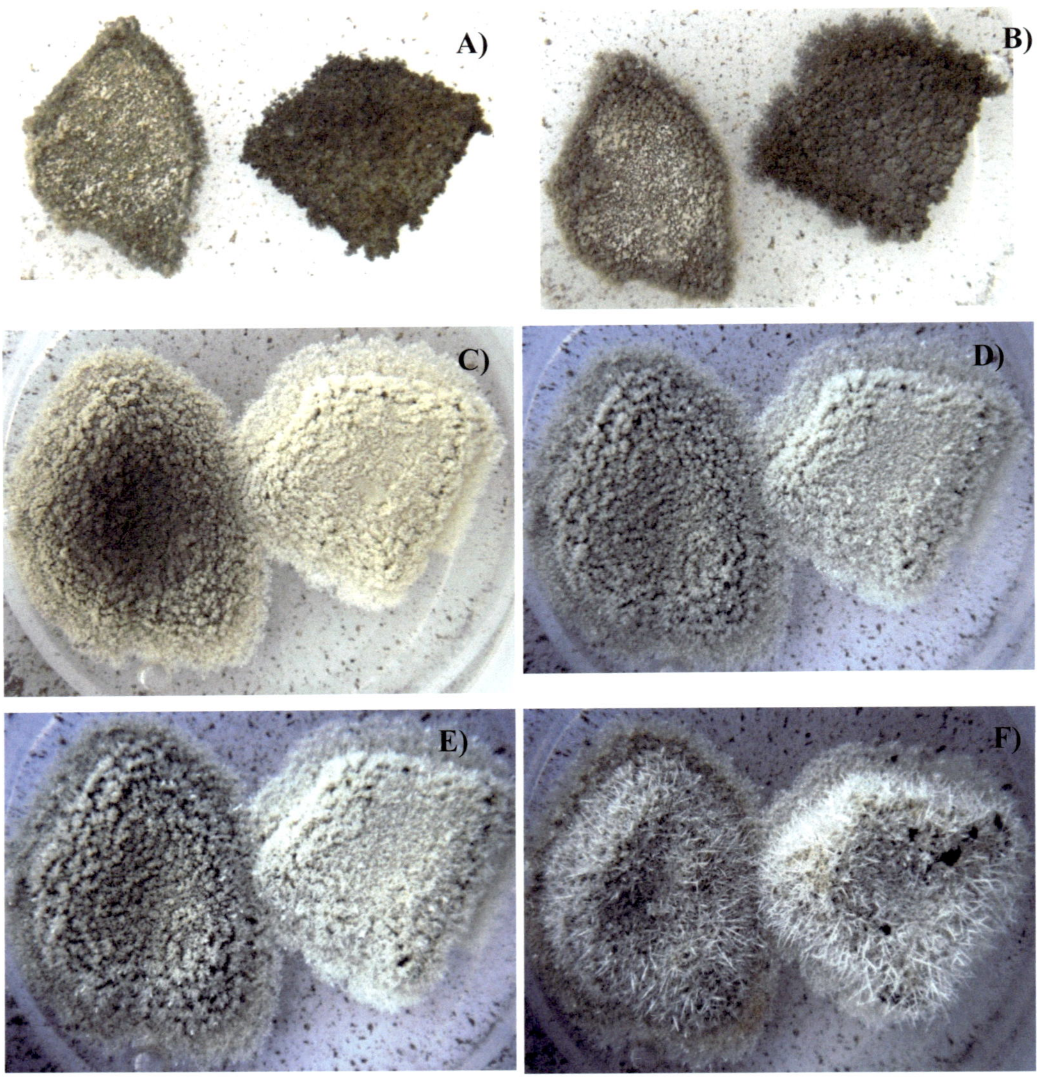

Fig. 13. Sequence of silver deposits growth on the system manganese-manganese oxide. The white appearance was a highly crystalline form of needle-dendritic silver. The dark appearance was dendritic silver.

covering the bottom of the container. Its size surpassed the original Mn piece, even reaching inches away from the substrate and was several millimeters thick. At some point, this layer reached its maximum extent but it is of low density with an algae-like formation and, with time, it increased its density and stiffness.

Fig. 13 shows the sequence of images of silver growth using two manganese pieces. The dendrites on the border of the pieces grew faster. The white appearance was due to the formation of shiny silver needles (Fig 13C). The deposition ended when bubbles, formed underneath, breaking the electrical contact between the active reduction sites and the metallic substrate. In Fig. 13F, the black spots, over the silver growths of the right piece, were the apertures made by the bubbles ejection.

The third stage was directly associated with bubble formation, which in turn may depend on the thickness of the first layer. A first thick dendritic formation may subsequently promote a rapid bubble generation. On the other hand, a first thin dendritic formation fosters the growth of crystals of silver, which started as needle-like formations but then grow in dendritic shapes. Higher concentrations may cause a short first stage, which indeed may reduce the probability of bubble formation and a corresponding abrupt interruption of the deposition. The silver ions concentration in the solution plays an important role in the kinetics of forming deposits. High concentrations and overpotentials propitiate such dendritic configurations (Herlach, 2004).

When considering the whole silver deposition process, the growth was in dendritic form, and there were different dendrite types in each stage. Upon removal of the deposition from the solution, there were two forms, an algae-like mass and long single silvery needles (some millimeters long).

Fig. 14. Another silver dendritic growth with needle-like formation. The black spots on top were holes left by the bubbles ejection.

As in the case of the previous section with lead, the dendritic growths were bigger at the borders of the Mn pieces. Also, exposing the deposits to air over the Mn piece leads to a manganese oxide coverage (Mn_3O_4). This not happened when the Mn piece was removed. In any circumstance, the silvery needles with single monocrystals appearance remained unchanged.

Figure 15 shows the status achieved by the growth of silver deposit on manganese when it is allowed to continue for several weeks without external disturbance. The same manganese piece, with a clean surface, can restart the process even in the same solution whether starting with 100 ppm or 1000 ppm. The latter contains enough silver ions to repeat the procedure several times. In the case of those silver growths, the substrate easily lost contact with the depositions. An easy cleaning of the surface made the Mn piece available for a new deposition.

Fig. 15. Two silver deposits, after a month, on Mn pieces located at the center of the formation.

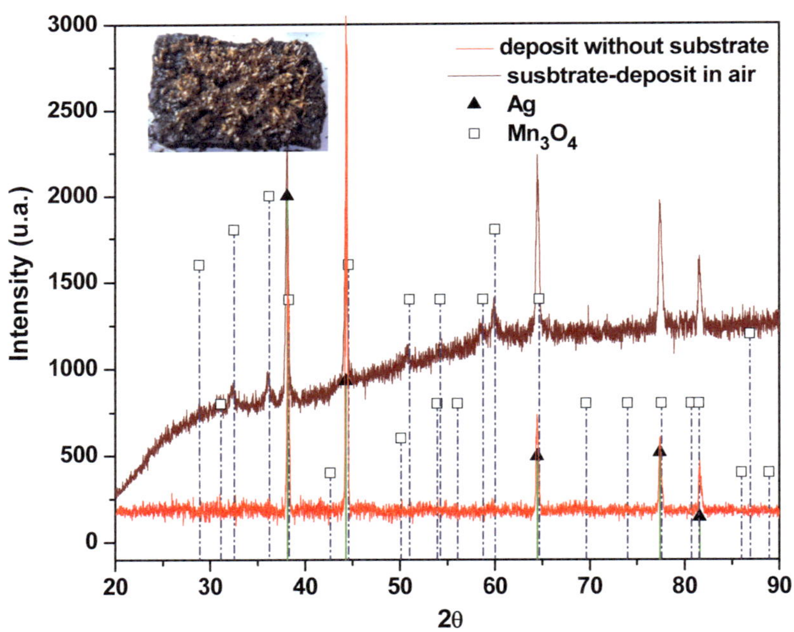

Fig. 16. Diffractograms of the bare silver deposit and on the Mn piece.

The electroless deposition process using manganese-manganese oxide usually stops by removing the metallic substrate. As mentioned previously, and similarly to the case of lead, by the exposure of deposits to air that remained on the substrate got a Mn_3O_4 coverage (Fig. 16 – inserted picture). Fig. 16 shows the diffractograms of a sample with and without the Mn substrate. The coverage of the deposits by a manganese oxide changes the shiny silver finish to a brownish colour, which happened in a matter of minutes. The reason for this change was not associated to a reaction of the silver with environmental sulfur, which usually constitutes the dark patina (Ag_2S).

On the other hand, if the silver deposit lost contact with the manganese substrate, a manganese oxide layer was not formed. This behaviour seems related to a Mn migration to the outer surface, which was interrupted. These are explained by redox reactions that require the transfer of electrons between the surface where oxidation takes place and the cathode, where deposits grew. Figures 13 and 14 show the deposited material corresponding to the three stages of the deposit, where highlighted the prominent needles and the foamy material generated in the third stage. A simply water cleaning allowed appreciate the state of the manganese surface with few firmly attached material.

From those tests, we can infer that Mn metal parts with its oxide may result with the potential for water remediation as well as for recovering some precious metals.

3.3 Electroless deposition from an aqueous solution containing Aluminum ions

In the study of a manganese-manganese oxide surface, immersed into an aluminum ion solution, a distinctive dense deposit, nearly continuous and homogeneous, was obtained. In this case, the deposit did not present dendritic formations and a high number of bubbles were formed on the surface. The proposed explanation refers to the potential difference between the areas acting as anodes and cathodes, which had favourable results the potential water electrolysis forming hydrogen and oxygen on the substrate.

The kinetic was fast using a solution with 500 ppm of aluminum. It was possible to observe a deposit and bubbles few seconds after immersion. However, the deposition rate decreased faster possibly attributed to two facts: a) the bubbles covering the surface (each measuring up to 2mm in diameter) and b) a thick and viscous layer of hydroxides forming above the surface. In order to permit deposited layer growth, some physical methods were used such as: a) magnetic stirring, b) ultrasound, or c) direct friction with other surface. Those methods were effective to displace both, the bubbles and the hydroxides. This procedure was contrary to the requirements in the previous case for silver deposition.

The deposited layer became perceptible through colour changes on the surface. The substrate acquired a lustrous golden metallic tonality, which appeared homogeneously and was perceptive to the naked eye and cameras without amplification. This effect was attributed to the interference of the light reflecting on the surface due to the thickness of the layer and its refraction index. Subsequently, as the layer became thicker, the surface took a lustrous metallic white tone.

The deposit from an aluminum containing solution did not show dependence on the initial surface condition, regardless of starting with a polished or oxide surface. Also, it differed from silver and lead susceptibility to a cover of manganese oxide. The final cleaned surface remained unchanged when exposed to air for long periods.

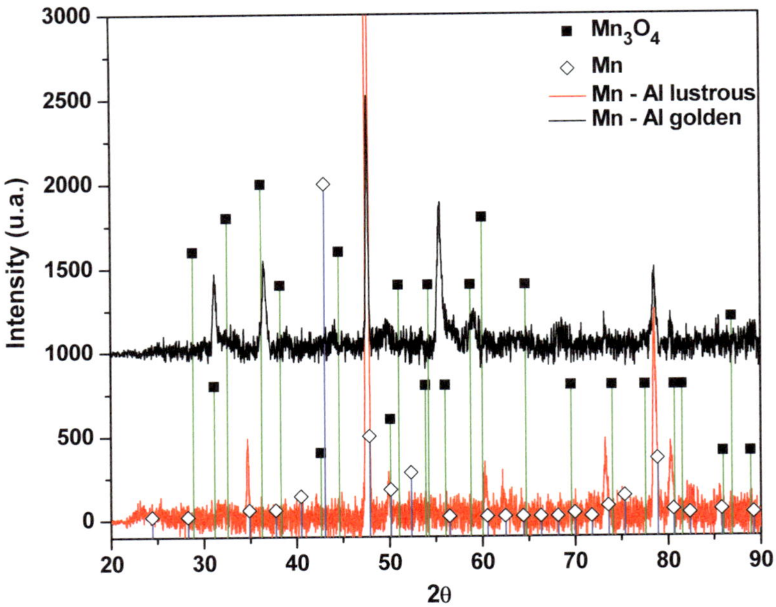

Fig. 17. Difractogram of aluminum deposit on manganese substrate.

Fig. 18. Aluminum deposits on manganese at different times in the process. The white metallic finishing remained unchanged with exposition to air.

The x-ray diffraction analysis of deposits, from an aluminum containing aqueous solution, did not allow the identification of the metallic aluminum phase. Fig. 17 shows a diffractogram that identifies only the manganese metallic substrate and, in some cases as this one, Mn_3O_4. Presumably, the X-ray diffraction was not able to identify the aluminum due to the small thickness of the layer.

The Fig. 18 shows the deposition from a solution containing aluminum on manganese substrate at different time during the process. It is possible to observe the bubbles forming on the substrate with a metallic deposit appearance.

Fig. 19 shows a SEM micrograph of the Aluminum deposit using 1000X (the scale bar indicate 25μm). The first notorious characteristic of this kind of deposits is a flat surface devoid of dendritic growths. An EDS microanalysis shows Mn, O and Al on surface. The signals for Aluminum and oxygen were very weak. The semi-quantitative calculus was: Oxygen 1.71wt% (5.6 atom%), Aluminium 0.42 (0.82 atom%) and Manganese 97.87 wt% (93.58 atom%). This can be attribute to a thin layer of Aluminum deposit, as in the case of the difractogram. The composition distribution maps on surface corresponding to Al, O, and Mn, respectively. There was a small distribution of Aluminum detected on surface.

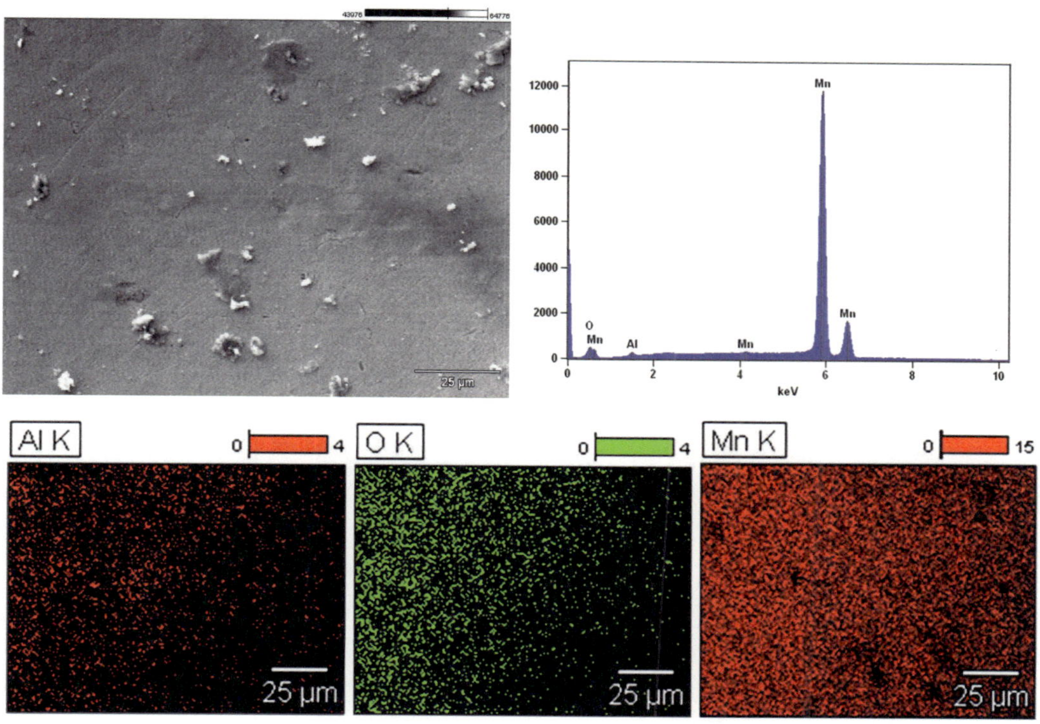

Fig. 19. SEM micrograph of the Aluminum deposit using 1000X (the scale bar indicate 25μm). Also, EDS microanalysis shows Mn, O and Al on surface. In the second line, the composition distribution maps on surface corresponding to Al, O, and Mn, respectively.

The Figure 20a shows the absorbance and reflectance spectra for a deposit from an aluminum solution on a manganese-manganese oxide substrate. The spectrum of the substrate is presented in order to compare the effect of the aluminum deposit on the absorbance and reflectance. In black, associated with the left **y** axis, the reflectance is shown. Correspondingly, associated with the right **y** axis, the absorbance is shown. The band presented in the reflectance spectrum for the bright aluminum deposit corresponds to the lamp used for illumination. Despite having used of an integration sphere in contact with the sample surface, the sample maintains a difference in reflectance in relationship to the reference that causes a lamp spectrum replication. Additionally, in Figure 20b, it is possible to quantify the surface appearance with the chromaticity diagrams.

Table 3 shows the color parameters, according to CIE (Commission Internationale de l'Eclairage): lightness (L*), red-green (a*) and yellow-blue (b*). The dominant wavelength is included, which evoke colour perception as given by a complicated light mixture. The purity perceived was thought to be in the range of 0-1 from white illumination to pure colour.

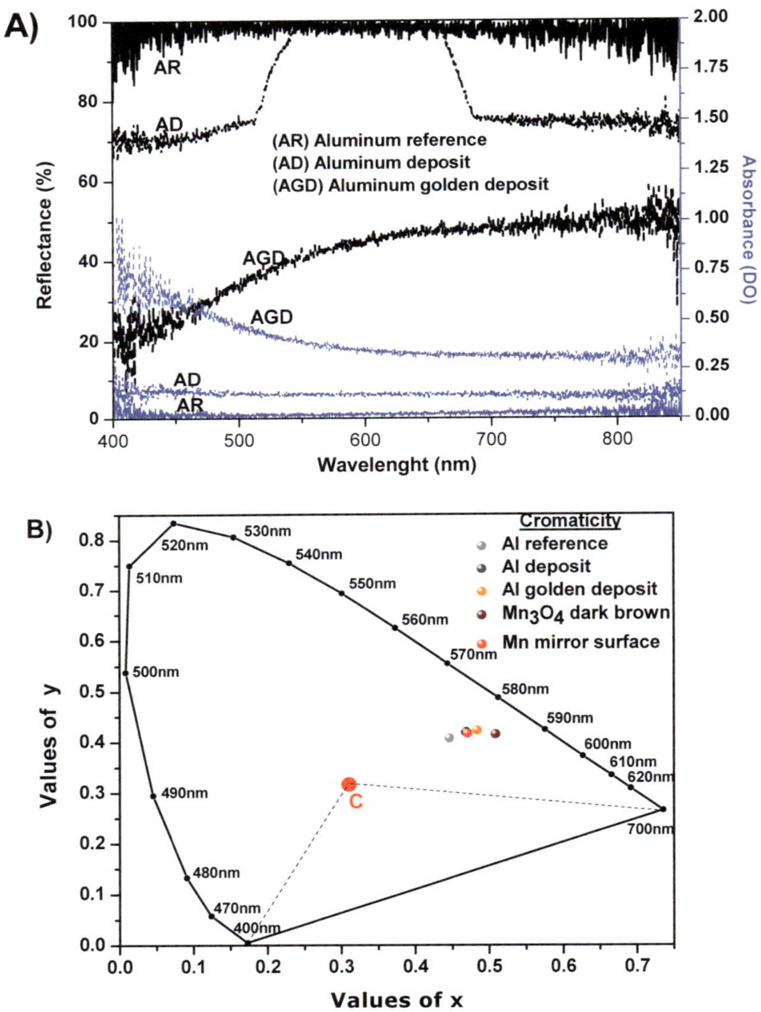

Fig. 20. A) Absorbance and reflectance spectra for the aluminum deposit. B) Chromaticity diagram for five samples of aluminum deposits.

Using Figure 20 and Table 3, it is possible to evaluate the obtained deposit finishing. It was possible to observe a bigger difference comparing the deposit and a manganese-manganese oxide substrate, which had a dark brown color. On the other hand, the bright deposit obtained was brighter than the deposit on the polished manganese.

The aluminum deposits on manganese- manganese oxide have a high reflectance. The profilometry analyses allowed quantify the surface roughness.

The images in Figure 21 were acquired by a video camera integrated to the profilometry apparatus, in the magnification range up to 70. A white LED was used as an illumination

	Aluminum	Al bright Deposit	Golden deposit	Dark brawn Mn_3O_4	Mn polished
CIE L*	99.5	98.4	76.2	44	97
CIE a*	-0.5	3	5.5	9.5	4.5
CIE b*	-0.4	18.1	22.9	20.6	17.2
$\lambda_{dominant}$ (nm)	583.0	580.6	589.3	595.0	589.2
purity	0.014	0.231	0.352	0.478	0.226

Table 3. Colour parameters for the five samples represented in Figure 16B.

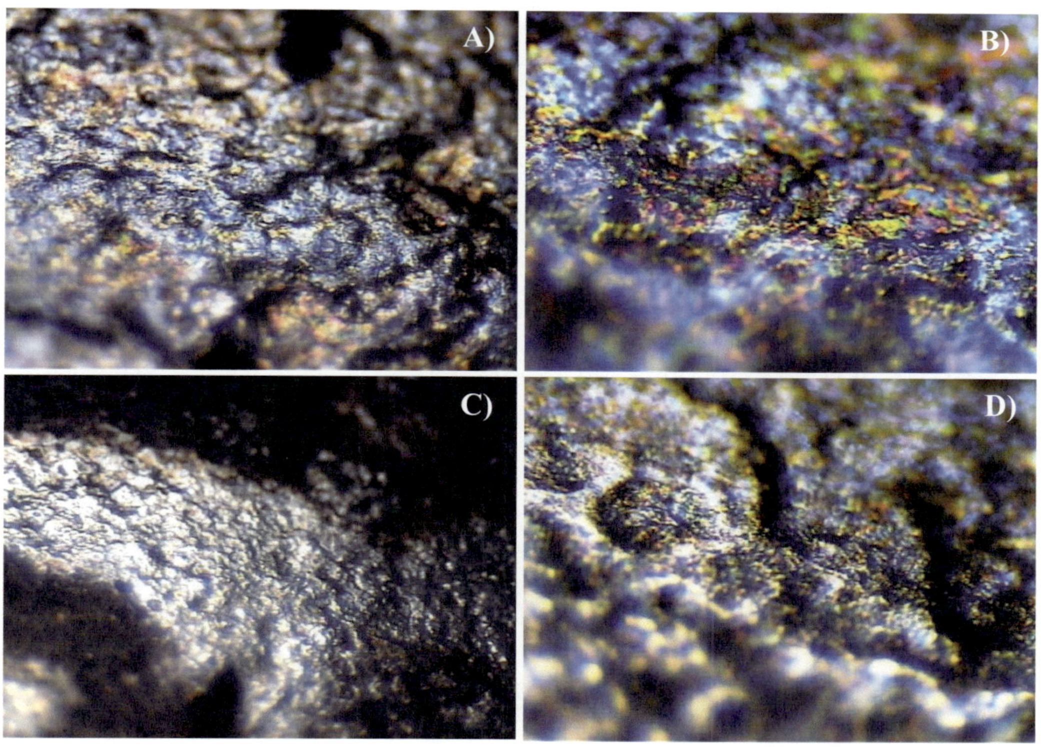

Fig. 21. Images of aluminum deposit on manganese-manganese oxide. A) and B) corresponds to a deposit on mangense-manganese oxide as susbtrate at 70 and 700x, respectively; C) and D) corresponds to a deposit on manganese with an ititially polished surface at 70 and 700x, respectively.

source which was put at the opposite side of the camera. It was possible to observe only specular light reflecting on surface (according to reflexion law).

Fig. 21 shows marked irregularities on the surface of deposits from an Al containing solution. This kind of surface is characteristic for aluminum deposit in contrast to the silver and lead dendritic deposits.

Table 4 shows the average roughness (Ra), the standard deviation (Rq), the number of peaks (Pc) and the distance between peaks (Sm). For references, the values for manganese substrate polished (using SiC #600) and a manganese-manganese oxide are presented. The measurements of aluminum surfaces are presented duplicated (A and B), the number (1) refers to the aluminum deposit on manganese-manganese oxide substrate and (2) refers to the polished manganese substrate. The unit of Ra or Rq is Armstrong. Pc shows, in those cases, only bigger picks, which made a great contrast between the substrate with oxide and the other surfaces. Also, the Ra of the polished surface had a great contrast compared with the others. The roughness of the deposits possessed relatively high values.

	Mn substrate with manganese oxide	Al (1A)	Al (1B)	Mn substrate polished (600)	Al (2A)	Al (2B)
Ra	13,213	61,908	47,460	4,459	48,332	14,467
Rq	18,169	77,715	63,501	5,681	58,349	18,086
Pc	119.98	12.00	17	30.00	6.00	21.43
Sm	113	1,137	753	630	1,869	519

Table 4. Roughness parameters profilometry obtaines for aluminun deposits.

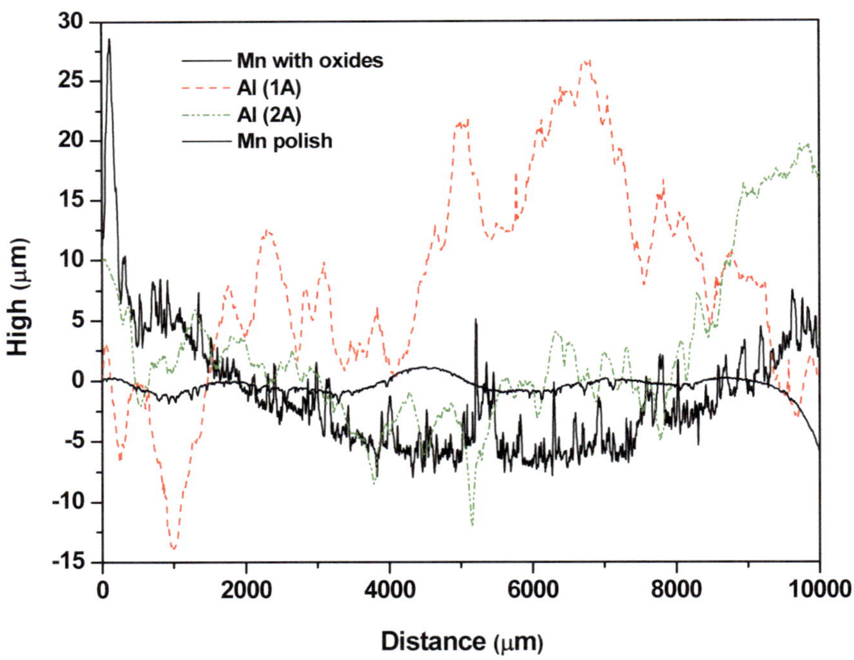

Fig. 22. Aluminum deposits profilometry from samples starting with polished and oxide surfaces.

Fig. 22 presents the graphs associated to the Table 4. The four profilometries correspond to samples prior (continuum lines) and after (doted lines) deposition in an aluminum containing solution. The original polished surface was smooth and the one with oxide had a roughness characteristic to it. After deposition, Al 1A that corresponds to a surface with oxide had wide and high picks. Meanwhile, Al 2A had a slightly inferior roughness.

4. Conclusion

This work has addressed the issue of water treatment contaminated with metal ions by deposition using the system manganese-manganese oxide. The research identified the case as a new alternative for water treatment and even the concentration of noble metals. Through the use of manganese-manganese oxide, electroless depositions were achieved for three cases under study: lead, silver and aluminum. Those deposits range from the fast dendritically structured silver with metallic crystals; passing through the lead oxides mixing with dendrites, hexagonal platelets and nucleations; and the extremely negative standard electrode potential of aluminum, which possessed a highly reflective layered growth.

Unlike the limited manganese oxides capacity for metal adsorption, the system manganese-manganese oxide promotes redox reactions leading to a massive deposition. Basically, the process continued as long as ions in solution and electrons transfer were kept. In the case of lead, the reduction of the deposition rate was closely related to the electrical conductivity drop due to growing oxide and the multiple grain frontiers between the active sites and the substrate.

After deposition and with exposure to air, a notorious effect was that the deposition growths were covered by Mn_3O_4.

5. Acknowledgements

The authors thank CONACYT for support through the Basic Science Project CB-2009-01: 133157. Especially appreciated was the help of M.C. Maria Guadalupe Martinez Almanza in conducting some laboratory tasks. We are also grateful to Guadalupe Olvera and Guillermo Carranza for performing ICP tests, as well as to Federico Manriquez Guerrero and M.C. Juan Manuel Alvarado for their support in the acquisition of SEM images.

6. References

Dries, J., Bastiaens, L., Springael, D., Kuypers, S., Agathos, S. N., & Diels, L. (2005). Effect of humic acids on heavy metal removal by zero-valent iron in batch and continuous flow column systems. *Water Research*, Vol. 39, pp. (3531–3540)

Herlach D. D., Funke O., Phanikumar G. & Galenko P., (2004), Rapid dendrite growth in undercooled melts: experiments and modelling. In: *Solidification Processes and Microstructures*. Eds: M. Rappaz, C. Beckermann, R. Trivedi. Warrendale, pp. (277-288), Wiley, Pennsylvania, USA

Kishimoto, N., Iwano, S. & Narazaki, Y. (2011) Mechanistic Consideration of Zinc Ion Removal by Zero-Valent Iron. Water Air Soil Pollut, pp. (1-7), DOI 10.1007/s11270-011-0781-1Kontyi, o.I., Zozulya, G.I., Kurrilets, O.G. (2007). Cementation of gold by Magnesium in Cyanide Solutions, Russian Journal of Non-Ferrous Metals, Vol. 48, No. 6, pp. (413-417). ISSN 1067-8212.

Noubactep, C. (2010). Elemental metals for environmental remediation: Learning from cementation process. Journal of Hazardous Materials. Vol. 181, pp. (1170–1174)

Noubactep, C. (2009). An Analysis of the evolution of reactive species in Fe0/H2O systems, Journal of Hazardous Materials, Vol. 168, pp. (1626-1631)

Oh, B., Lee, J., & Yoon, J. (2007). Removal of contaminants in leachate from landfill by waste steel scrap to converter slag. Environmental Geochemistry and Health, Vol. 29, pp. (331–336).

Peng, K., Yan, Y., Gao, S., Zhu J. (2003). Dendrite-Assited Growth of Silicon Nanowires in Electroless Metal Deposition, Advance Functional Materials, Vol. 13, No.2, (February, 2003), pp. (127-132).

Zouboulis, A. I., Lazaridis, N. K., & Grohmann, A. (2002). Toxic metals removal from waste waters by upflow filtration with floating filter medium. I. The case of zinc. Separation Science and Technology, Vol. 37, pp. (403–416)